CISM Courses and Lectures

The series presents lecture notes, monographs, edited works and proceedings in the field of Mechanics, Engineering, Computer Science and Applied Mathematics.
Purpose of the series is to make known in the international scientific and technical community results obtained in some of the activities organized by CISM, the International Centre for Mechanical Sciences.

International Centre for Mechanical Sciences

Courses and Lectures Vol. 557

For further volumes:
www.springer.com/series/76

Hans Irschik · Alexander K. Belyaev
Editors

Dynamics of Mechanical Systems with Variable Mass

Springer

Editors

Hans Irschik
University of Linz, Austria

Alexander K. Belyaev
Russian Academy of Sciences, St. Petersburg, Russia

ISSN 0254-1971
ISBN 978-3-7091-1808-5 ISBN 978-3-7091-1809-2 (eBook)
DOI 10.1007/ 978-3-7091-1809-2
Springer Wien Heidelberg New York Dordrecht London

All contributions have been typeset by the authors

Printed on acid-free paper

Springer is part of Springer Science+Business Media (www.springer.com)

PREFACE

This Book is a result of the Advanced School entitled "Dynamics of Mechanical Systems with Variable Mass", which took place in the International Centre for Mechanical Sciences (CISM), Udine, Italy, in September 2012.

The fundamental equations of classical mechanics were originally formulated for the situation that mass is conserved in the mechanical system under consideration. Mass is generally not conserved when a supply of mass is present, or when open systems with a flow of mass through their surface are to be considered. Mass of the mechanical system then is said to be variable. In such a situation, the general methodological approaches of mechanics have to be properly modified. In fluid mechanics, open systems are encountered when studying a non-material control volume. In solid mechanics, systems with a variable mass appear as the result of a problem-oriented modelling, e.g., when mass is expelled or captured by a structure or machine. This again leads to the treatment as an open system, or to the assumption that mass is explicitly dependent on the position. In solid mechanics as well as in fluid mechanics, it is often appropriate to model the exchange of mass between the system under consideration and the environmental world by means of a supply of mass in the interior. This is of particular interest in the continuum theory of mixtures, for which mass and other entities are exchanged between the various components.

It is the goal of the book to present up-to-date and unifying formulations for treating the dynamics of different types of mechanical systems with variable mass. The book is opened with an overview of the continuum mechanics relations of balance and jump for open systems. Afterwards, at the level of analytical mechanics, extended Lagrange and Hamiltonian formulations approaches are presented. The dynamics of multi-body mechanical systems with a continual and discontinual variation of mass is treated in detail. Another chapter is devoted to axially moving structures, like belts and chains, and on pipes with an axial flow of fluid, where the stability of motion is also considered. Constitutive relations appearing in the dynamics of mechanical systems with variable mass are studied with particular reference to the modelling of multi-component mixtures, in which damage of steel structures in the form of hydrogen diffusion is also addressed. Last not least, fluid-structural coupling is studied, where novel finite element formulations for open systems are presented. As a whole, the book should be of interest not only in the fields of civil and mechanical engineering, but also in mechatronics.

In more detail, the material presented in the book is organized as follows. In the introductory Chapter by H. Irschik and A. Humer, a rational treatment of the relations of balance and jump for mechanical systems with a time-variable mass and other non-classical supplies is presented. This presentation starts from the proper extensions of the fundamental relation of balance of mass, linear and angular momentum, and total energy. Relations for balance of moment of momentum, intrinsic spin, translational and rotational kinetic energy, and internal energy are derived afterwards, as mathematical consequences of the former fundamental relations. First, a single mass point with an intrinsic spin is treated, which then is extended to a deformable body with a finite extension, where special emphasis is laid upon the consequences of balance of mass and linear momentum. Center-of-mass oriented formulations are presented in some detail, a formulation for non-classical supplies in extension of the so-called Meshchersky reaction force is given, and a specialization to rigid bodies with a surface growth is studied. Continual variations of mass due to both, sources of mass in the interior and a flow of mass through the surface of a non-material control volume are studied, and analogies between these two cases are worked out. Jump relations for the case that a singular surface is travelling through the variable mass system are presented in an extended form, which takes into account concentrated surface supplies. The Cayley example of a chain hanging over the edge of the table and set into motion is used as an illustrative example.

In the following Chapter by C.P. Pesce and L. Casetta, the important case of variable mass systems, in which mass explicitly depends on position, and thus is variable in the course of the motion, is studied from the point of view of analytical mechanics. It is pointed out that for this case the Lagrange equation has to be carefully re-interpreted, since an extra non-conservative generalized force term, linearly proportional to the mass gradient and quadratic on velocities, turns out to emerge from first variational principles. In this chapter, a cor-

rect extended form of the Lagrange equation is derived through the Lagrangean and Hamiltonian approaches, and is discussed in detail. Various illustrative and practical examples from offshore engineering and civil engineering are presented, such as the reel laying operation of marine cables, the dynamics of a water column inside a free surface piercing open pipe, and the hydrodynamic impact of a solid body against a free surface of water. Also, the governing equation of motion of vertically collapsing towers is properly derived in the framework of the extended Lagrange equation.

The Chapter by L. Cveticanin addresses the dynamics of the body with time variable mass and time variable moment of inertia. The discontinual and the continual cases of variations of mass and moment of inertia are considered. The basic laws of dynamics are extended to the case in which the mass is varying in time. The principles of momentum and angular momentum are applied to obtain the velocity and angular velocity of the body after discontinual mass variation. The dynamics of mass addition is treated as the plastic impact. In addition to the reactive force, the reactive torque is introduced for the case of the continual time-variation of the mass and the moment of inertia. The free motion and the vibration of the mass variable body are treated as special cases. The influence of the reactive force on the vibration properties of the body is analyzed.

The Chapter prepared by D. Indeitsev and Yu. Mochalova deals with the dynamics of the material with complex internal structure. A two-component continuum model is utilised to this end. The approach allows one to describe the internal evolution processes in materials basing on the Euler equations and the mass balance equations containing the source terms is proposed. The influence of exchange mass between the components on the internal structure of the materials is investigated. The source terms determining the mass transfer between material components are defined. The following examples are delivered: the structured liquids in nanochannels, the metals with impurities and dissolved hydrogen.

The Chapter by A. Zilian is devoted to mechanics of coupled systems with mass which is dependent on structural motion and deformation; the effects of added mass, damping, stiffness; the models for fluid-structure interaction of discrete and distributed mass systems. In this contribution, concepts for modelling the interaction of structures and fluids are presented. Starting from excitation mechanisms and associated classifications, various model depth approaches are compared. Among them, the use of added coefficients for quasi-steady problems is discussed. On the basis of potential flow theory, the different approaches for determining fluid-induced additional mass are established and illustrated by using an analytical example. Given the limitations of simplifying the engineering models, the second part of the chapter provides a brief overview on computational methods for fluid-structure interaction and presents a monolithic modelling approach using space-time finite elements for discretisation of both fluid and structure. Applications from aero- and hydro-elasticity show the applicability of computational methods for problems involving flow-induced added mass, damping, and stiffness.

The Chapter by A.K. Belyaev is predominantly concerned with a special case of dynamics of the engineering systems with variable mass, namely open systems with moving continua. A characteristic feature of these systems is that a material enters the system and leaves the system. The transported material itself is assumed to be deformable and under some conditions the entire system exhibits unstable behaviour in the transverse direction. The intent of the present chapter is to demonstrate that the systems with axially moving material are inherently unstable. Dynamics and stability are known to be strongly related to each other, for this reason the study of dynamics and stability for each engineering system under consideration is carried out in the framework of the same approach. The chapter is opened with a general discussion and notion of the static and dynamics stability. A number of special cases important for mechanical engineering are considered. Some of these belong to the class of problems of fluid-structure interaction, in particular, dynamics and stability of the fluid conveying pipes and the shaft rotating in oil film plain bearings. The dynamics and stability of belts and chains are studied in detail, too.

It is a great pleasure for the editors to acknowledge the significant contributions made to the CISM Advanced School "Dynamics of Mechanical Systems with Variable Mass" and to this book by Professors Livija Cveticanin from Serbia, Dmitry Indeitsev from Russia, Celso Pesce from Brazil and Andreas Zilian from Luxembourg. They delivered excellent lectures in Udine in September 2012, and they contributed chapters to this book making our joint project a truly international effort. We also wish to thank them for their invaluable contributions.

The school brought together of about 40 participants from 7 countries. We are grateful to all participants for their interest and the numerous discussions that took place during and after the lectures. We are particularly thankful to the Scientific Council of CISM for supporting this Advanced School and recognizing the importance of the topic. The continuous support of the co-operation of the undersigned Editors in the framework of the Austrian Comet K2-Center of Excellence in Mechatronics ACCM is gratefully acknowledged.

Hans Irschik and Alexander K. Belyaev

CONTENTS

A rational treatment of the relations of balance for mechanical systems with a time-variable mass and other non-classical supplies

Hans Irschik* and Alexander Humer*,†

* Institute of Technical Mechanics, Johannes Kepler University Linz, Austria
† Linz Center of Mechatronics, Austria

Abstract This contribution intends to present a rational methodology for mechanical systems with a variable mass, represented by a supply of mass. Special emphasis is given to the relations of balance and jump for such systems. In these relations, we also allow for other types of additional, non-classical supplies, e.g., supplies of linear and angular momentum. In doing so, we aim at completing and substantially extending formulations laid down in the famous article by Truesdell and Toupin (1960), who stated local relations of balance of mass and linear momentum in the presence of sources of mass, and, among other formulations with relevance to the present article, gave fundamental formulations for the case that a flow of mass through the surface of the system is present in the global relations of balance.

Our presentation is organized as follows: We remain in the framework of non-relativistic mechanics, referring to a common inertial frame. Throughout the Chapter, we formulate our relations in the Euler or spatial description, in which every entity is understood as a function of the instantaneous place of the material particles under consideration, and of time. In Section 1, the general equation of balance is stated and is applied to the model of a single mass point with a variable mass. This general equation is specified for the fundamental relations of balance of mass, linear momentum, angular momentum and total energy first. The variable mass is associated with a supply of mass. Afterwards, as mathematical consequences of the fundamental statements, we derive the statements of balance of moment of momentum, intrinsic spin, kinetic energy and internal energy for the single mass point. As a rational procedure for formulating the additional, non-classical supplies that are present in the relations of balance, we assume that the single mass point is gaining or losing differential masses by means of continuous impacts, which

H. Irschik, A. K. Belyaev (Eds.), *Dynamics of Mechanical Systems with Variable Mass*,
CISM International Centre for Mechanical Sciences DOI 10.1007/ 978-3-7091-1809-2_1

are again studied in the framework of the general equation of balance. The outcomes of this procedure include a Seeliger-Meschersky type additional supply of linear momentum.

In Section 2, theorems on balance of mass, linear momentum, moment of momentum and kinetic energy for deformable bodies of finite extension with a variable mass are presented. Among these, the first two can be considered as fundamental, while balance of moment of momentum and kinetic energy are derived from balance of mass and linear momentum as mathematical consequences. The supply of mass is associated with distributed sources of mass attached to the material particles, which we call material sources of mass. Both global and local relations of balance are considered, including global and local non-classical supplies of mass and linear momentum. The supplies of moment of momentum and kinetic energy follow as mathematical consequences. A Seeliger-Meschersky type local model for the non-classical supply of linear momentum is presented. Due to limited space, the fundamental relations of balance of angular momentum and total energy for bodies of finite extension are not considered. However, useful global relations concerning the notion of center-of-mass are given, introducing the notions of center-of-mass linear momentum and relative linear momentum, center-of-mass moment of momentum and relative moment of momentum, as well as center-of-mass kinetic energy and relative kinetic energy. Our relations extend some formulations that are well-known for bodies in the absence of a supply of mass. The corresponding relations of balance again follow as mathematical consequences of the fundamental ones, including non-classical supply terms related to the non-classical supplies of mass and linear momentum. In Section 3, global relations of balance for open systems are studied, and are set into analogy to the results of Section 2. An open system is represented by a non-material control volume, the surface of which moves at a velocity different from the velocity of the material particles instantaneously located on that surface, such that a flow of mass takes place. Supplies of mass and linear momentum due to this flow of mass are shown to be analogous to the supplies introduced in Section 2. The theoretically as well as practically important special case of a rigid body that experiences a surface growth is exemplarily treated. Section 4 deals with extended relations of jump for systems with a variable mass. Relations of jump are needed, when certain entities suffer considerable changes across some region of transition. This region of transition is replaced by an equivalent singular surface, for which relations of jump are formulated by including additional non-classical surface supply terms, such as surface supply of mass and linear momentum. Other surface supply terms are derived as mathematical consequences of the latter. As an example for the formulations presented in Sections 3 and 4, the problem of a chain heaped up on a table, the hanging part of the chain being set into motion, is considered in Section 5. It is believed that the corresponding formulation can explain some seemingly controversial results from the literature.

It should be mentioned that our emphasis lies on a rational treatment of the topics under consideration. While our methodology has many important predecessors, but appears to be novel in the systematic manner here presented, we do not intend to give a historical review on the topic due to the limited space available. For the latter, the reader is referred to reviews by Mikhailov (1975), historical presentations to be found, e.g., in the important works of Eke (1998) and Cveticanin (1998), as well as to a review by Irschik and Holl (2004) on balance of mass and momentum for systems with a variable mass. (An extended review by the latter authors concerning balance of moment of momentum and kinetic energy for variable mass systems is being prepared since and hoped to be finished soon). Last but not least, the reader is referred to the other chapters of the present book.

1 An introductory example: The single mass point with a continual time-variation of mass.

1.1 The general and the differential relations of balance

In this Section, we present introductory material on the relations of balance for mechanical systems with a variable mass. The general equation of balance for any system can be written as:

$$Q(t+\Delta t) - Q(t) = \int_{t}^{t+\Delta t} R\,(\tau)\, d\tau \tag{1}$$

In (1), Q denotes some physical meaningful quantity that properly describes the system within the time interval $t \leq \tau \leq t+\Delta t$, and R is the physical cause that is responsible for a time change of Q, also denoted as the source of that time change. When $Q(t+\Delta t) = Q(t)$, i.e., when the integral at the right hand side of (1) does vanish, then $Q(t)$ is said to be conserved with respect to the time instant $t+\Delta t$. When the right hand side of (1) does not vanish, Q is said to be balanced by the integral of the source R over the time interval under consideration.

Eq. (1) represents the most general statement of balance; in order to make physical sense, it is only necessary to require that Q has unique values at the instants t and $t + \Delta t$, and that the integral over the source R does exist, but the source R itself does not need to be continuous in time.

A large part of the history of mechanics (and of physics as a whole) can be understood as an intense struggle for finding physically meaningful quantities Q and corresponding physically meaningful sources R that do satisfy (1). In the Newtonian theory of mechanics, which is used subsequently, time t and quantities Q are defined with respect to an inertial frame, and Q basically stands for the notions of mass, momentum and energy. The notion of energy directly connects the fields of mechanics and thermodynamics, the latter bringing into the play the additional notion of entropy, which, however, will not be addressed in the following due to limited space.

In case R is continuous and bounded, it makes sense to consider an infinitesimal time interval, $\Delta t \to dt$. Then the finite statement in (1) can be replaced by the following differential relation of balance:

$$Q(t + dt) - Q(t) = dQ = R\, dt \tag{2}$$

In the following Subsections 1.2–1.5, we apply the differential statement (2) to the model of a single mass point, also denoted as a point mass, a problem that is elementary for the dynamics of mechanical systems. In the present context, (2) is specified for several relations of balance for the single mass point with a time-varying mass, $m = m(t)$. The source that is responsible for the change in mass will be denoted as a supply of mass. We start with the fundamental relations of balance of mass, and discuss the relations of linear momentum, angular momentum and total energy afterwards. Into these relations, we incorporate further supply terms, additional to classical formulations. For a comprehensive representation of the classical balance statements of mechanics, see, e.g., Ziegler (1998). To add non-classical supply terms to the classical statements can be motivated, e.g., by the theory of multiphase mixtures, in which it is assumed that a particle of a single constituent exchanges mass, momentum and energy with the particles of the other constituents of the mixture, and thus is being supplied with the latter entities, see, e.g., the book by Hutter and Jöhnk (2004).

A continuous impact model then is presented in order to express the non-classical supplies of linear momentum, angular momentum and energy. This model assures that mass, linear momentum and energy is continuously gained from differential masses at an own velocity. As a special case, the Seeliger-Meshchersky formulation for the supply of linear momentum is contained therein.

Having stated the fundamental relations of balance of mass, linear momentum, angular momentum and total energy for a single mass point with a supply of mass and other non-classical supplies, we proceed to consequences of the latter fundamental statements, and derive the statements of balance of moment of momentum, intrinsic spin, kinetic energy and internal energy. All the above mentioned relations of balance will be referred to a common inertial frame without further reference.

1.2 Balance of mass

For balance of mass, (2) reads

$$Q(t+dt) - Q(t) = dQ = dm, \quad R = s[m] \tag{3}$$

The supply of mass from or to the environment is denoted as $s[m]$. If mass is added to the point mass, then $s[m] > 0$, and if mass is ejected from the point mass, there is $s[m] < 0$. Note that the notation $s[\lambda]$, say, reads *"the supply of* λ", where the physical dimension of $s[\lambda]$ is the dimension of λ per dimension of time. Using (2) and (3), balance of mass becomes:

$$\frac{dm}{dt} = s[m] \tag{4}$$

1.3 Balance of linear momentum

The vector of linear momentum is defined as

$$j = m\,v \tag{5}$$

where the absolute velocity vector is $v = dp/dt$, and p is the position vector of the single mass point with respect to the origin of the inertial frame. For linear momentum j, the quantities in (2) are:

$$dQ = dj, \quad R = F + s[j] \tag{6}$$

The resultant of imposed and restoring forces that acts upon the single mass point is the vector F, and the vector $s[j]$ stands for a non-classical (additional) supply of momentum. Utilizing balance of mass (4), the following relation of balance of linear momentum for a single mass point is obtained from (2) and (6):

$$\frac{dj}{dt} = m\frac{dv}{dt} + s[m]\,v = F + s[j] \tag{7}$$

1.4 Balance of angular momentum

The vector of angular momentum is defined as

$$\alpha = p \times m\,v + l \tag{8}$$

where the vector product $p \times m\, v$ is called the moment of (linear) momentum, and the vector l denotes an intrinsic spin of the point mass. For angular momentum, the quantities in (2) are:

$$dQ = d\alpha, \quad R = p \times F + M + s\,[\alpha] \tag{9}$$

The resultant of imposed and restoring couples that act upon the mass point is abbreviated by the (free) vector M. The notions of intrinsic spin l and the resultant couple M, while often not taken into account in the mechanics of single mass points, have been introduced in (9) in order to render balance of angular momentum a relation of balance in its own right, bringing additional physical notions into the play that are not present in balance of linear momentum (7). Motivations for introducing l and M can be taken, e.g., from the model of material particles in a continuous polar medium. For a recent comprehensive representation of the theory of micro-polar media, see the book by Eremeyev et al. (2012). Another example is the case that a single mass point shall be used to model the rotational motion of a rigid body of finite extension. From these examples, it makes sense to set

$$l = J \cdot \omega \tag{10}$$

where J is the symmetric second order tensor of inertia, and ω is the angular velocity vector of the mass point. Note that we use the simple single dot product operation that has been introduced in the exposition on tensor fields by Ericksen (1960). Introducing a non-classical (additional) supply of angular momentum $s\,[\alpha]$, using balance of mass (4), and substituting (10), the relation of balance of angular momentum is obtained from (2) and (9) as

$$\frac{d\alpha}{dt} = p \times m \frac{dv}{dt} + J \cdot \frac{d\omega}{dt} + p \times s\,[m]\, v + \frac{dJ}{dt} \cdot \omega = p \times F + M + s\,[\alpha] \tag{11}$$

1.5 Balance of total energy

The total energy E of the mass point is defined as the sum

$$E = E_{kin} + E_{int} \tag{12}$$

where the kinetic energy of the mass point is given by

$$E_{kin} = \frac{1}{2}\left(m\, v \cdot v + \omega \cdot (J \cdot \omega)\right) \tag{13}$$

The internal energy, a notion that stems from thermodynamics, is abbreviated by E_{int}. For total energy, the quantities in (2) are:

$$dQ = dE, \quad R = F \cdot v + M \cdot \omega + r + s\,[E] \tag{14}$$

where r stands for non-mechanical sources of total energy, and $s[E]$ is a non-classical (additional) supply of energy. Using balance of mass (4), and substituting (10) and (13), the relation of balance of total energy is obtained from (2) and (12) as

$$\frac{dE}{dt} = m\,v\cdot\frac{dv}{dt} + \omega\cdot\left(J\cdot\frac{d\omega}{dt}\right) + \frac{1}{2}\left(s[m]\;v\cdot v + \omega\cdot\left(\frac{dJ}{dt}\cdot\omega\right)\right) + \frac{dE_{int}}{dt}$$
$$= F\cdot v + M\cdot\omega + r + s[E] \quad (15)$$

Balance of total energy is also known as the *First Law of Thermodynamics*; the relation in (14) accounts for the case of a variable mass and an additional supply of total energy.

1.6 A rational procedure for obtaining expressions for the additional supplies

Additional modeling is necessary in order to describe the supply of mass $s[m]$. The same is true for the non-classical supplies $s[j]$, $s[\alpha]$ and $s[E]$. In the present subsection, we introduce a rational model, in which the latter supplies can be expressed using the former. This model may be called an extension of the continuous impact model by Cayley (1856), who assumed that the single mass under consideration *"is continually taking into connexion with itself particles of infinitesimal mass [...], so as not itself to undergo any abrupt change of velocity, but to subject to abrupt changes of velocity the particles so taken into connexion."*

Hence, consider a mass point with an infinitesimal mass of amount $d\bar{m} = dm = s[m]\,dt$, with $s[m] > 0$. Assume that, at the time instant t, the linear momentum of the infinitesimal mass is $dm\,u$, the intrinsic spin is $dJ\cdot\Omega$, and the internal energy is $d\bar{E}_{int}$, where u is the velocity vector of the infinitesimal mass, and Ω is the angular velocity vector of the infinitesimal mass at time t. Assume further that this infinitesimal mass $d\bar{m}$ is absorbed during the time-interval dt by the mass point with finite mass m, where the linear momentum, angular momentum and energy of the infinitesimal mass are completely transferred to the mass m at time $t+dt$. In accordance with (2), the equations of balance for $d\bar{m}$ become:

Balance of linear momentum

$$dQ = Q(t+dt) - Q(t) = 0 - dm\,u = -dm\,u, \qquad R = s[\bar{j}] \quad (16)$$

$$\Rightarrow \quad \frac{dm}{dt}\,u = s[m]\;u = -s[\bar{j}] \quad (17)$$

Balance of angular momentum

$$dQ = -p \times dm\, u - dJ \cdot \Omega, \quad R = s\,[\bar{\alpha}] \tag{18}$$

$$\Rightarrow\ p \times \frac{dm}{dt} \cdot u + \frac{dJ}{dt} \cdot \Omega = p \times s\,[m] \cdot u + \frac{dJ}{dt} \cdot \Omega = -s\,[\bar{\alpha}] \tag{19}$$

Balance of energy

$$dQ = -\frac{1}{2}\left(dm\, u \cdot u + \Omega \cdot (dJ \cdot \Omega)\right) - d\bar{E}_{int}, \quad R = s\left[\bar{E}\right] \tag{20}$$

$$\Rightarrow\ \frac{1}{2} s\,[m]\, u \cdot u + \frac{1}{2}\Omega \cdot \left(\frac{dJ}{dt} \cdot \Omega\right) + d\bar{E}_{int} = -s\left[\bar{E}\right] \tag{21}$$

Above, no imposed or reactive forces and couples or non-mechanical sources of energy have been considered for the infinitesimal mass for the sake of brevity. Requiring that the supplies for m and $d\bar{m}$ must be mutual, we obtain:

$$s\,[j] = -s\,[\bar{j}] = s\,[m]\, u \tag{22}$$

$$s\,[\alpha] = -s\,[\bar{\alpha}] = p \times s\,[m]\, u + \frac{dJ}{dt} \cdot \Omega \tag{23}$$

$$s\,[E] = -s\left[\bar{E}\right] = \frac{1}{2} s\,[m]\, u \cdot u + \frac{1}{2}\Omega \cdot \left(\frac{dJ}{dt} \cdot \Omega\right) + d\bar{E}_{int} \tag{24}$$

Substituting into the fundamental equations of balance of linear momentum, angular momentum and total energy stated in Subsections 1.3–1.5, these relations become:

$$\frac{dj}{dt} = m\frac{dv}{dt} + s\,[m]\; v = F + s\,[m]\, u \tag{25}$$

$$\begin{aligned}\frac{d\alpha}{dt} &= p \times m\frac{dv}{dt} + J \cdot \frac{d\omega}{dt} +\ p \times s\,[m]\, v + \frac{dJ}{dt} \cdot \omega \\ &= p \times F + M + p \times s\,[m]\, u + \frac{dJ}{dt} \cdot \Omega\end{aligned} \tag{26}$$

$$\begin{aligned}\frac{dE}{dt} &= m\, v \cdot \frac{dv}{dt} + \omega \cdot \left(J \cdot \frac{d\omega}{dt}\right) + \frac{1}{2}\left(\, s\,[m]\, v \cdot v + \omega \cdot \left(\frac{dJ}{dt} \cdot \omega\right)\right) + \frac{d}{dt}E_{int} \\ &= F \cdot v + M \cdot \omega + r + \frac{1}{2}\left(s\,[m]\, u \cdot u + \Omega \cdot \left(\frac{dJ}{dt} \cdot \Omega\right)\right) + \frac{d}{dt}\bar{E}_{\mathrm{int}}\end{aligned} \tag{27}$$

For the sake of comparison with the literature, the relation of balance of linear momentum is re-written as

$$m\frac{dv}{dt} = F + s\,[m]\,(u - v) \tag{28}$$

The term $s\,[m]\,(u - v)$ dates back to Seeliger (1890) and Meshchersky (1897); it is called the Meshchersky reactive force, while (28) as a whole is denoted as the Tsiolkovsky-Meshchersky rocket equation. Tsiolkovsky in 1897 independently derived the solution for the case of a constant relative velocity $u-v$, cf. Kosmodemyansky (2000). For historical expositions, see Mikhailov (1975) and Irschik and Holl (2004). The results in (13)–(17) also hold for the case $s\,[m] < 0$. We then assume that the infinitesimal mass $d\bar{m} = -dm > 0$ is ejected during the time-interval dt from the mass point m, where momentum, angular momentum and energy of the infinitesimal mass have been completely released from m at time instant $t + dt$. E.g., balance of linear momentum for the infinitesimal mass reads, cf. (2):

$$dQ = Q\,(t + dt) - Q\,(t) = d\bar{m}\;u - 0 = -dm\;u, \qquad R = s\,[\bar{j}] \tag{29}$$

which again yields (17).

It must be emphasized that the above continuous impact model does result in balance relations that are invariant with respect to a change of the common inertial frame. In these relations, we may add a constant position vector to the vector p, a constant velocity vector to v and u, etc., and then subtract the results from the original formulations without obtaining any discrepancies. E.g., adding a constant position vector to p in (26), the difference of the result with respect to the original relation (26) vanishes, since (25) holds. Of course, the supply of mass $s\,[m]$ itself and the internal energies must also be formulated accordingly.

1.7 Consequences of the fundamental relations of balance

The fundamental relations of balance can be mathematically manipulated in order to obtain further relations of balance. This will be done in the following Subsection using the continuous impact model presented in Subsection 1.6 above, i.e., by studying mathematical consequences of the relations stated in (25)–(27). Particularly, we derive balance relations for moment of momentum, intrinsic spin, translational and rotational kinetic energy and internal energy from the fundamental relations. The additional supply terms in the latter derived relations are expressed by the supply terms in the fundamental relations of balance, as it should be. For a systematic treatment of relations between non-classical supply or growth terms in the framework of continuum mechanics, the reader is referred to Irschik (2005, 2007).

Balance of moment of momentum Performing the vector product of the balance of linear momentum (25) with the position vector p yields

$$p \times m\frac{dv}{dt} + p \times s\,[m]\,v = \frac{d}{dt}\,(p \times m\,v) = p \times F + p \times s\,[m]\,u \tag{30}$$

Introducing the moment of momentum as

$$\alpha^* = p \times m\,v = \alpha - l \tag{31}$$

we obtain:

$$\frac{d\alpha^*}{dt} = p \times m\,\frac{dv}{dt} + p \times s\,[m]\;v = p \times F + s\,[\alpha^*] \tag{32}$$

with the non-classical supply of moment of momentum

$$s\,[\alpha^*] = p \times s\,[m]\;u = p \times s\,[j] \tag{33}$$

Balance of intrinsic spin Subtracting (32) from balance of angular momentum (26) gives:

$$\frac{dl}{dt} = J \cdot \frac{d\omega}{dt} + \frac{dJ}{dt} \cdot \omega\; = M + s\,[l] \tag{34}$$

with the non-classical supply of intrinsic spin

$$s\,[l]\; = \frac{dJ}{dt} \cdot \Omega \tag{35}$$

Balance of translational kinetic energy Performing the scalar product of the balance of linear momentum (25) with the velocity v yields

$$v \cdot m\frac{dv}{dt} + v \cdot s\,[m]\,v = \frac{d}{dt}\left(\frac{1}{2}m\,v \cdot v\right) + s\,[m]\,\frac{1}{2}v \cdot v = F \cdot v + s\,[m]\,u \cdot v \tag{36}$$

Introducing the translational kinetic energy as

$$E^{tr}_{kin} = \frac{1}{2}m\,v \cdot v \tag{37}$$

we obtain:

$$\frac{dE^{tr}_{kin}}{dt} = v \cdot m\;\frac{dv}{dt} + \frac{1}{2}s\,[m]\,v \cdot\; v = F \cdot v + s\left[E^{tr}_{kin}\right] \tag{38}$$

with the non-classical supply

$$s\left[E^{tr}_{kin}\right] = s\,[m]\left(u - \frac{1}{2}v\right) \cdot v \tag{39}$$

Balance of rotatory kinetic energy Performing the scalar product of the balance of intrinsic spin (34) with the angular velocity ω yields

$$\begin{aligned}\omega \cdot \frac{dl}{dt} &= \omega \cdot \left(J \cdot \frac{d\omega}{dt}\right) + \omega \cdot \left(\frac{dJ}{dt} \cdot \omega\right) \\ &= \frac{d}{dt}\left(\frac{1}{2}\omega \cdot (J \cdot \omega)\right) + \frac{1}{2}\omega \cdot \left(\frac{dJ}{dt} \cdot \omega\right) = M \cdot \omega + \omega \cdot \left(\frac{dJ}{dt} \cdot \Omega\right) \qquad (40)\end{aligned}$$

Introducing the rotational kinetic energy as

$$E_{kin}^{rot} = \frac{1}{2}\omega \cdot (J \cdot \omega) \qquad (41)$$

we obtain:

$$\frac{dE_{kin}^{rot}}{dt} = \omega \cdot \left(J \cdot \frac{d\omega}{dt}\right) + \frac{1}{2}\omega \cdot \left(\frac{dJ}{dt} \cdot \omega\right) = M \cdot \omega + s\left[E_{kin}^{rot}\right] \qquad (42)$$

with the non-classical supply of rotatory kinetic energy

$$s\left[E_{kin}^{rot}\right] = \omega \cdot \left(\frac{dJ}{dt} \cdot \left(\Omega - \frac{\omega}{2}\right)\right) \qquad (43)$$

Balance of internal energy Subtracting the relations of balance for kinetic energy, (38) and (42), from the relation of balance of total energy in the form of (27), we obtain:

$$\frac{dE_{int}}{dt} = r + s\left[E_{int}\right] \qquad (44)$$

with the non-classical supply of internal energy

$$s\left[E_{int}\right] = \frac{1}{2}s\left[m\right](u - v)\cdot(u - v) + \frac{1}{2}(\Omega - \omega)\cdot\left(\frac{dJ}{dt} \cdot (\Omega - \omega)\right) + \frac{d}{dt}\bar{E}_{int} \qquad (45)$$

2 Balance relations for bodies of finite extension with a variable mass

So far, we have dealt with the relations of balance for a single mass point with a variable mass. In the following, we extend these considerations to the case of a material body of finite size. Hence, we subsequently deal with a mechanical system that consists of an infinite set of continuously distributed material particles, for which mass is not conserved. Due to limited space, we remain in the framework of purely mechanical notions in the present section, dealing with balance of mass, linear momentum, moment of momentum and kinetic energy, the latter two as consequences of mathematical manipulations of the former ones. This restriction is also motivated by a remark of Truesdell and Toupin (1960), who stated that the concepts of linear momentum, moment of momentum and kinetic energy of a finite body are *"the stuff of which classical mechanics is made"*, and that they *"deserve the most minute analysis"*. It is the scope of the present section to present a rational methodology for formulating relations of balance for these quantities, considering the case of a body with variable mass and a finite extension.

The relations of balance in their most basic form, see (2), are to be referred to the total of a system under consideration, in the present case by integrating over the finite volume of the body. In order to highlight this fact, the corresponding balance equations are also denoted as global relations of balance. Considering the so-called localization argument, i.e., requiring that the relations of balance must hold for any sub-volume of the body, the global relations can be localized under obvious continuity conditions to so-called local relations of balance.

We start our considerations by dealing with global and local balance of mass. For writing global relations, we consider balance with respect to a material volume at first, in other words, for a finite volume with a surface that moves together with the particles instantaneously located in it. This material volume constitutes the material body under consideration. As the reason for a variable mass of the so defined material body, we assume that the elementary masses carried by the particles of the body do change in time. In order to characterize this situation, we say that material sources of mass are attached to the particles. In our subsequent balance formulations, these sources are associated with global and local supplies of mass, the latter being spatially distributed. Since these supplies are absent in classical formulations of continuum mechanics, we talk about non-classical supplies.

In a next step, the notion of the center of mass is introduced in the presence of material sources of mass, which, to a certain extent, connects the formulation for the single mass point in Section 1 above with the problem of a material body of finite spatial extension. Some useful global relations in connection with the notion of center of mass are presented, introducing the notions of center-of-mass linear momentum and relative linear momentum, center-of-mass moment of momentum and relative moment of momentum, as well as center-of-mass kinetic energy and relative kinetic energy. The corresponding relations extend some formulations well-known for bodies in the absence of a supply of mass. The global and local relations of balance of linear momentum are then discussed, where non-classical supplies of linear momentum are additionally introduced. The global relation of balance of linear momentum can be replaced by a center-of-mass oriented form, from which a relation of balance of relative linear momentum follows. The global relations of balance of moment of momentum, center-of-mass moment of momentum and relative moment of momentum are afterwards derived as consequences of mathematical manipulations of the local relation of linear momentum. The same strategy is used for deriving the global relations of balance of kinetic energy, center-of-mass kinetic energy and relative kinetic energy. In all of these derived relations of balance, additional, non-classical supply terms emerge. These, however, are expressed by the non-classical supplies of mass and linear momentum, and which vanish when the latter two are absent. From this, one can conclude that the notion of a non-classical supply as such is necessary in order to ensure consistency of the various relations of balance.

A problem oriented constitutive modeling is needed, in order to properly formulate the non-classical supply terms for mass and linear momentum. In order to provide a rational formulation for the global and local relations of balance of linear momentum, a simple local model for the non-classical supply of linear momentum is presented, which, in a continuum mechanics framework, was suggested by Irschik (2005) for bodies with a growing mass. In this model, mass is locally added to (or removed from) the particles at an own velocity and at a rate equal to the local non-classical supply of mass. For a system consisting of several distinct single mass points, an analogous model was considered by Federhofer (1922). The model assures the invariance of the global and local relations of balance of linear momentum with respect to a Galilean transformation of the inertial frame. Due to the apparent analogy to the relation of balance of linear momentum for the single mass point (28), it can be called a model of the Seeliger-Meshchersky type, cf. Section 1. Using this particular model, the relations of balance of global and center-of-mass moment of momentum and kinetic energy are exemplarily re-formulated.

2.1 Variable mass due to material sources of mass in the interior of a material volume

The instantaneous total mass m of a deformable body of finite extension that occupies the volume V in the current configuration is denoted as

$$m = \int_V \rho \, dV \tag{46}$$

In (46), ρ denotes the current mass density, i.e., the local mass per unit volume in the current configuration, and V is assumed to be a material volume, i.e., a volume that moves together with the particles instantaneously located in it. The case of a flow of mass through the surface of V, i.e., the case of a non-material volume, will be studied in Section 3 below. As the reason for a non-vanishing time-rate of the total mass, for the moment being we assume that distributed sources (sinks) of mass are assigned to the particles in V, where the change of mass is not understood as a change in the number of particles in V, but as a change of the elementary mass of the particles contained in V. We therefore talk about material sources of mass, which result in a non-classical supply of mass, the latter being absent in classical formulations dealing with mechanical systems with a conserved mass. We start our considerations with balance of mass.

2.2 Global and local relations of balance of mass

Assume that the material body under consideration is subjected to a continual, non-classical supply of mass due to material sources of mass. From the general balance relation (2), and in analogy to the relation (4), which has been stated above for a single point mass, we write the global statement of balance of mass as

$$\frac{dm}{dt} = \frac{d}{dt} \int_V \rho \, dV = s\,[m] \tag{47}$$

where $s\,[m]$ is the total supply of mass of the body due to the material sources of mass. We now introduce a local supply of mass $s\,[1]$ by setting

$$s\,[m] = \int_V s\,[1]\, \rho \, dV \tag{48}$$

Interchanging the time derivative and the integral in (47), and requiring that this relation must hold for any sub-volume of the material volume V, the following relation for the time rate of the elementary mass carried by a particle is obtained:

$$\frac{d}{dt}\left(\rho \, dV\right) = s\,[1]\, \rho \, dV \tag{49}$$

An example for the material frame indifferent constitutive modeling of $s\,[1]$ can be found in the chapter written by D. Indeitsev in the present book, where the case of a single constituent of a binary mixture is studied.

The left hand side of (49) can be transformed to

$$\frac{d}{dt}\left(\rho \, dV\right) = \frac{d\rho}{dt} dV + \rho \frac{d}{dt}\left(dV\right) \tag{50}$$

The place of the particles is characterized by their position vectors p with respect to an inertial frame. Particularly, for the mass density we write $\rho = \rho\,(p,t)$. The material time derivative of the mass density in (50) then becomes

$$\frac{d}{dt}\rho\,(p,t) = \frac{\partial \rho}{\partial t} + v \cdot \operatorname{grad} \rho \tag{51}$$

where the absolute velocity of the particles with respect to the inertial frame is denoted as $v = dp/dt$. The divergence and gradient operators with respect to the place of the particles in the current configuration are written as "div" and "grad", respectively. Utilizing the Euler expansion formula for the time rate of the elementary volume carried by a particle,

$$\frac{d}{dt}\,(dV) = (\operatorname{div} v)\; dV \tag{52}$$

together with the following vector identity,

$$\operatorname{div}\,(\rho\, v) = \rho \operatorname{div} v + v \cdot \operatorname{grad} \rho \tag{53}$$

we can put (50) into the form

$$\frac{d}{dt}\,(\rho\, dV) = \left(\frac{\partial \rho}{\partial t} + \operatorname{div}\,(\rho\, v)\right) dV \tag{54}$$

Substituting (54) into (49) yields the local equation of balance of mass in the presence of material sources of mass:

$$\frac{\partial \rho}{\partial t} + \operatorname{div}\,(\rho\, v) = s\,[1]\,\rho \tag{55}$$

Note that the term $s\,[1]\,\rho$ in (55) is denoted as J in D. Indeitsev's Chapter of the present book.

2.3 Some useful relations involving the center of mass

The position vector c of the instantaneous center of mass from the origin of the inertial frame is defined by

$$c\,m = \int_V p\,\rho\, dV \tag{56}$$

where p is the position vector of a particle currently having the elementary mass $\rho\, dV$. Setting

$$p = c + p' \tag{57}$$

with the position vector p' relative to the center of mass, it follows from (56) with (46) that

$$\int_V p' \rho \, dV = 0 \tag{58}$$

Time-wise differentiation of (56) gives, recall that $v = dp/dt$ and $s[m] = dm/dt$, and see (49):

$$\begin{aligned} \frac{d}{dt}(m\,c) &= m\frac{dc}{dt} + s\,[m]\,c = \frac{d}{dt}\left(\int_V p\,\rho\,dV\right) \\ &= \int_V v\,\rho\,dV + \int_V p\frac{d}{dt}(\rho\,dV) = \int_V v\,\rho\,dV + \int_V p\,s[1]\,\rho\,dV \end{aligned} \tag{59}$$

Note that the center of mass defined in (56) is not to be confused with the place of some material particle, which instantaneously might coincide with it. Indeed, the center of mass might not even be situated within the body, think, e.g., of a hollow sphere. Hence, the velocity dc/dt of the center of mass in general will be different from the velocity of a material particle that instantaneously might be located at the place of the center of mass. (One may ask, why dp/dt in (59) has been identified as the absolute velocity v of a material particle. The reason is that the position of the center of mass of the elementary mass $\rho\,dV$ carried by that material particle is given by its position vector p, which, however, does not move relative to the particle. Hence, the time derivatives are consistently applied in the various expressions presented in (59).)

Now, a further characteristic place $\tilde{c}$ is introduced by defining

$$s\,[m]\,\tilde{c} = \int_V p\frac{d}{dt}(\rho\,dV) = \int_V p\,s\,[1]\,\rho\,dV \tag{60}$$

Substituting into (59) yields

$$\int_V v\,\rho\,dV = m\frac{dc}{dt} + s\,[m]\,(c - \tilde{c}) = m\frac{dc}{dt} - \int_V p' s[1]\,\rho\,dV \tag{61}$$

The left hand side of (61) represents the total linear momentum j of the body, being defined as

$$j = \int_V v\,\rho\,dV = m\frac{dc}{dt} + \int_V v'\rho\,dV = m\frac{dc}{dt} + s\,[m]\,(c - \tilde{c}) \tag{62}$$

In (62), use has been made of the relation

$$v = \frac{dp}{dt} = \frac{dc}{dt} + v' \tag{63}$$

The velocity $v' = dp'/dt$ in (63) represents the difference between the velocity of some material particle and the velocity of the center of mass, dc/dt. With (63), the total linear momentum in (62) can be decomposed into

$$j = j^c + j' \tag{64}$$

with the pseudo- or center-of-mass linear momentum

$$j^c = m\frac{dc}{dt} \tag{65}$$

and the relative linear momentum

$$j' = \int_V v' \rho \, dV = s\,[m]\,(c - \tilde{c}) = -\int_V p' s[1]\, \rho \, dV \tag{66}$$

which follows by comparison of (64) with (61). Hence, the relative linear momentum in a body with a non-vanishing supply of mass $s\,[m]$ in general does not vanish. The relation in (66) thus extends a theorem on the center of mass by Thomson and Tait (1867), who treated bodies without a supply of mass. Note from (48), (57), (58) and (60) that $c = \tilde{c}$, when $s\,[1]$ is uniformly distributed over the body, i.e., $j' = 0$ in (66).

The total moment of momentum of the body with respect to the origin of the inertial frame is defined as

$$\alpha^* = \int_V p \times v \, \rho \, dV \tag{67}$$

Substituting (57), (58), (63) and (66), we get

$$\begin{aligned} \alpha^* &= \int_V p \times \frac{dc}{dt} \rho \, dV + \int_V p \times v' \rho \, dV \\ &= c \times m\frac{dc}{dt} + c \times \int_V v' \rho \, dV + \int_V p' \times v' \rho \, dV = \alpha^c + \alpha' + c \times s\,[m]\,(c - \tilde{c}) \end{aligned} \tag{68}$$

The center-of-mass moment of momentum is

$$\alpha^c = c \times j^c = c \times m\frac{dc}{dt} \tag{69}$$

and the relative moment of momentum reads

$$\alpha' = \int_V p' \times v' \rho \, dV \tag{70}$$

Hence, the total moment of momentum of a body with a non-vanishing supply of mass $s\,[m]$ in general is not represented by the sum of the center-of-mass moment of momentum and the relative moment of momentum only. Instead, the last term in (68) needs to be taken into account, in which $c \times s[m]\, c = 0$ holds, of course, but which is kept for formal reasons.

The total kinetic energy of the of the body in the absence of an intrinsic spin of the particles is defined as

$$E_{kin} = \frac{1}{2} \int_V v \cdot v \, \rho \, dV \tag{71}$$

Substituting (57), (63) and (66), we obtain

$$\begin{aligned} E_{kin} &= \frac{1}{2} m \frac{dc}{dt} \cdot \frac{dc}{dt} + \frac{dc}{dt} \cdot \int_V v' \rho \, dV + \frac{1}{2} \int_V v' \cdot v' \rho \, dV \\ &= E^c_{kin} + E'_{kin} + \frac{dc}{dt} \cdot s\,[m]\,(c - \tilde{c}) \end{aligned} \tag{72}$$

with the center-of-mass kinetic energy

$$E^c_{kin} = \frac{1}{2} m \frac{dc}{dt} \cdot \frac{dc}{dt} \tag{73}$$

and the relative kinetic energy

$$E'_{kin} = \frac{1}{2} \int_V v' \cdot v' \rho \, dV \tag{74}$$

Hence, the total kinetic energy of a body with a non-vanishing supply of mass $s\,[m]$ in general is not given by the sum of the center-of-mass kinetic energy and the relative kinetic energy, but the last term in (72) must be taken into account. This represents an extension of a theorem on the center of mass by König (1751), who treated bodies without a supply of mass.

2.4 Global and local balance of linear momentum

We now specialize the general equation of balance (2) for the linear momentum defined in (62). This yields

$$dQ = dj = d\left(\int_V v \, \rho \, dV \right), \quad R = F + s\,[j] \tag{75}$$

where

$$F = \int_V b\, dV + \int_S \sigma_n\, dS \tag{76}$$

The imposed body forces per unit current volume are denoted by b, and the surface tractions are given by the stress vector σ_n. The latter is connected to the stress tensor by Cauchy's fundamental law, which describes the stress vector at the current material surface S of V as the linear mapping

$$\sigma_n = n \cdot \Sigma \tag{77}$$

The unit outer normal vector at S is denoted by n, and the Cauchy stress tensor is written as Σ. The resultant of the body forces and the surface tractions is the vector F. In (75), the total non-classical supply of linear momentum is defined as

$$s\,[j] = \int_V s\,[v]\, \rho\, dV \tag{78}$$

with the local non-classical supply of linear momentum per unit mass denoted by $s\,[v]$. We thus deduce from (75) that the relation of global balance of linear momentum reads

$$\frac{dj}{dt} = \frac{d}{dt}\left(\int_V v\, \rho\, dV\right) = F + s\,[j] = \int_V b\, dV + \int_V (\operatorname{div} \Sigma)\, dV + \int_V s\,[v]\, \rho\, dV \tag{79}$$

Noting from (49) that

$$\begin{aligned} \frac{d}{dt}\left(\int_V v\, \rho\, dV\right) &= \int_V \frac{d}{dt}\,(v\, \rho\, dV) \\ &= \int_V \frac{dv}{dt}\, \rho\, dV + \int_V v \frac{d}{dt}\,(\rho\, dV) = \int_V \left(\frac{dv}{dt} + s\,[1]\, v\right) \rho\, dV \end{aligned} \tag{80}$$

the global statement (79) localizes to

$$\rho \frac{dv}{dt} + s\,[1]\, \rho\, v = b + \operatorname{div} \Sigma + s\,[v]\, \rho \tag{81}$$

This equation represents the local relation of balance of linear momentum in the presence of a supply of mass and a non-classical supply of linear momentum. Note that in the Chapter by D. Indeitsev, the term $s\,[1]\, \rho\, v$ is denoted as $J\, v$, and $s\,[v]\, \rho$ is written as R.

2.5 Balance of center-of-mass linear momentum

As a center-of-mass oriented formulation of balance of linear momentum, using (64) and (66), in (2) we can also write

$$dQ = dj = d\left(j^c + s\,[m]\,(c - \tilde{c})\right) \tag{82}$$

With (65), the relation of balance of linear momentum (79) can be replaced by the following center-of-mass oriented form:

$$\frac{dj^c}{dt} = m\,\frac{d^2c}{dt^2} + s\,[m]\,\frac{dc}{dt} = F + s\,[j^c] \tag{83}$$

with the non-classical supply of center-of-mass linear momentum

$$s\,[j^c] = -\frac{d}{dt}\left(s\,[m]\,(c - \tilde{c})\right) + s\,[j] = -s\,[m]\left(\frac{dc}{dt} - \frac{d\tilde{c}}{dt}\right) - \frac{d^2m}{dt^2}\,(c - \tilde{c}) + s\,[j] \tag{84}$$

The relation in (83) provides a formulation of balance of center-of-mass linear momentum. Subtracting (83) from (79) with (64), it is found that

$$\frac{dj'}{dt} = \frac{dj}{dt} - \frac{dj^c}{dt} = s\,[m]\left(\frac{dc}{dt} - \frac{d\tilde{c}}{dt}\right) + \frac{d^2m}{dt^2}\,(c - \tilde{c}) \tag{85}$$

2.6 Balance of relative linear momentum

Using (64), a relation of balance of relative moment of momentum, see (66), is directly obtained from (85):

$$\frac{dj'}{dt} = s\,[j'] \tag{86}$$

with the non-classical supply term

$$s\,[j'] = s\,[j] - s\,[j^c] = s\,[m]\left(\frac{dc}{dt} - \frac{d\tilde{c}}{dt}\right) + \frac{d^2m}{dt^2}\,(c - \tilde{c}) \tag{87}$$

Note that the time rate of the relative linear momentum in general does not vanish if a supply of mass is present. Only if m is conserved, or in the exceptional case of $c = \tilde{c}$, $dj'/dt = 0$ holds.

2.7 Balance of moment of momentum

Performing the vector product of the local form of balance of linear momentum (81) with the position vector p and integrating over the volume

V gives

$$\int_V p \times \frac{d}{dt}(v\,\rho\,dV) = \frac{d}{dt}\int_V p \times \rho\,v\,dV = \int_V p \times (b + \operatorname{div}\Sigma + s\,[v]\,\rho)\,dV \tag{88}$$

Substituting (67) and (80), and using an extended divergence theorem, this can be re-written as:

$$\begin{aligned}\frac{d\alpha^*}{dt} &= \int_V p \times \left(\rho\frac{dv}{dt} + s\,[1]\,\rho\,v\right) dV \\ &= \int_V p \times (b - \Sigma_\times)\,dV + \int_S p \times \sigma_n dS + s\,[\alpha^*]\end{aligned} \tag{89}$$

The so-called Gibbsian cross-vector of the Cauchy stress tensor is denoted as $\Sigma_\times$. It is twice the axial vector of the skew-symmetric part of the Cauchy stress tensor Σ, and thus vanishes if Σ is symmetric. With (57), the non-classical supply of moment of momentum in (89) becomes

$$s\,[\alpha^*] = \int_V p \times s\,[v]\,\rho\,dV = c \times s\,[j] + \int_V p' \times s\,[v]\,\rho\,dV \tag{90}$$

Note that (89) and (90) represent pure consequences of a mathematical manipulation of the local relation of balance of linear momentum (81). Now, in the absence of an intrinsic spin and of applied body and surface couples, the fundamental relation of balance of angular momentum in principle does coincide with balance of moment of momentum (89), compare the analogous relation (32) for the angular momentum of a single mass point. The only exceptions are that the term with $\Sigma_\times$ is not present in the relation of balance of angular momentum, and that the non-classical supply of moment of momentum $s\,[\alpha^*]$ is to be replaced by a possibly different non-classical supply of angular momentum. Hence, if the non-classical supplies of moment of momentum and of angular momentum can be assumed to be equal, it follows that the volume integral over $\Sigma_\times$ in (89) vanishes. The localization argument then leads to the conclusion that the stress tensor must be symmetric, such that $\Sigma_\times = 0$ in (89). This can be considered as an extension of an axiom for non-polar bodies without a supply of mass, see Ziegler (1998)

2.8 Balance of center-of-mass moment of momentum

Performing the vector product of the center-of-mass form of balance of linear momentum (83) with c gives

$$c \times \frac{dj^c}{dt} = \frac{d}{dt}\left(c \times m\frac{dc}{dt}\right) = c \times F + c \times s\left[j^c\right] \tag{91}$$

Substituting (69), this can be re-written into the following relation of balance for the center-of-mass moment of momentum:

$$\frac{d\alpha^c}{dt} = c \times m\frac{d^2c}{dt^2} + c \times s\left[m\right]\frac{dc}{dt} = c \times F + s\left[\alpha^c\right] \tag{92}$$

with the non-classical supply, see (84),

$$s\left[\alpha^c\right] = c \times s\left[j^c\right] \tag{93}$$

Subtracting (92) from (89), cf. (84) and (85), gives

$$\begin{aligned}\frac{d\alpha^*}{dt} - \frac{d\alpha^c}{dt} = &\int_V p' \times b\,dV + \int_S p' \times \sigma_n dS - \int_V p \times \Sigma_\times dV \\ &+ c \times \frac{d}{dt}\left(s\left[m\right]\left(c - \tilde{c}\right)\right) + \int_V p' \times s\left[v\right]\rho\,dV\end{aligned} \tag{94}$$

2.9 Balance of relative moment of momentum

Utilizing (68), a relation of balance of relative moment of momentum, see (70), is obtained from (94):

$$\frac{d\alpha'}{dt} = \int_V p' \times b\,dV + \int_S p' \times \sigma_n dS - \int_V p \times \Sigma_\times dV + s\left[\alpha'\right] \tag{95}$$

with the non-classical supply term

$$s\left[\alpha'\right] = -\frac{dc}{dt} \times s\left[m\right]\left(c - \tilde{c}\right) + \int_V p' \times s\left[v\right]\rho\,dV \tag{96}$$

2.10 Balance of kinetic energy

Performing the scalar product of the local relation of balance of linear momentum (81) with the absolute velocity vector v, and integrating over the material volume V yields

$$\int_V v \cdot \frac{dv}{dt} \rho \, dV + \int_V s[1] \, v \cdot v \, \rho \, dV$$
$$= \frac{d}{dt}\left(\int_V \frac{1}{2} v \cdot v \, \rho \, dV\right) + \int_V \frac{1}{2} s[1] \, v \cdot v \, \rho \, dV$$
$$= \int_V (b \cdot v + v \cdot \operatorname{div} \Sigma + \rho \, s[v] \cdot v) dV \quad (97)$$

Using (71) and an extended divergence theorem, we obtain the relation of balance of kinetic energy as

$$\frac{dE_{kin}}{dt} = \int_V v \cdot b \, dV + \int_S v \cdot \sigma_n dS - \int_V \operatorname{tr}(\Sigma \cdot \operatorname{grad} v) \, dV + s[E_{kin}] \quad (98)$$

The trace of a second order tensor is denoted by "tr". Using (63), the relation in (98) can be re-written to

$$\frac{dE_{kin}}{dt} = F \cdot \frac{dc}{dt} + \int_V v' \cdot b \, dV + \int_S v' \cdot \sigma_n dS - \int_V \operatorname{tr}(\Sigma \cdot \operatorname{grad} v') \, dV + s[E_{kin}] \quad (99)$$

The non-classical supply of kinetic energy in (98) and (99) turns out to be

$$s[E_{kin}] = -\frac{1}{2} \int_V v \cdot v \, s[1] \, \rho \, dV + \int_V v \cdot s[v] \, \rho \, dV \quad (100)$$

Again with (63), this becomes

$$s[E_{kin}] = -\frac{1}{2} s[m] \frac{dc}{dt} \cdot \frac{dc}{dt} - \frac{dc}{dt} \cdot \int_V v' s[1] \, \rho \, dV$$
$$- \int_V \frac{1}{2} v' \cdot v' s[1] \, \rho \, dV + \frac{dc}{dt} \cdot s[j] + \int_V v' \cdot s[v] \, \rho \, dV \quad (101)$$

2.11 Balance of center-of-mass kinetic energy

The scalar product of the center-of-mass form of balance of linear momentum (83) with the absolute velocity of the center of mass, dc/dt, gives

$$\frac{dc}{dt} \cdot m \frac{d^2c}{dt^2} + s\,[m]\, \frac{dc}{dt} \cdot \frac{dc}{dt} = \frac{d}{dt} \left(\frac{1}{2} m \frac{dc}{dt} \cdot \frac{dc}{dt} \right) + \frac{1}{2}\, s\,[m]\, \frac{dc}{dt} \cdot \frac{dc}{dt}$$
$$= F \cdot \frac{dc}{dt} + s\,[j^c] \cdot \frac{dc}{dt} \quad (102)$$

Using (73), we obtain the following relation of balance for the center-of-mass kinetic energy:

$$\frac{dE_{kin}^c}{dt} = F \cdot \frac{dc}{dt} + s\,[E_{kin}^c] \quad (103)$$

with the non-classical supply of the center-of-mass kinetic energy

$$s\,[E_{kin}^c] = -\frac{1}{2}\, s\,[m]\, \frac{dc}{dt} \cdot \frac{dc}{dt} + s\,[j^c] \cdot \frac{dc}{dt} \quad (104)$$

Substituting (84), this can be re-written to

$$s\,[E_{kin}^c] = -s\,[m] \left(\frac{3}{2} \frac{dc}{dt} - \frac{d\tilde{c}}{dt} \right) \cdot \frac{dc}{dt} - \frac{d^2m}{dt^2} (c - \tilde{c}) \cdot \frac{dc}{dt} + s\,[j] \cdot \frac{dc}{dt} \quad (105)$$

Subtracting (103) from (99), cf. (101) and (105), it is found that

$$\frac{dE_{kin}}{dt} - \frac{dE_{kin}^c}{dt} = \int_V v' \cdot b\, dV + \int_S v' \cdot \sigma_n dS$$
$$- \int_V \mathrm{tr}\,(\Sigma \cdot \mathrm{grad}\, v')\, dV + s\,[m] \left(\frac{dc}{dt} - \frac{d\tilde{c}}{dt} \right) \cdot \frac{dc}{dt}$$
$$+ \frac{d^2m}{dt^2} (c - \tilde{c}) \cdot \frac{dc}{dt} - \frac{dc}{dt} \cdot \int_V v' s\,[1]\, \rho\, dV$$
$$- \int_V \frac{1}{2} v' \cdot v' s\,[1]\, \rho\, dV + \int_V v' \cdot s\,[v]\, \rho\, dV \quad (106)$$

2.12 Balance of relative kinetic energy

Using (72), a relation of balance of relative kinetic energy, see (74), is obtained from (106):

$$\frac{dE'_{kin}}{dt} = \int_V v' \cdot b\, dV + \int_S v' \cdot \sigma_n dS - \int_V \mathrm{tr}\,(\Sigma \cdot \mathrm{grad}\, v')\, dV + s\,[E'_{kin}] \quad (107)$$

with the non-classical supply of relative kinetic energy

$$s\left[E'_{kin}\right] = -s\left[m\right]\frac{d^2c}{dt^2}\cdot(c-\tilde{c}) - \frac{dc}{dt}\cdot\int\limits_V v' s\left[1\right]\rho\, dV \\ - \int\limits_V \frac{1}{2} v'\cdot v' s\left[1\right]\rho\, dV + \int\limits_V v'\cdot s\left[v\right]\rho\, dV \quad (108)$$

2.13 Material volume with distributed material sources of mass of the Seeliger-Meshchersky type

Motivated by the procedure for obtaining a rational expression for the additional, non-classical supply of linear momentum presented in Subsection 1.3 above, see (17), we set

$$s\left[v\right] = s\left[1\right]u \quad (109)$$

where the velocity u, at which mass is locally gained or lost by the particles, in general will be different from the particle velocity v. The local relation of balance (81) then can be written as

$$\rho\frac{dv}{dt} = b + \operatorname{div}\Sigma + s\left[1\right]\rho\,(u-v) \quad (110)$$

The relation of balance of global linear momentum (79) follows to

$$\int\limits_V \frac{dv}{dt}\,\rho dV = \int\limits_V b\, dV + \int\limits_S \sigma_n\, dS + \int\limits_V s\left[1\right](u-v)\,\rho\, dV \quad (111)$$

The continuum mechanics based model for the additional supply of linear momentum in (109) was suggested and embedded into the literature by Irschik (2005) in the framework of growing materials. Due to the apparent analogy to the relation of balance of linear momentum for the single mass point, (27), we denote it as a model of the Seeliger-Meshchersky type. Note that this model meets the requirements of the so-called Galileian invariance of the relations of balance of linear momentum in (110) and (111). In other words, seen from an observer, who moves with a constant velocity relative to the global inertial frame, the relations given in (110) and (111) do remain invariant. The same remains true, if a translational rigid-body motion with constant velocity is superimposed upon the actual motion. Moreover, would we add to the model in (109) some vector, which does not depend on the velocity but acts like a body force, this term could be treated as an additional body force and would not interfere with the Galileian invariance.

Introducing (109) into (90), the relation of balance of moment of momentum (89) becomes

$$\int_V p \times \rho \frac{dv}{dt} \, dV = \int_V p \times (b - \Sigma_\times) \, dV + \int_S p \times \sigma_n dS + \int_V p \times s\,[1]\,(u - v) dV \quad (112)$$

In order to write down further global statements, we now introduce another characteristic velocity w as

$$s\,[m]\,w = s\,[j] = \int_V s\,[v]\,\rho\,dV = \int_V s\,[1]\,u\,\rho\,dV \quad (113)$$

With the definition stated in (113), the various relations of balance, which have been presented above, reduce to the following useful forms:

Balance of total linear momentum, see (79):

$$\frac{dj}{dt} = F + s\,[m]\,w \quad (114)$$

This coincides with the Seeliger-Meshchersky relation (25) for the single point mass.

Balance of center-of-mass linear momentum, see (83) and (84):

$$\frac{dj^c}{dt} = F + s\,[m] \left(w - \frac{dc}{dt} + \frac{d\tilde{c}}{dt} \right) + \frac{d^2 m}{dt^2} \left(\tilde{c} - c \right) \quad (115)$$

With (65), this becomes

$$m \frac{d^2 c}{dt^2} = F + s\,[m] \left(w - 2\frac{dc}{dt} + \frac{d\tilde{c}}{dt} \right) - \frac{d^2 m}{dt^2} \left(\tilde{c} - c \right) \quad (116)$$

Note that this relation is Galilei-invariant. For $c = \tilde{c}$, (116) reduces to (27) for the single mass point. For a mechanical system consisting of a set of discrete mass points, (116) is due to Federhofer (1922).

Balance of moment of momentum, see (89) and (90):

$$\frac{d\alpha^*}{dt} = c \times F + \int_V p' \times (b - \Sigma_\times)\, dV + \int_S p' \times \sigma_n dS + c \times s\,[m]\, w + \int_V p' \times s\,[1]\, u\, \rho\, dV \quad (117)$$

If the material volume V shrinks to a single mass point, such that the relative position vectors p' shrink to zero, this Galilei-invariant form reduces to (32).

Balance of center-of-mass moment of momentum, see (92), (93) and (84):

$$\frac{d\alpha^c}{dt} = c \times F + c \times s\,[m] \left(w - \frac{dc}{dt} + \frac{d\tilde{c}}{dt} \right) + c \times \frac{d^2 m}{dt^2} (\tilde{c} - c) \quad (118)$$

or, substituting (69),

$$c \times m \frac{d^2 c}{dt^2} = c \times F + c \times s\,[m] \left(w - 2\frac{dc}{dt} + \frac{d\tilde{c}}{dt} \right) + c \times \frac{d^2 m}{dt^2} (\tilde{c} - c) \quad (119)$$

When $c = \tilde{c}$, this reduces to (32).

Balance of kinetic energy, see (99)–(101):

$$\frac{dE_{kin}}{dt} = F \cdot \frac{dc}{dt} + \int_V v' \cdot b\, dV + \int_S v' \cdot \sigma_n dS - \int_V \mathrm{tr}\,(\Sigma \cdot \mathrm{grad}\, v')\, dV + s\,[m] \left(w - \frac{1}{2}\frac{dc}{dt} \right) \cdot \frac{dc}{dt} - \frac{dc}{dt} \cdot \int_V v' s\,[1]\, \rho\, dV + \int_V v' \cdot s\,[1] \left(u - \frac{1}{2} v' \right) \rho\, dV \quad (120)$$

When V shrinks towards a single mass point, this reduces to the balance of translational kinetic energy (38).

Balance of center-of-mass kinetic energy, see (103)–(105):

$$\frac{dE^c_{kin}}{dt} = F \cdot \frac{dc}{dt} + s\,[m] \left(w - \frac{3}{2}\frac{dc}{dt} + \frac{d\tilde{c}}{dt} \right) \cdot \frac{dc}{dt} - \frac{d^2 m}{dt^2} (c - \tilde{c}) \cdot \frac{dc}{dt} \quad (121)$$

When $c = \tilde{c}$, this reduces to (38).

3 Global relations of balance written for a non-material control volume

3.1 The Reynolds transport theorem

We now release the assumption of a material volume V that moves together with the material particles contained in it. Instead, we consider a so-called open system with a non-material control volume V, the surface S of which moves at a velocity u that is different from the velocity v of the material particles instantaneously located on that non-material control surface S. In this situation, a flow of mass through S will be present implying that the total mass contained in V in general will not be conserved. As is shown below, the above methodology for taking into account material sources of mass inside a material volume can be adopted in the present case by analogy. Moreover, the presence of material sources of mass as well as of non-classical supply terms for other entities can be additionally taken into account, when there is a flow of mass through the non-material control surface S. In demonstrating this, we use the transport theorem by Reynolds (1903), cf. Truesdell and Toupin (1960). As has been discussed in some detail by Irschik and Holl (2004), the transport theorem can be put into a form that involves both, the rate of the total of some entity contained in a non-material control volume, as well as the rate of this entity contained in the material volume that instantaneously coincides with the non-material control volume.

In a generalized form, this version of the Reynolds transport theorem can be stated as follows. Consider a scalar or vector quantity Ψ, which is the total of a local entity $\psi\,\rho$ carried by the particles in the volume V under consideration:

$$\Psi = \int_V \psi\,\rho\,dV \tag{122}$$

Then the transport theorem can be written as

$$\frac{d_u}{dt}\Psi = \frac{d\Psi}{dt} + s_u\left[\Psi\right] \tag{123}$$

with the non-classical supply

$$s_u\left[\Psi\right] = \int_S n \cdot (u - v)\psi\,\rho\,dS \tag{124}$$

The operator d_u/dt in (123) indicates that the time-rate refers to the motion of the non-material control volume V, while d/dt means the time rate considering the motion of the material volume that instantaneously coincides with the latter. The surface integral in (124) vanishes, when V is a material volume, since then $u = v$. Various useful versions of (123) and (124) are listed below, where the rates with respect to the motion of the material volume, $d\psi/dt$ in (123), can be substituted directly from the equations of balance presented above in Subsections 2.2–2.12. Note that additional considerations are needed in the presence of a singular surface, on which $\psi\,\rho$ takes on different values at the two sides of this surface. This case will be discussed in Section 4 below.

3.2 Balance of mass

For balance of mass, there is $\psi = 1$, $\Psi = m$, see (46). With (47), we obtain from (123) and (124) that

$$\frac{d_u}{dt} m = s\,[m] + s_u\,[m] \tag{125}$$

with the additional, non-classical supply of mass

$$s_u\,[m] = \int\limits_S n \cdot (u - v)\rho\, dS \tag{126}$$

3.3 Balance of linear momentum

With the definition of total linear momentum j stated in (62), we set $\psi = v$, $\Psi = j$ in (123) and (124). Substituting (79), this yields

$$\frac{d_u}{dt} j = \frac{d}{dt} j + s_u\,[j] = F + s\,[j] + s_u\,[j] \tag{127}$$

with the additional non-classical supply of linear momentum

$$s_u\,[j] = \int\limits_S n \cdot (u - v)\rho\, v\, dS \tag{128}$$

The apparent analogy between the global equations of balance of mass and linear momentum for a material volume with material sources of mass and for a non-material control volume with a flow of mass trough its surface along with various applications, particularly in fluid mechanics and rocketry, has been discussed in the review by Irschik and Holl (2004). Note that the cases of material sources of mass in the interior and of a flow of mass through the surface have been treated separately in Irschik and Holl (2004). In the present case of taking into account both, material sources in the interior and a flow of mass through the surface, in the equations of balance presented above in Subsections 2.2–2.12, it is only necessary to replace the time rate $d\Psi/dt$ by $d_u\Psi/dt$, in order to indicate that there is a flow of mass through the surface S, and to replace $s\,[\Psi]$ by the sum $s\,[\Psi] + s_u\,[\Psi]$. This is subsequently shown for various relations of balance. A further remark seems to be in order: It is evident from (127) that a relation of balance for a non-material control volume with a flow of mass through the surface can be brought into the form valid for the material volume that instantaneously coincides with the former non-material volume. This follows from the canceling of the non-classical supply $s_u[\Psi]$ in (127). However, in many practical applications it is necessary to compute the time rate $d_u\Psi/dt$, following the motion of the volume with a flow of mass through the surface, instead of $d\Psi/dt$.

3.4 Balance of moment of momentum

We set $\psi = p \times v$ and $\Psi = \alpha^*$ in (123) and (124). Using (89), this gives

$$\begin{aligned}\frac{d_u}{dt}\alpha^* &= \frac{d}{dt}\alpha^* + s_u[\alpha^*] \\ &= \int\limits_V p \times (b - \Sigma_\times)\, dV + \int\limits_S p \times \sigma_n dS + s\,[\alpha^*] + s_u\,[\alpha^*] \end{aligned} \quad (129)$$

with the additional non-classical supply of moment of momentum

$$s_u\,[\alpha^*] = \int\limits_S n \cdot (u - v)\, p \times \rho\, v\, dS \quad (130)$$

3.5 Balance of kinetic energy

Setting $\psi = v{\cdot}v/2$ and $\Psi = E_{kin}$ in (123) and (124) and substituting (98) gives:

$$\begin{aligned}&\frac{d_u}{dt}E_{kin} = \frac{d}{dt}E_{kin} + s_u\,[E_{kin}] \\ &= \int\limits_V v \cdot b\, dV + \int\limits_S v \cdot \sigma_n dS - \int\limits_V \mathrm{tr}\,(\Sigma \cdot \mathrm{grad}\, v)\, dV + s\,[E_{kin}] + s_u\,[E_{kin}]\,, \end{aligned} \quad (131)$$

with the non-classical additional supply of kinetic energy

$$s_u\,[E_{kin}] = \int\limits_S n \cdot (u - v)\, \frac{1}{2}\rho\, v \cdot v\, dS \quad (132)$$

Similar extensions can be written down for the center-of-mass oriented relations of balance stated above. Particularly, balance of center-of-mass linear momentum has been treated in some detail in Irschik and Holl (2004), but separately from the case of sources of mass in the interior. In the following, we give extended and additional relations.

3.6 Some useful relations involving the center of mass

Subsequently, we repeatedly utilize balance of mass in the form (123)–(124). The definition of the position vector c of the instantaneous center of mass from the origin of the inertial frame stated above in (56), the split of the position vector in (57) and its consequence (58) remain unchanged.

Time-wise differentiation of (56) in the presence of a flow of mass through the surface gives

$$\begin{aligned}\frac{d_u}{dt}(m\,c) &= \frac{d_u}{dt}\int\limits_V c\,\rho\,dV = m\,\frac{dc}{dt} + (s\,[m] + s_u\,[m])\,c \\ &= \frac{d_u}{dt}\left(\int\limits_V p\,\rho\,dV\right) = \frac{d}{dt}\left(\int\limits_V p\,\rho\,dV\right) + \int\limits_S n\cdot(u-v)\,\rho\,p\,dS \\ &= \int\limits_V v\,\rho\,dV + \int\limits_V p\,s\,[1]\,\rho\,dV + \int\limits_S n\cdot(u-v)\,\rho\,p\,dS \qquad (133)\end{aligned}$$

The transport theorem (125) and (126) has been used for expressing the time-rate of the volume integral in (133). Analogous to (60), a further characteristic place $\tilde{c}$ is introduced by defining

$$\begin{aligned}(s\,[m] + s_u\,[m])\,\tilde{c} &= \int\limits_V p\,s\,[1]\,\rho\,dV + \int\limits_S n\cdot(u-v)\,\rho\,p\,dS \\ &= (s\,[m] + s_u\,[m])\;c + \int\limits_V p'\,s\,[1]\,\rho\,dV + \int\limits_S n\cdot(u-v)\,\rho\,p'dS \qquad (134)\end{aligned}$$

see (49). Substituting into (133) yields

$$\begin{aligned}\int\limits_V v\,\rho\,dV &= m\frac{dc}{dt} + (s\,[m] + s_u\,[m])\,(c - \tilde{c}) \\ &= m\frac{dc}{dt} - \int\limits_V p'\,s\,[1]\,\rho\,dV - \int\limits_S n\cdot(u-v)\,\rho\,p'dS \qquad (135)\end{aligned}$$

Recalling that the split of the absolute velocity in (63), the resulting representation of the linear momentum in (62), as well as the definition of the center-of-mass linear momentum in (65) and the split of the linear momentum in (64) remain unchanged, we now have, in extension of (62):

$$j = \int_V v\,\rho\,dV = m\frac{dc}{dt} + \int_V v'\rho\,dV$$
$$= m\frac{dc}{dt} + (s[m] + s_u[m])\,(c - \tilde{c}) = j^c + j' \quad (136)$$

The relative linear momentum becomes

$$j' = \int_V v'\rho\,dV = (s[m] + s_u[m])\,(c - \tilde{c})$$
$$= -\int_V p's[1]\,\rho\,dV - \int_S n\cdot(u - v)\,\rho\,p'dS \quad (137)$$

Analogously, we get for the total moment of momentum of the body with respect to the origin of the inertial frame, see (67):

$$\alpha^* = \alpha^c + \alpha' + c \times (s[m] + s_u[m])\,(c - \tilde{c}) \quad (138)$$

The definitions of the center-of-mass and the relative moment of momentum in (69) and (70) remain unchanged.

The total kinetic energy of the body in the absence of an intrinsic spin of the particles, see (71), becomes

$$E_{kin} = E^c_{kin} + E'_{kin} + \frac{dc}{dt}\cdot(s[m] + s_u[m])\,(c - \tilde{c}) \quad (139)$$

with the center-of-mass and the relative kinetic energy as given in (73) and (74).

3.7 Balance of center-of-mass linear momentum

Using the definition of the center-of-mass linear momentum in (65) and substituting (136), the relation of balance of linear momentum (127) can be reformulated to the following form:

$$\frac{d_u}{dt}j = \frac{d_u}{dt}\left(j^c + (s[m] + s_u[m])\,(c - \tilde{c})\right) = F + s[j] + s_u[j] \quad (140)$$

Performing the operation d_u/dt, we obtain the following center-of-mass oriented formulation of balance of linear momentum:

$$\frac{d_u}{dt}j^c = m\frac{d^2c}{dt^2} + (s[m] + s_u[m])\frac{dc}{dt} = F + s[j^c] + s_u[j^c] \tag{141}$$

with the non-classical supply of center-of-mass linear momentum

$$s_u[j^c] = -s_u[m]\left(\frac{dc}{dt} - \frac{d\tilde{c}}{dt}\right) - \left(\frac{d}{dt}s_u[m]\right)(c - \tilde{c}) + s_u[j] \tag{142}$$

For the case without a flow of mass through the surface, see (84). Having derived (141)–(142) in some detail, we subsequently list additional results by direct analogy to the equations of balance, which have been presented above in Subsections 2.2–2.12 for the case without a flow of mass through the surface.

3.8 Balance of relative moment of momentum

$$\frac{d_u}{dt}j' = s[j'] + s_u[j'] \tag{143}$$

$$s_u[j'] = s_u[j] - s_u[j^c] = s_u[m]\left(\frac{dc}{dt} - \frac{d\tilde{c}}{dt}\right) + \left(\frac{d}{dt}s_u[m]\right)(c - \tilde{c}) \tag{144}$$

with $s_u[j^c]$ of (142).

3.9 Balance of center-of-mass moment of momentum

$$\frac{d_u}{dt}\alpha^c = c \times m\frac{d^2c}{dt^2} + c \times (s[m] + s_u[m])\frac{dc}{dt} = c \times F + s[\alpha^c] + s_u[\alpha^c] \tag{145}$$

$$s_u[\alpha^c] = c \times s_u[j^c] \tag{146}$$

with $s_u[j^c]$ of (142).

3.10 Balance of relative moment of momentum

$$\frac{d_u}{dt}\alpha' = \int_V p' \times b\, dV + \int_S p' \times \sigma_n dS - \int_V p \times \Sigma_\times dV + s[\alpha'] + s_u[\alpha'] \tag{147}$$

$$s_u[\alpha'] = -\frac{dc}{dt} \times (s_u[m](c - \tilde{c})) + \int_S n \cdot (u - v)\, p' \times \rho v\, dS \tag{148}$$

3.11 Balance of center-of-mass kinetic energy

$$\frac{d_u}{dt} E^c_{kin} = F \cdot \frac{dc}{dt} + s\left[E^c_{kin}\right] + s_u\left[E^c_{kin}\right] \tag{149}$$

$$s_u\left[E^c_{kin}\right] = -s_u\left[m\right]\left(\frac{3}{2}\frac{dc}{dt} - \frac{d\tilde{c}}{dt}\right) \cdot \frac{dc}{dt} - \left(\frac{d}{dt} s_u\left[m\right]\right)\left(c - \tilde{c}\right) \cdot \frac{dc}{dt} + s_u\left[j\right] \cdot \frac{dc}{dt} \tag{150}$$

3.12 Balance of relative kinetic energy

$$\begin{aligned} \frac{d_u}{dt} E'_{kin} = \int\limits_V v' \cdot b\, dV + \int\limits_S v' \cdot \sigma_n dS \\ - \int\limits_V \operatorname{tr}\left(\Sigma \cdot \operatorname{grad} v'\right) dV + s\left[E'_{kin}\right] + s_u\left[E'_{kin}\right] \end{aligned} \tag{151}$$

$$\begin{aligned} s_u\left[E'_{kin}\right] = -s_u\left[m\right] \frac{d^2 c}{dt^2} \cdot \left(c - \tilde{c}\right) \\ - \int\limits_S n \cdot \left(u - v\right) \rho v' dS - \int\limits_S n \cdot \left(u - v\right) \rho \left(\frac{1}{2} v' \cdot v'\right) dS \end{aligned} \tag{152}$$

3.13 Application to rigid bodies

The above formulations do hold irrespective of the specific deformation behavior of the bodies under consideration. As a theoretically as well as practically important special case, which also allows an exemplary comparison of our formulations with results from the literature, we consider a rigid body in what follows. An interesting case is represented, e.g., by the problem of a rigid body that experiences a surface growth, see Ong and O'Reilly (2004). In this case, the surface of the body moves at a velocity u that in general is different from the velocity v of the particles instantaneously located on it. Thus, our above methodology is applicable. For a rigid body, the velocity of a material particle is given by the Euler velocity formula, see Ziegler (1998):

$$v = v_c + \omega \times p' \tag{153}$$

Here, v_c denotes the velocity vector of the material particle, which *"instantaneously coincides with the position vector c of the center of mass"*. Of course, it is a delicate matter to talk about such a particle, since the center of mass may not even be situated inside the body, see the remarks given above. However, for a rigid body it is quite straightforward to consider a fictitious rigid extension of the body towards the instantaneous location of the center of mass. In general, the velocity v_c is different from the velocity of the center of mass, dc/dt. The angular velocity of the rigid body in (153) is denoted as ω.

We first present some useful formulas involving the velocity of the material particle that instantaneously coincides with the center of mass c. Substituting (153) into (62), we obtain for the linear momentum of the rigid body

$$j = \int_V v\,\rho\,dV = m\,v_c + \omega \times \int_V p'\rho\,dV = m\,v_c \tag{154}$$

which coincides with (4.2) of Ong and O'Reilly (2004).

Moreover, substituting (153) into (67), the moment of momentum for the rigid body becomes

$$\alpha^* = \int_V p \times v\,\rho\,dV = c \times m\,v_c + \int_V p' \times (\omega \times p')\,\rho\,dV = c \times j + J' \cdot \omega \tag{155}$$

with

$$J' = \int_V ((p' \cdot p')\,I - p' \otimes p')\,\rho\,dV \tag{156}$$

The tensor of inertia relative to the center of mass is denoted as J', see Ziegler (1998), compare (10) for the single mass point. Note that (155) coincides with (4.4) and (4.6) of Ong and O'Reilly (2004).

For kinetic energy, substituting (153) into (71) yields

$$\begin{aligned} E_{kin} = \frac{1}{2}\int_V v \cdot v\,\rho\,dV =& \frac{1}{2} m\,v_c \cdot v_c + v_c \cdot \omega \times \int_V p'\rho\,dV \\ &+ \frac{1}{2}\int_V (\omega \times p') \cdot (\omega \times p')\,\rho\,dV = \frac{1}{2} m\,v_c \cdot v_c + \frac{1}{2}\,(J' \cdot \omega) \cdot \omega \end{aligned} \tag{157}$$

see (4.8) of Ong and O'Reilly (2004).

It is to be noted that the relations (154)–(157) are well-known for rigid bodies without sources of mass in the interior and without a flow of mass trough the surface, cf. Ziegler (1998). In order to establish a connection to the more general relations stated above, we equate (153) and (63), from which it is found that

$$v' = v_c + \omega \times p' - \frac{dc}{dt} \tag{158}$$

Substituting into the relation for the relative linear momentum (137) yields

$$j' = \int\limits_V v' \rho \, dV = m \left(v_c - \frac{dc}{dt} \right) = (s\,[m] + s_u\,[m])\,(c - \tilde{c}) \tag{159}$$

from which it follows that

$$m\, v_c = m \frac{dc}{dt} + (s\,[m] + s_u\,[m])\,(c - \tilde{c}) \tag{160}$$

This is also obtained from a direct comparison of (136) and (154).

Having stated the useful expressions (154)–(157) for linear momentum, moment of momentum and kinetic energy, which are based on the entities v_c, J' and ω of the rigid body, we proceed with formulating the relations of balance taking into account the assumption of rigidity.

Balance of linear momentum With (154), the time rate of linear momentum in the presence of sources of mass in the interior and a flow of mass through the surface can be formulated as

$$\begin{aligned} \frac{d_u}{dt} j &= \frac{d_u}{dt} (m\, v_c) = \frac{d_u}{dt} \int\limits_V v_c \, \rho \, dV \\ &= \frac{d}{dt} \int\limits_V v_c \, \rho \, dV + \int\limits_S n \cdot (u - v)\ \rho \, v_c \, dV = m \frac{dv_c}{dt} + \ (s\,[m] + s_u\,[m])\, v_c \end{aligned} \tag{161}$$

where the transport theorem (123), (124) has been utilized. Equating (161) and (127), we obtain

$$m \frac{dv_c}{dt} + (s\,[m] + s_u\,[m])\, v_c = F + s\,[j] + s_u\,[j] \tag{162}$$

Substituting the Euler velocity formula (153), the non-classical supplies of momentum $s_u\,[j]$, see (128), for the rigid body becomes:

$$\begin{aligned} s_u\,[j] &= v_c \int\limits_S n \cdot (u - v)\, \rho \, dS + \omega \times \int\limits_S n \cdot (u - v)\, \rho \, p' dS \\ &= v_c \, s_u\,[m] + \omega \times \int\limits_S n \cdot (u - v)\, \rho \, p' dS \end{aligned} \tag{163}$$

Relation (162) thus reduces to

$$m\frac{dv_c}{dt} + s\,[m]\,v_c = F + s\,[j] + \omega \times \int\limits_S n \cdot (u - v)\,\rho\, p' dS \tag{164}$$

This coincides with (7.3) and (7.4) of Ong and O'Reilly (2004) for the case of $s\,[m] = 0$ and $s\,[j] = 0$. Now, since v_c is the velocity of the material particle that instantaneously coincides with the center of mass, and since the latter in general is moving with respect to the rigid body, the time derivative dv_c/dt in general can not be expected to coincide with the acceleration a_c of the material point that instantaneously is located at the center of mass. A further clarification of the mechanical meaning of the derivative dv_c/dt therefore is deemed desirable. For that sake, we use the Euler formula for the acceleration a of a material particle of a rigid body, see Ziegler (1998):

$$a = \frac{dv}{dt} = a_c + \beta \times p' + \omega \times (\omega \times p') \tag{165}$$

The angular acceleration of the body is denoted as β, and a_c is the absolute acceleration of the material particle that instantaneously is situated at the place of the center of mass. Utilizing (80), the relation of balance of linear momentum (127) becomes:

$$\frac{d_u}{dt} j = \frac{d}{dt} j + s_u\,[j] = \int\limits_V \frac{dv}{dt}\,\rho\, dV + \int\limits_V v\, s\,[1]\,\rho\, dV + s_u\,[j] = F + s\,[j] + s_u\,[j] \tag{166}$$

Cancelling out the non-classical supply $s_u\,[j]$, substituting (166) and (153) and noting (58) yields

$$m\,a_c + s\,[m]\,v_c = F + s\,[j] - \omega \times \int\limits_V p' s\,[1]\,\rho\, dV \tag{167}$$

Comparing (164) and (167), the following illustrative explanation of the difference between a_c and dv_c/dt is obtained:

$$\begin{aligned} m\,a_c - m\frac{dv_c}{dt} &= -\omega \times \left(\int\limits_V p' s\,[1]\,\rho\, dV + \int\limits_S n \cdot (u - v)\,\rho\, p' dS \right) \\ &= \omega \times j' = \omega \times (s\,[m] + s_u\,[m])\,(c - \tilde{c}) \end{aligned} \tag{168}$$

see (137) for the relative linear momentum j'. Only when $\omega \times j'$ vanishes, this difference disappears.

A further remark seems to be in order. As already discussed above, a relation of balance can be brought into a form, in which the non-classical supply due to the flow of mass through the surface formally cancels out. This is demonstrated for balance of relative moment of momentum next.

Balance of relative moment of momentum For comparison's sake, we study balance of moment of momentum relative to the canter of mass. From (155), it is immediately seen that the relative moment of momentum (70) for the rigid body is

$$\alpha' = \int_V p' \times (\omega \times p')\, \rho\, dV = J' \cdot \omega \tag{169}$$

This follows by substituting (156) into (70), and noting (58). Using $\Psi = p' \times (\omega \times p')$ in (123)–(124), the transport theorem yields the relation of balance of relative moment of momentum as

$$\frac{d_u}{dt}\alpha' = \int_V p' \times b\, dV + \int_S p' \times \sigma_n dS - \int_V p \times \Sigma_\times dV \\ + s\,[\alpha'] + \int_S n \cdot (u - v)\rho\, (p' \times (\omega \times p'))\, dS \tag{170}$$

where the non-classical supply due to a flow of mass through the surface in case of a rigid body is given by

$$s_u[\alpha'] = \int_S n \cdot (u - v)\rho\, (p' \times (\omega \times p'))\, dS \tag{171}$$

For this relation, compare (6.10) of Ong and O'Reilly (2004), and the literature cited there. For the rigid body, we moreover obtain

$$\frac{d}{dt}\alpha' = \frac{d}{dt}(J' \cdot \omega) = \left(\frac{d}{dt}J'\right) \cdot \omega + J' \cdot \dot{\omega}, \tag{172}$$

see (156) and (169). Taking into account (49), the time derivative of the tensor of relative inertia J' can be written as

$$\frac{d}{dt}J' = \frac{d}{d}\int_V ((p' \cdot p')\, I - p' \otimes p')\, \rho\, dV \\ = \omega \times J' + \omega \cdot \int_V ((p' \cdot p')\, I - p' \otimes p')\, s[1]\rho\, dV \tag{173}$$

We thus can re-write (170) in the following form, which no longer contains $s_u[\alpha']$:

$$\left(\frac{d}{dt}J'\right)\cdot\omega +' \cdot\dot{\omega} = \int_V p'\times b\,dV + \int_S p'\times\sigma_n dS - \int_V p\times\Sigma_\times dV + s[\alpha'] \quad (174)$$

In case of $\Sigma_\times = 0$, $s[1] = 0$ and $s\,[\alpha'] = 0$, this coincides with (7.7) of Ong and O'Reilly (2004). Analogous derivations can be performed for the other relations of balance, e.g., for balance of kinetic energy.

Balance of kinetic energy For the rigid body, using (153) and (58) and some vector algebra, the relation of balance of kinetic energy in (131) reduces to

$$\frac{d_u}{dt}E_{kin} = \frac{d}{dt}E_{kin} + s_u[E_{kin}]$$
$$= v_c\cdot F + \omega\cdot\left(\int_V p'\times b\,dV + \int_S p'\times\sigma_n dS\right) + s[E_{kin}] + s_u[E_{kin}] \quad (175)$$

Setting $s[E_{kin}] = 0$, and substituting $s_u[E_{kin}]$ of (130), this coincides with (8.3) and (8.5) of Ong and O'Reilly (2004).

4 Presence of a singular surface in a material volume

So far, we have tacitly assumed that the entities under consideration are distributed continuously throughout the volume V. Now assume that a (smooth) singular surface $\bar{S}$ is present within V, such that $\bar{S}$ subdivides V into two non-material sub-volumes. At a singular surface $\bar{S}$, certain local entities $\rho\,\psi$, e.g., mass and linear momentum, and their supplies may take on different values at the two sides of the singular surface $\bar{S}$. Moreover, the points of $\bar{S}$ may move at a velocity w, which in general is different from the velocities of the particles that instantaneously are located at the two sides of $\bar{S}$. In order to connect the local forms of the balance relations at the two sides of the singular surface, jump relations are needed. A classical strategy for deriving jump relations was presented by Truesdell and Toupin (1960): The Reynolds transport theorem is applied to the two non-material sub-volumes, taking into account the motion of $\bar{S}$ relative to the material particles and adding the results. The volume V then is shrunken down to the singular surface $\bar{S}$, assuming that the integrands of the respective volume integrals remain bounded, and letting the surface $\bar{S}$ be finite in the limit. The result of this procedure is the jump relation of Kotchine (1926). The Kotchine jump relation takes into account the jump of the entity $\rho\psi$ across $\bar{S}$ in a straightforward manner, however, this procedure is at the cost of not considering the particles that instantaneously are passing the singular surface $\bar{S}$, as well as some non-classical concentrated supplies that may travel with $\bar{S}$. This conceptual drawback can be removed by applying the following more general strategy, which was described in some detail by Irschik (2003). This strategy is discussed and adopted for variable mass systems in the following.

4.1 On a generalized form of the general relation of jump at a singular surface

The methodology proposed by Irschik (2003) consists of several steps. First, instead of starting with the assumption of a singular surface, one studies a non-material shell-type layer of transition, within which the entity $\rho\psi$ and its sources are subjected to considerable changes in their spatial distribution. One then replaces the shell-type layer of transition by an equivalent singular surface $\bar{S}$, where several terms in the transport theorem for the non-material layer are represented by surface supply terms that equivalently describe the behavior of the respective quantity contained in the non-material shell-type layer. Motivated by fundamental studies of Slattery (1990) on interfacial transport theorems, a rational mechanical and thermodynamic formulation for this general strategy was presented in Irschik (2003), see also Irschik (2004) and Irschik (2007). This formulation contains the classical Kotchine jump relation as a special case. The additional equivalent surface supply terms have been denoted as surface growth terms by Irschik (2003). In the following, we shortly review the latter rational formulation with special emphasis on the presence of both, a surface supply of mass and linear momentum at the equivalent singular surface $\bar{S}$, and we demonstrate that and show how a surface supply of kinetic energy must be introduced for the sake of consistency, even when the surface supplies of mass and linear momentum do vanish. In the subsequent Section 5, we will exemplarily apply these extended jump relations to the case of a chain heaped up on the edge of a table, the hanging part of the chain being set into motion. In this study, the transition from the heaped part to the moving part of the chain is described by the extended jump relations, while the equation of motion for the moving part is obtained by the relations of balance written for a non-material control volume, which have been stated in Section 3 above.

4.2 Jump of mass

The extended relation of jump of mass at an equivalent singular surface $\bar{S}$ as derived by Irschik (2003) can be written as

$$n_{\bar{S}} \cdot \left(i_{\bar{S}} \left[\rho \right] + \llbracket (v - w)\, \rho \rrbracket \right) = 0 \tag{176}$$

The outer unit normal vector at the equivalent singular surface is $n_{\bar{S}}$, and the vector $i_{\bar{S}}\left[\rho\right]$ represents the non-classical equivalent surface supply of mass at $\bar{S}$. Note that the dimension "dim" of any non-classical surface supply $i_{\bar{S}}\left[k\right]$, say, is given by:

$$\dim\left(i_{\bar{S}} \left[k \right] \right) = \frac{\dim\left(k\right)}{\dim\left(time\right)\ \dim\left(area\right)} \tag{177}$$

The jump operator at $\bar{S}$ is defined as the difference of the entities k at the two sides of:

$$\llbracket k \rrbracket = k^{+} - k^{-} \tag{178}$$

In the following, the unit outer normal vectors at the two sides of $\bar{S}$ are taken such that $n^{+} = n_{\bar{S}}$ and $n^{-} = -n_{\bar{S}}$.

4.3 Jump of linear momentum

The extended relation of jump of linear momentum reads

$$n_{\bar{S}} \cdot \left(i_{\bar{S}} \left[\rho\, v \right] + \llbracket (v - w) \otimes \rho\, v - \Sigma \rrbracket \right) = 0 \tag{179}$$

where the second-order tensor $i_{\bar{S}}\left[v\,\rho\right]$ represents the non-classical equivalent surface supply of linear momentum at $\bar{S}$.

4.4 Jump of kinetic energy

The extended relation of jump of kinetic energy reads

$$n_{\bar{S}} \cdot \left(i_{\bar{S}} \left[\rho\, \frac{v \cdot v}{2} \right] + \left\llbracket (v - w)\, \rho\, \frac{v \cdot v}{2} - v \cdot \Sigma \right\rrbracket \right) = 0 \tag{180}$$

The non-classical surface supply of kinetic energy at $\bar{S}$ is a vector denoted as $i_{\bar{S}}\left[\rho\, v \cdot v/2\right]$. It has been shown by Irschik (2003) that the non-classical surface supplies of mass, linear momentum and kinetic energy are not independent. Rather, $i_{\bar{S}}\left[\rho\, v \cdot v/2\right]$ must obey the following relation:

$$\begin{aligned} n_{\bar{S}} \cdot i_{\bar{S}} \left[\rho\, \frac{v \cdot v}{2} \right] = {} & \left(\frac{\langle v \cdot v \rangle}{2} - \langle v \rangle \cdot \langle v \rangle \right) n_{\bar{S}} \cdot i_{\bar{S}} \left[\rho \right] \\ & + \langle v \rangle \cdot \left(n_{\bar{S}} \cdot i_{\bar{S}} \left[\rho\, v \right] \right) + \llbracket v \rrbracket \cdot \left(n_{\bar{S}} \cdot \langle \Sigma \rangle \right) \end{aligned} \tag{181}$$

The mean value operator of some entity k across $\bar{S}$ is given by

$$\langle k \rangle = \frac{1}{2}\left(k^{+} + k^{-}\right) \tag{182}$$

It is seen from (181) that, even if the non-classical equivalent surface supplies of mass and linear momentum are absent, it is generally necessary to consider a non-classical equivalent surface supply of kinetic energy. Subsequently, we give an example, in which a non-classical equivalent surface supply of mass must be introduced for modeling reasons.

5 Example for the formulations in Sections 3 and 4: Caley's chain set into motion

The example of a chain heaped up on a table with the hanging part being set into motion has been chosen by Cayley (1856) in a fundamental contribution on the dynamics of what he called continuous impact problems, see Fig. 1 for a sketch.

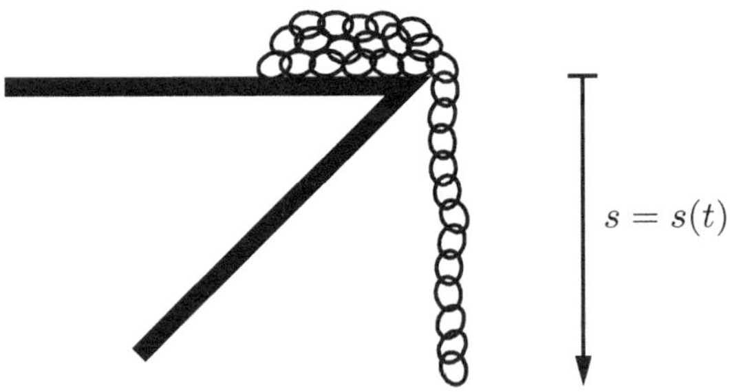

Figure 1. Chain hanging over the edge of a table.

Cayley wrote: *"A problem of the sort arises when a portion of a heavy chain hangs over the edge of the table, the remainder of the chain being coiled or heaped up close to the edge of the table, the part hanging over constitutes the moving system, and in each element of time the system takes into connexion with itself, and sets into motion with a finite velocity an infinitesimal length of the chain."* Cayley used this problem in order to demonstrate the application of a novel variational formulation, as well as of a corresponding extended form of Lagrange's equations. For a contemporary discussion on Cayley (1856), see Irschik (2012).

Subsequently, we use Cayley's example of a chain set into motion in order to apply the formulations stated in Sections 3 and 4. For extended forms of Lagrange's equations for mechanical systems with a variable mass, see Irschik and Holl (2002), Pesce (2003) and the Chapter by Pesce in the present book.

5.1 The hanging part of the chain as a system with a time-varying mass

The chain is assumed to be inextensible and to be coiled up loosely at the table, the hanging part having the instantaneous length $s = s(t)$. In a first step, the hanging part of the chain is described as an open system. For that sake, it is enclosed by a non-material control surface S, see the dashed surface in Fig. 2. For this surface, the formulas of Section 3 above do apply. The upper horizontal part of S is located immediately under the edge of the table and is fixed in space, such that there is $u = 0$; the particles of the inextensible chain enter there the control volume at the velocity $v = (ds/dt)e_x = \dot{s}\, e_x$, see Fig. 2. The outer unit normal vector n at this location is opposite to the global x-direction, $n = -e_x$; a superimposed dot denotes the time derivative. Hence, there is $n \cdot (u - v) = \dot{s}$ in (124). Note that the chain in general is stressed at this upper horizontal location of the control surface S. Denoting the mass of the chain per unit length as μ, and the tensile force in the chain at the upper horizontal part of the control surface as N, Cauchy's fundamental theorem on stresses yields $n \cdot \Sigma = -N(\rho/\mu)e_x$. The lower horizontal part of the control surface moves together with the tip of the chain, $u = v = \dot{s}\, e_x$, such that there is $u - v = 0$ in (124); also, the chain is free of stress there, $n \cdot \Sigma = 0$, see Fig. 2. The vertical portions of S do not contribute to the relations of balance, since no material is present there.

The instantaneous mass of the hanging part of the chain is $m = \mu\, s$, such that

$$\frac{d_u}{dt} m = \mu\, \dot{s} \tag{183}$$

Recall that the operator d_u/dt means that we consider the motion of the non-material control volume; in other words, $s = s(t)$ in $m = \mu\, s$ is not kept fixed, which leads to (183). The additional, non-classical supply of mass due to the flow of mass through the control surface becomes, see (126),

$$s_u[m] = \int_S n \cdot (u - v)\rho\, dS = \mu\, \dot{s} \tag{184}$$

Further recall that only the upper part of the control surface in Fig. 2 contributes to the surface integral. Substituting (183) and (184), it is seen

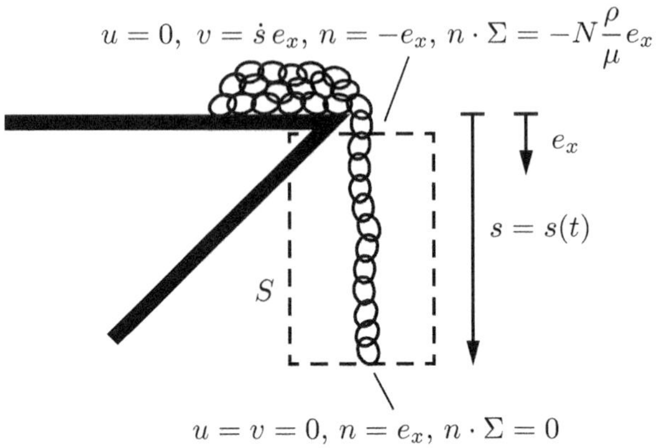

Figure 2. Non-material control volume S enclosing the hanging part of the inextensible chain.

that balance of mass (125) indeed is satisfied. No sources of mass in the interior are present, $s[m]=0$.

We now proceed to balance of linear momentum (127) which we apply for the open system depicted in Fig. 2. The instantaneous linear momentum of the hanging part of the chain in x-direction is

$$j=\mu\,s\,\dot{s}\,e_x \tag{185}$$

such that

$$e_x\cdot\frac{d_u}{dt}j=\mu\,s\,\ddot{s}+\mu\,\dot{s}^2 \tag{186}$$

For the additional, non-classical supply of linear momentum due to the flow of mass through the control surface, see (128), one obtains analogous to (184) that

$$e_x\cdot s_u[j]=e_x\cdot\int_S n\cdot(u-v)\rho\,v\,dS=\mu\,\dot{s}^2 \tag{187}$$

The resultant force acting upon the hanging part of the chain is

$$e_x\cdot F=-N+\mu\,g\,s \tag{188}$$

where the last term represents the instantaneous weight of the hanging part and g denotes the gravitational acceleration. With $s[j]=0$, the relation of balance of linear momentum (127) thus yields the following equation of motion:

$$\mu\, s\, (\ddot{s} - g) + N = 0 \tag{189}$$

Under the assumptions of an inextensible homogeneous chain in a vertical motion due to its own weight, this relation is to be considered as exact; however the normal force N in the chain at the upper horizontal part of the control surface in Fig. 2 is yet unspecified.

5.2 Modeling of the region of transition between the moving and resting parts of the chain by an equivalent singular surface

In order to set up an additional relation for N, the region of transition between the part of the chain in motion and the heaped up part at rest now is modeled by means of an equivalent singular surface $\bar{S}$, see the dashed line in Fig. 3, where the region of transition with the two outer surfaces S^+ and S^- is also sketched.

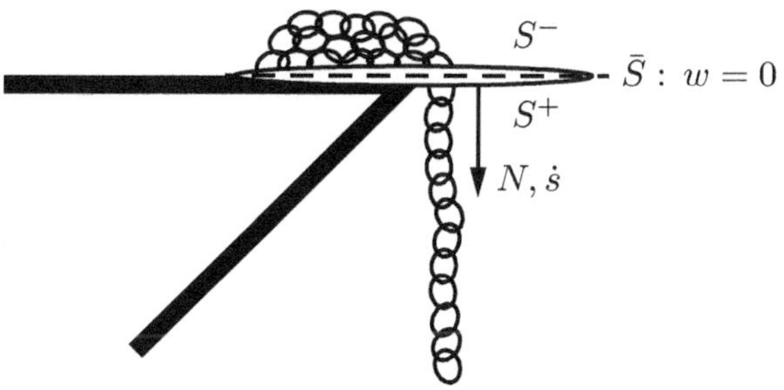

Figure 3. Equivalent singular surface modeling the region of transition between the moving and the resting parts of the chain.

Since the equivalent singular surface $\bar{S}$ is at rest, there is $w = 0$ in the formulas of Section 4 above. Outer unit normal vectors, velocities and stresses of the particles at the two sides of $\bar{S}$ are as follows, see also Fig. 4:

$$\begin{aligned} &\bar{S}: && w = 0, && n_{\bar{S}} = n^+ = e_x, && \\ &S^-: && v^- = 0, && n^- = -e_x, && n^- \cdot \Sigma^- = 0, \\ &S^+: && v^+ = \dot{s}\, e_x, && n^+ \cdot \Sigma^+ = N \frac{\rho}{\mu} e_x && \end{aligned} \tag{190}$$

The heaped part of the chain, which is at rest, can be taken as unstressed for the present purpose, $N = 0$ at S^-, while the force in the chain at the side

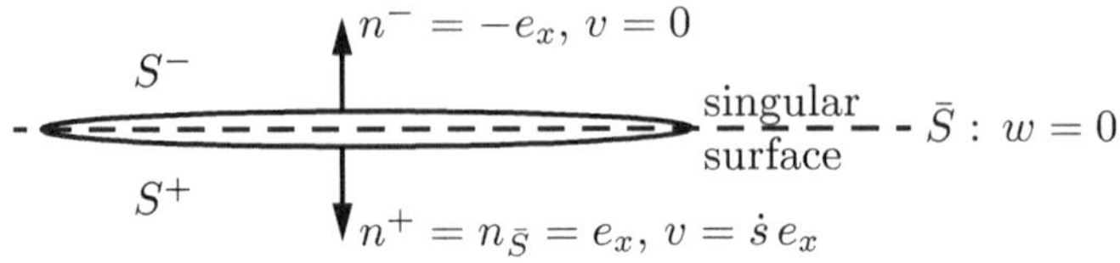

Figure 4. Details of equivalent singular surface.

S^+ is N. For the situation sketched in Fig. 4, (190) yields for the relation of jump of mass (176) that

$$n_{\bar{S}} \cdot i_{\bar{S}}[\rho] = -\rho\, \dot{s} \tag{191}$$

Note that $n_{\bar{S}} \cdot i_{\bar{S}}[\rho]$ only vanishes, if the hanging part is at rest. This example clearly demonstrates the necessity of extending the classical Kotchine-type relations of jump by non-classical equivalent surface supply terms, as proposed by Irschik (2003).

The next step is the relation of jump of linear momentum. (179). Substituting (190), one obtains for the non-classical equivalent surface supply of linear momentum

$$e_x \cdot i_{\bar{S}}[\rho\, v] = \left(-\rho\, \dot{s}^2 + N\, \frac{\rho}{\mu}\right) e_x \tag{192}$$

Analogously, the relation of jump of kinetic energy (180) gives for the non-classical equivalent surface supply of kinetic energy that

$$e_x \cdot i_{\bar{S}} \left[\rho\, \frac{v \cdot v}{2}\right] = -\frac{\rho\, \dot{s}^3}{2} + \frac{N\, \rho}{\mu}\, \dot{s} \tag{193}$$

It is easily checked that (181), which states the relation between the surface supplies of mass, linear momentum and kinetic energy derived by Irschik (2003), indeed is satisfied, since, at the equivalent singular surface $\bar{S}$, there is

$$\bar{S}: \quad [\![v]\!] = \dot{s}\, e_x, \quad \langle v \rangle = \frac{\dot{s}}{2} e_x, \quad \langle v \cdot v \rangle = \frac{\dot{s}^2}{2}, \quad [\![v]\!] \cdot (n_{\bar{S}} \cdot \langle \Sigma \rangle) = \frac{N\, \rho}{\mu}\, \frac{\dot{s}}{2} \tag{194}$$

We now may distinguish two special cases: The first is obtained by assuming that there is no non-classical surface supply of linear momentum,

$$e_x \cdot i_{\bar{S}}\, [\rho\, v] = 0 \tag{195}$$

in (192), from which the chain force N becomes

$$N = N_{nc} = \mu \, \dot{s}^2 \tag{196}$$

Substituting into the equation of motion (189) gives

$$\mu \, s_{nc} \, (\ddot{s}_{nc} - g) + \mu \, \dot{s}_{nc}^2 = 0 \tag{197}$$

The solution with homogeneous intial conditions is

$$s_{nc} = \frac{g \, t^2}{6} \tag{198}$$

The solution stated in (197) and (198) has been obtained by Cayley (1856) in the framework of another methodology. In the literature, this type of solution has been called a non-conservative solution, see, e.g., Wong and Yasui (2006). From this naming, the index "nc" has been introduced in (196)–(198). Indeed, substituting (196), from (193) we obtain the following equivalent surface supply of kinetic energy:

$$e_x \cdot i_{\bar{S}_{nc}} \left[\rho \, \frac{v \cdot v}{2} \right] = \rho \, \dot{s}^3 \tag{199}$$

On the other hand, if we assume that the equivalent surface supply of kinetic energy vanishes,

$$e_x \cdot i_{\bar{S}} \left[\rho \, \frac{v \cdot v}{2} \right] = 0 \tag{200}$$

then (193) yields

$$N = N_c = \mu \, \frac{\dot{s}^2}{2} \tag{201}$$

Substituting into the equation of motion (197), we now obtain

$$\mu \, s_c \, (\ddot{s}_c - g) + \frac{1}{2} \, \mu \, \dot{s}_c^2 = 0 \tag{202}$$

The solution for homogeneous initial conditions this time becomes

$$s_c = \frac{g \, t^2}{4} \tag{203}$$

This solution, which predicts a fall of the chain faster than Cayley's non-conservative solution (198), has been called a conservative solution in the literature, see Wong and Yasui (2006). Note, however, that this conservative solution is associated with a non-classical supply of linear momentum, which follows by substituting (201) into (192):

$$e_x \cdot i_{\bar{S}_c} [\rho \, v] = \frac{\rho \, \dot{s}^2}{2} \, e_x \tag{204}$$

The discrepancy between conservative and non-conservative solutions in Cayley's problem, as well as in related problems of falling chains, ropes, cables and strings, has given raise to a long-term controversy in the literature. Here, we mention the more recent theoretical and/or experimental works by Tomaszewski et al. (2006), Wong and Yasui (2006), Wong et al. (2007), Grewal et al. (2011) and Irschik (2012).

It is hoped that the rational methodology given in the present section, which is based on relations of balance of mass, linear momentum and kinetic energy for open systems, and on generalized corresponding relations of jump, will contribute to a further clarification of the different results in the literature. It should have become clear from our reasoning that a more detailed modeling of the region of transition between the heaped and the moving parts of the chain in general will lead to both, a non-vanishing equivalent surface supply of linear momentum, as well as a non-vanishing equivalent surface supply of kinetic energy, and thus will give raise to solutions different from the above discussed two cases, cf. O'Reilly and Varadi (1999) for a study on shocks.

Acknowledgments

This work has been supported by the Linz Center of Mechatronics (LCM) in the framework of the Austrian COMET-K2 program.

Bibliography

A. Cayley. On a class of dynamical problems. *Proceedings of the Royal Society of London*, 8:506–511, 1856.

L. Cveticanin. *Dynamics of Machines with Variable Mass.* Gordon and Breach Science Publishers, 1998.

F.O. Eke. Dynamics of variable mass systems. Technical report, UC Davis, 1998.

V.A. Eremeyev, L.P. Lebedev, and H. Altenbach. *Foundations of micropolar mechanics.* Springer-Verlag, Berlin Heidelberg, 2012.

J.L. Ericksen. Tensor fields. In S. Flügge, editor, *Prinzipien der Klassischen Mechanik und Feldtheorie*, volume III/1 of *Handbuch der Physik (Encyclopedia of Physics)*, pages 794–858. Springer-Verlag, Berlin Göttingen Heidelberg, 1960.

K. Federhofer. Dynamik sich ändernder Massen. *Mitteilungen des Deutschen Ingenieur-Vereins in Mähren, Hauptvereines deutscher Ingenieure in der tschechoslowakischen Republick*, 11(6–7):83–86, 115–118, 1922.

A. Grewal, P. Johnson, and A. Ruina. A chain that speeds up, rather than slows, due to collisions: how compression can cause tension. *American Journal of Physics*, 79(7):723–729, 2011.

K. Hutter and K.D. Jöhnk. *Continuum Methods of Physical Modeling.* Springer-Verlag, Berlin Heidelberg, 2004.

H. Irschik. On the necessity of surface growth terms for the consistency of jump relations at a singular surface. *Acta Mechanica*, 162(1-4):195–211, 2003.

H. Irschik. A treatise on the equations of balance and on the jump relations in continuum mechanics. In H. Irschik and K. Schlacher, editors, *Advanced Dynamics and Control of Structures and Machines*, volume 444 of *CISM Courses and Lectures*, pages 9–20. Springer-Verlag, Wien New York, 2004.

H. Irschik. Über Wachstumsterme in den Bilanzgleichungen der Kontinuumsmechanik, speziell beim Wachstum von biologischen Materialien. *Sitzungsberichte der Österreichischen Akademie der Wissenschaften, Mathematisch-naturwissenschaftliche Klasse Abt. II*, 213:95–111, 2005.

H. Irschik. On rational treatments of the general laws of balance and jump, with emphasis on configurational formulations. *Acta Mechanica*, 194 (1-4):11–32, 2007.

H. Irschik. The Cayley variational principle for continuous-impact problems: a continuum mechanics based version in the presence of a singular surface. *Journal of Theoretical and Applied Mechanics*, 50(3):717–727, 2012.

H. Irschik and H.J. Holl. The equations of Lagrange written for a non-material volume. *Acta Mechanica*, 153(3-4):231–248, 2002.

H. Irschik and H.J. Holl. Mechanics of variable-mass systems—part 1: Balance of mass and linear momentum. *Applied Mechanics Reviews*, 57(2): 145, 2004.

J.S. König. De universali principio æquilibrii & motus, in vi viva reperto, deque nexu inter vim vivam & actionem, utriusque minimo, dissertatio. *Nova Acta Eruditorum*, pages 125–135, 162–176, 1751.

A. Kosmodemyansky. *Konstantin Tsiolkovsky His Life and Work.* University Press of the Pacific, 2000.

N.E. Kotchine. Sur la théorie des ondes de choc dans un fluide. *Rendiconti del Circolo Matematico di Palermo*, 50(2):305–344, 1926.

I.V. Meshchersky. Dinamika tochki peremennoj massji. *Magistarskaja disertacija, Peterburgski Universitet, Peterburg*, 1897.

G.K. Mikhailov. On the history of variable-mass system dynamics. *Mechanics of Solids*, 10(5):32–40, 1975.

J.J. Ong and O.M. O'Reilly. On the equations of motion for rigid bodies with surface growth. *International Journal of Engineering Science*, 42 (19):2159–2174, 2004.

O.M. O'Reilly and P.C. Varadi. A treatment of shocks in one-dimensional thermomechanical media. *Continuum Mechanics and Thermodynamics*, 11(6):339–352, 1999.

C.P. Pesce. The application of Lagrange equations to mechanical systems with mass explicitly dependent on position. *Journal of Applied Mechanics*, 70(5):751–756, 2003.

O. Reynolds. *The sub-mechanics of the universe.* Cambridge University Press, 1903.

H. Seeliger. Über Zusammenstösse und Theilungen planetarischer Massen. *Abhandlungen der Königlichen Bayerischen Akademie der Wissenschaften, II. CL.*, XVII(II):459–490, 1890.

J.C. Slattery. *Interfacial transport phenomena.* Springer-Verlag, New York, 1990.

W. Thomson and P.G. Tait. *Treatise on Natural Philosophy.* Clarendon Press, Oxford, 1867.

W. Tomaszewski, P. Pieranski, and J.-C. Geminard. The motion of a freely falling chain tip. *American Journal of Physics*, 74(9):776–783, 2006.

C.A. Truesdell and R.A. Toupin. The classical field theories. In S. Flügge, editor, *Prinzipien der Klassischen Mechanik und Feldtheorie*, volume III/1 of *Handbuch der Physik (Encyclopedia of Physics)*, pages 226–793. Springer-Verlag, Berlin Göttingen Heidelberg, 1960.

C.W. Wong and K. Yasui. Falling chains. *American Journal of Physics*, 74 (6):490–496, 2006.

C.W. Wong, S.H. Youn, and K. Yasui. The falling chain of Hopkins, Tait, Steele and Cayley. *European Journal of Physics*, 28(3):385, 2007.

F. Ziegler. *Mechanics of Solids and Fluids.* Springer-Verlag, New York, second edition, 1998.

Systems with mass explicitly dependent on position

Celso Pupo Pesce* and Leonardo Casetta†
Escola Politécnica, University of São Paulo, São Paulo, Brazil

Abstract This chapter addresses an interesting type of variable mass systems. Those in which mass may be explicitly written as function of position. Two perspectives can be followed: systems with a material type of source, attached to particles continuously gaining or loosing mass and systems for which the variation of mass is of a "control volume type", mass trespassing a control surface. This is the case if, for some theoretical or practical reason, partitions into sub-systems are considered. Whenever mass depends explicitly on position, the Lagrange equation has to be carefully re-interpreted. As a matter of fact, an extra non-conservative generalized force term, linearly proportional to the mass gradient and quadratic on velocities, emerges from first variational principles. Ignoring this term has been the cause of misleading derivations of equations of motions and even of many misinterpretations, not rarely provoking claims of false paradoxes. The present chapter derives such an extended form of the Lagrange equation, through Lagrangean and Hamiltonian approaches. Illustrative and practical examples are taken from two engineering fields, offshore engineering and civil engineering. In the first category are included: (i) the reel laying operation of marine cables; (ii) the dynamics of a water column inside a free surface piercing open pipe (and the analogous moon pool problem) and (iii) the hydrodynamic impact of a solid body against a free surface of water. In the second category, the governing equation of motion of vertically collapsing towers is properly derived.

1 Introduction

According to Mikhailov (1975), in his work on the history of variable mass system dynamics, early in the nineteenth century the Czech scientist Buquoy

*Professor of Mechanical Sciences.

†Dr. in Mechanics and Research Fellow.

H. Irschik, A. K. Belyaev (Eds.), *Dynamics of Mechanical Systems with Variable Mass*,
CISM International Centre for Mechanical Sciences DOI 10.1007/ 978-3-7091-1809-2_2

proposed a generalization of the laws of dynamics to account for what he named "possibility of continuous variation of the masses of parts of the system"[1]. Buquoy noted that the classical form of Newton's law is only valid for the case of constant mass, the consideration of an extra term related to the time rate of change of momentum of the system being necessary when dealing with the case of mass variation. In the specialized literature, this term is known as Meshcherski's reactive force. In particular, Meshcherski's Master Thesis (Meshcherski (1897)), and his subsequent work Meshcherski (1904), have been ever since recognized in the Russian technical literature as the limestone in the study of variable mass systems within the context of Classical Mechanics; see, e.g., Targ (1976), page 394 or Starjinski (1986), page 498.

From Buquoy to Meshcherski, important period when the mechanics of systems with variable mass started to be recognized as a particular branch of mechanics, other important names, as Poisson (1819), Tait & Steele (1856), Cayley (1857), contributed to the comprehension of the physical effects that were involved in the process of mass variation of a system; see Mikhailov (1975). Those authors independently suggested possible modifications and interpretations in the classical formalism of mechanics in order to have variable mass effects coherently taken into account. In fact, this is the prime difficult in the proposition of a proper formalism for variable mass systems, since the classical form of the fundamental principles and equations of mechanics were originally conceived to account for well defined bodies and particles of constant mass. However, although this may be seen as an interesting and challenging aspect, apart from few others, as Levi-Civita (1928), and Agostinelli (1936), the subject did not attract too much attention of their foregoers[2].

A renewed interest in this subject strongly reappeared in the 1950's and 1960's, motivated by the "rocket problem" and the emerging Aerospace Engineering. As a matter of fact, an interesting debate took place among American scholars on how to properly interpret and teach variable mass system dynamics in engineering education; see, e.g., Meriam (1960), Thorpe (1962), Arons & Bork (1964), Bork & Arons (1964), Pomeranz (1964), Van de Akker (1964), Gadsden (1966), Copeland (1982), Calkin & March (1989a,b), Keifer (2001), Eke & Mao (2002), Chicon (2003), Wong & Yasui (2006), Grewal et al (2011); see also Plastino & Muzzio (1992), Pesce & Casetta (2007).

[1] In Buquoy's terminology.

[2] The superb book by Dugas (1988) does not give too much emphasis on the variable mass problem.

The theoretical formalism concerning variable mass systems mechanics regained vigor in the 1970's, after Truesdell & Toupin (1960) and the work by McIver (1973), followed, from the 90's on, by a number of fundamental articles, as those by Cvetićanin (1993, 2009), Irschik & Holl (2002, 2004), Mušicki (1999, 2000, 2004), Shao-kai & Feng-xiang (1992) and Shan-jun et al (2005). See also the recent contributions by Irschik (2012), Casetta & Pesce(2012, 2013a,b).

Such a renewed interest followed the analysis and solution of practical problems, which do represent important aspects of technology advances as well as industrial necessities. This scenario can be featured, for example, by the study of industrial coiling processes and variable mass machinery dynamics, Cvetićanin (1998),Cveticanin & Kovacic (2007), Holl et al (2006) and Holl & Hammelmuller (2011); or by the aerospace industry, in particular motivated by tethered satellites dynamics, Janssens et al (1995); Crellin et al (1997), Schwarzbart et al (2009). Applications and analogies within other fields of sciences and engineering are also encountered, as in surface growth mechanisms, Ong & O'Reilly (2004), or in the dynamics of electrical circuits, Jeltsema & Doria-Cerezo (2011), Jeltsema (2012).

However, the comprehension of the adequate extension of the fundamental principles of mechanics to variable mass systems has shown to be a process with many subtleties. Not surprisingly, controversial results and paradoxes, even concerning apparently simple problems, as is the case of the classic falling chain and rope problems, have been and still are discussed; see, e.g., Calkin & March (1989a,b), Casetta (2008), Casetta & Pesce (2013) Chicon (2003),Grewal et al (2011), Jimenez et al (2012), Keifer (2001), Prato & Gleiser (1982), Schagerl et al (1997), Šima & Podolsky (2005), Tomaszewski et al (2006), Wong & Yasui (2006), Wong et al (2007). Analogously, this is also the case of the vertically collapsing towers problem; see, e.g., Bažant & Verdure (2007), Bažant et al (2008), Bažant & Le (2008), Beck (2007, 2008), Pesce et al (2012), Seffen (2008). All those problems are usually formulated under a special form of mass variation: the explicit dependence on position.

However, whenever mass depends explicitly on position, the application of the Lagrange equation has to be carefully re-interpreted. As a matter of fact, an extra non-conservative generalized term, linearly proportional to the mass partial derivative with respect to generalized coordinates and quadratic on velocities, emerges from first variational principles; Cvetićanin (1993), Pesce (2003). Ignoring this term may be attributed as the main cause of misleading derivations and of many misinterpretations, Šima & Podolsky (2005), Tomaszewski et al (2006), usually provoking claims of apparent paradoxes, as those contained in Wong & Yasui (2006), later proven false by the same authors, in a meritorious scientific attitude; see Wong et al (2007).

The present chapter derives such an extended form of the Lagrange equation, through Lagrangean and Hamiltonian approaches. Illustrative and practical examples are addressed in two engineering fields, offshore and civil engineering. In the first category are included: (i) the reel laying operation of marine cables; (ii) the dynamics of a water column inside a free surface piercing open pipe (and the analogous moon pool problem) and (iii) the hydrodynamic impact of a solid body against a free surface of water. In the second category, the governing equation of motion of vertically collapsing towers is properly derived.

2 The Extended Lagrange Equations for Variable Mass Systems

In this section, not only explicitly dependence on time and position but, for completeness, a possible mass variation dependence on velocities is also considered. A previous and independent derivation, restricted to the case of explicit mass dependences on position and time may be seen in Cvetićanin (1993, 1998). Two perspectives are followed: Lagrangean and Hamiltonian. For completeness sake, the prologue and the following first subsection, corresponding to the Lagrangean approach, are essentially recovered from Pesce (2003) - almost in its totality[3]. The second subsection addresses the problem from a Hamiltonian point of view for dissipative systems and is, indeed, new material.

Prologue

Kinetic energy, $T = T(q_j; \dot{q}_j; t)$, is, by definition, at least a bi-linear form of generalized velocities, $\dot{q}_j$, and also, in many cases, a function of the generalized coordinates, . However, physically speaking, there is a rather large distinction between being the kinetic energy an explicit or an implicit function of the kinematic state of the system, the latter through a possible dependence of mass, in the form $m_i(q_j; \dot{q}_j; t)$. Whenever this is the case, the mechanical system does not obey the usual form of the classical Euler-Lagrange equations $\frac{d}{dt}\left(\frac{\partial T}{\partial \dot{q}_j}\right) - \frac{\partial T}{\partial q_j} = Q_j$, *unless non-conservative terms associated to fluxes of mass are already considered included in* Q_j , as those usually referred to as Mescherski's reactive forces. Otherwise, the derived equations of motion take an erroneous form. It could be argued that the

[3]Under kind permission of ASME - the American Society of Mechanical Engineers.

usual Lagrange equations would be suitable to the case for which $m = m(t)$, mass varying solely as an explicit function of time. This is true, however, only if mass is gained or lost at zero-velocity with respect to an inertial frame of reference, as assumed by Agostinelli (1936), page 257, who concluded being the Lagrange equations invariant for holonomic variable-mass systems. In that particular work, Levi-Civita's special form of the momentum equation, $\dot{\mathbf{p}} = \mathbf{f}$, was used[4], with no reference to any reactive force proportional to the velocity of the particle that is being expelled from or accreted to the system.

The reason for these subtle distinctions, concerning how mass changes, if as an explicit or an implicit function of time, is shown next. The answer hides behind the derivation of the most general form of the Lagrange equations, as presented, e.g., in Cvetićanin (1993). For now, it is pointed out that, in systems with mass explicitly dependent on position, a naive application of the usual Lagrange equations, without any special consideration on generalized forces, leads to erroneous equations of motions which lack terms of the form $\frac{1}{2}\frac{\partial m}{\partial q}\dot{q}^2$.

The Lagrangean approach[5]

Consider a system of N particles of mass $m_i; i = 1, \ldots, N$. Let P_i be the corresponding position in a given inertial frame of reference and $\mathbf{p_i} = m_i \mathbf{v_i}$ the momentum. By extending Levi-Civita's form of Newton's law to cases when mass is gained or lost with no null velocity, D'Alembert's Principle can be written

$$\sum_i (\dot{\mathbf{p}}_\mathbf{i} - \mathbf{F}_\mathbf{i}).\delta P_i = \mathbf{0} \tag{1}$$

where

$$\mathbf{F_i} = \mathbf{f_i} + \mathbf{h_i} \tag{2}$$

being $\mathbf{f_i}$ the sum of all active forces acting on P_i, and

$$\mathbf{h_i} = \dot{m}_i \mathbf{w_i} \tag{3}$$

a non conservative force, proportional to the rate of variation of mass with respect to time and to the velocity of the expelled (or gained) mass.

[4]Dot designates the total derivative with respect to time, as usual.

[5]From another perspective, the inverse problem of Lagrangian mechanics for Meshcherski's equation has been recently worked out in Casetta & Pesce (2013b), in which a principle of stationary action is constructed from the equations of motion, by applying the method of Darboux.

Eq.1 is the so-called *first form of the fundamental equation* for a system of discrete particles, which, according to Pars (1965), is a generalization both of the virtual work principle as well as of D'Alembert principle. Mikhailov (1975) attributes this form to Poisson. Concisely, it will be simply referred to as D'Alembert principle.

Note that the reactive force known as Meshcherski's is usually written as function of relative velocities, in the form

$$\mathbf{\Phi}_i = \dot{m}_i(\mathbf{w_i} - \mathbf{v_i}) = \mathbf{h_i} - \dot{m}_i\mathbf{v_i} \tag{4}$$

making the first form of the fundamental equation, Eq.1, to be equivalently written (see, e.g., Cvetićanin (1993))

$$\sum_i (m_i\mathbf{a_i} - (\mathbf{f_i} + \mathbf{\Phi_i})).\delta P_i = \mathbf{0} \tag{5}$$

with $\mathbf{a_i} = \dot{\mathbf{v}}_\mathbf{i}$, the acceleration. Consider virtual displacements δP_i , and a n-set of generalized coordinates q_j (for simplicity, the system is considered holonomic) such that.

$$\delta P_i = \sum_j \frac{\partial P_i}{\partial q_j}.\delta q_j \tag{6}$$

The velocities $\mathbf{v_i} = \mathbf{v_i}(q_j; \dot{q}_j; t); j = 1, \ldots, n$ are, as usual, considered as functions of generalized coordinates and derivatives, as well of time. The following common and straightforwardly derivable kinematic relations, see, e.g., Targ (1976), page 508, will be used as well:

$$\frac{\partial \mathbf{v_i}}{\partial q_j} = \frac{d}{dt}\left(\frac{\partial P_i}{\partial q_j}\right) \tag{7}$$

$$\frac{\partial \mathbf{v_i}}{\partial \dot{q}_j} = \frac{\partial P_i}{\partial q_j} \tag{8}$$

and

$$\mathbf{a_i}.\frac{\partial P_i}{\partial q_j} = \frac{d\mathbf{v_i}}{dt}.\frac{\partial P_i}{\partial q_j} = \frac{d}{dt}\left(\frac{1}{2}\frac{\partial \mathbf{v_i}^2}{\partial \dot{q}_j}\right) - \frac{\partial}{\partial q_j}\left(\frac{1}{2}\mathbf{v_i}^2\right) \tag{9}$$

We also define the generalized, non-conservative force, which already includes the non conservative force $\mathbf{h_i} = \dot{m}_i\mathbf{w_i}$, as

$$Q_j = \sum_i \mathbf{F_i}.\frac{\partial P_i}{\partial q_j} = \sum_i (\mathbf{f_i} + \mathbf{h_i}).\frac{\partial P_i}{\partial q_j} \tag{10}$$

The simplest case of systems of particles with constant mass

For systems of constant mass, the kinetic energy $T_i = \frac{1}{2}m_i\mathbf{v_i}^2$ of a given particle P_i is, apart the mass m_i, identified in both terms of equation 9, such that

$$m_i \frac{d\mathbf{v_i}}{dt} . \frac{\partial P_i}{\partial q_j} = \frac{d}{dt}\left(\frac{1}{2}\frac{\partial m_i \mathbf{v_i}^2}{\partial \dot{q_j}}\right) - \frac{\partial}{\partial q_j}\left(\frac{1}{2}m_i \mathbf{v_i}^2\right) = \frac{d}{dt}\left(\frac{\partial T_i}{\partial \dot{q_j}}\right) - \frac{\partial T_i}{\partial q_j} \quad (11)$$

Observing that, in this simplest case, $\dot{m_i} = 0$, such that $\dot{\mathbf{p_i}} = m_i \mathbf{a_i}$, substituting (6) and (111) in the fundamental equation (1), and observing that the generalized forces Q_j reduce, from (10), to the usual form

$$Q_j = \sum_i \mathbf{f_i} . \frac{\partial P_i}{\partial q_j} \quad (12)$$

one easily obtains the usual Lagrange equations

$$\frac{d}{dt}\frac{\partial T}{\partial \dot{q_j}} - \frac{\partial T}{\partial q_j} = Q_j; \quad j = 1, \ldots, n \quad (13)$$

for a system where all particles have invariant mass.

Systems of particles with mass as explicit function of time $m_i = m_i(t)$

Before discussing the more general case, where $m_i = m_i(q_j; \dot{q_j}; t)$, it is didactic to consider the case where mass is solely an explicit function of time, $m_i = m_i(t)$. The fundamental equation 1 reads,

$$\sum_i (\dot{\mathbf{p_i}} - (\mathbf{f_i} + \mathbf{h_i})).\delta P_i = \sum_j \sum_i (m_i \mathbf{a_i} + \dot{m_i}\mathbf{v_i} - (\mathbf{f_i} + \mathbf{h_i})). \frac{\partial P_i}{\partial q_j}\delta q_j = \mathbf{0} \quad (14)$$

The first term transforms as follows (see equation 9)

$$
\begin{aligned}
&m_i\mathbf{a_i}.\frac{\partial P_i}{\partial q_j}=m_i\frac{d\mathbf{v_i}}{dt}.\frac{\partial P_i}{\partial q_j}=\\
&=\frac{d}{dt}\left(\frac{1}{2}m_i\frac{\partial \mathbf{v_i}^2}{\partial \dot{q_j}}\right)-\frac{dm_i}{dt}\left(\frac{1}{2}\frac{\partial \mathbf{v_i}^2}{\partial \dot{q_j}}\right)-\frac{\partial}{\partial q_j}\left(\frac{1}{2}m_i\mathbf{v_i}^2\right)=\\
&=\frac{d}{dt}\left(\frac{1}{2}\frac{\partial m_i\mathbf{v_i}^2}{\partial \dot{q_j}}\right)-\frac{dm_i}{dt}\left(\frac{\partial}{\partial \dot{q_j}}\left(\frac{\partial}{\partial m_i}\left(\frac{1}{2}m_i\mathbf{v_i}^2\right)\right)\right)-\frac{\partial}{\partial q_j}\left(\frac{1}{2}m_i\mathbf{v_i}^2\right)=\\
&=\frac{d}{dt}\left(\frac{\partial T_i}{\partial \dot{q_j}}\right)-\frac{dm_i}{di}\left(\frac{\partial}{\partial \dot{q_j}}\left(\frac{\partial T_i}{\partial m_i}\right)\right)-\frac{\partial T_i}{\partial q_j}
\end{aligned}
\tag{15}
$$

Observing equation 8, the second term in equation 14 transforms as,

$$
\begin{aligned}
&\dot{m}_i\mathbf{v_i}.\frac{\partial P_i}{\partial q_j}=\frac{dm_i}{dt}\mathbf{v_i}.\frac{\partial P_i}{\partial q_j}=\frac{dm_i}{dt}\mathbf{v_i}\frac{\partial \mathbf{v_i}}{\partial t}=\\
&=\frac{1}{2}\frac{dm_i}{dt}\frac{\partial \mathbf{v_i}^2}{\partial \dot{q_j}}=\frac{dm_i}{dt}\left(\frac{\partial}{\partial \dot{q_j}}\left(\frac{\partial T_i}{\partial m_i}\right)\right)
\end{aligned}
\tag{16}
$$

This latter expression is the most general and concise form for the parcel that depends on the variation of mass in the momentum time derivative. Note that this form is exactly the opposite of the second term appearing on the right hand side of equation 15. They cancel each other when (15) and (16) are substituted into Eq. (14), leading to an equation of motion which has the same form as (13), with the generalized forces extended as given by (10).

This is a very subtle step which explains why a system of particles with variable mass, but given solely as an explicit function of time, $m_i = m_i(t)$, obey the same form of Lagrange Equations that govern a system of particles of invariant mass. This is essentially Agostinelli (1936) result, page 257, now having the generalized forces extended according to equation (10), by including the nonconservative forces defined by (3).

Systems of particles with variable mass as function of time, generalized coordinates and velocities, $m_i = m_i(q_j; \dot{q}_j; t)$

As before, use is made of the fundamental equation, in the form (14) besides equations (7) - (10). In this general case the first term in (14) is given as

$$m_i \frac{d\mathbf{v_i}}{dt} \cdot \frac{\partial P_i}{\partial q_j} = \frac{d}{dt}\left(\frac{1}{2} m_i \frac{\partial \mathbf{v_i}^2}{\partial \dot{q}_j}\right) - \frac{1}{2}\frac{dm_i}{dt}\left(\frac{\partial \mathbf{v_i}^2}{\partial \dot{q}_j}\right) -$$
$$- \frac{\partial}{\partial q_j}\left(\frac{1}{2} m_i \mathbf{v_i}^2\right) + \frac{1}{2}\frac{\partial m_i}{\partial q_j}\mathbf{v_i}^2 =$$
$$= \frac{d}{dt}\left(\frac{1}{2}\frac{\partial m_i \mathbf{v_i}^2}{\partial \dot{q}_j}\right) - \frac{1}{2}\frac{d}{dt}\left(\frac{\partial m_i}{\partial \dot{q}_j}\mathbf{v_i}^2\right) - \frac{1}{2}\frac{dm_i}{dt}\left(\frac{\partial \mathbf{v_i}^2}{\partial \dot{q}_j}\right) -$$
$$- \frac{\partial}{\partial q_j}\left(\frac{1}{2} m_i \mathbf{v_i}^2\right) + \frac{1}{2}\frac{\partial m_i}{\partial q_j}\mathbf{v_i}^2 =$$
$$= \frac{d}{dt}\left(\frac{\partial T_i}{\partial \dot{q}_j}\right) - \frac{1}{2}\frac{d}{dt}\left(\frac{\partial m_i}{\partial \dot{q}_j}\mathbf{v_i}^2\right) - \frac{dm_i}{dt}\left(\frac{\partial}{\partial \dot{q}_j}\left(\frac{\partial T_i}{\partial m_i}\right)\right) - \frac{\partial T_i}{\partial q_j} + \frac{1}{2}\frac{\partial m_i}{\partial q_j}\mathbf{v_i}^2 \tag{17}$$

This is the most general form for the parcel that depends on the acceleration in the momentum time derivative. Taking both most general forms, (16) and (17), and substituting in (14), with the generalized forces given by (10), one finally obtains[6]

$$\frac{d}{dt}\frac{\partial T}{\partial \dot{q}_j} - \frac{\partial T}{\partial q_j} = Q_j + \sum_i \left(\frac{1}{2}\frac{d}{dt}\left(\frac{\partial m_i}{\partial \dot{q}_j}\mathbf{v_i}^2\right) - \frac{1}{2}\frac{\partial m_i}{\partial q_j}\mathbf{v_i}^2\right); \quad j = 1, \ldots, n \tag{18}$$

These are the dynamic equations for a system of particles with variable mass in the form $m_i = m_i(q_j; \dot{q}_j; t)$. Or, simply, *the extended Lagrange* equations. The last two terms can be interpreted as additional parcels of the momentum time rate, caused by the changes in mass as functions of position and velocities. Alternatively, they could be interpreted as additional nonconservative generalized forces which take into account the variation of mass of each particle in the system. By properly defining a total nonconservative generalized force, $\widehat{Q}_j$, equation (18) can finally be written in the compact form,

$$\begin{aligned} &\frac{d}{dt}\frac{\partial T}{\partial \dot{q}_j} - \frac{\partial T}{\partial q_j} = \widehat{Q}_j; \quad j = 1, \ldots, n \\ &\widehat{Q}_j = \sum_i (\mathbf{f_i} + \dot{m}_i \mathbf{w_i}) . \frac{\partial P_i}{\partial q_j} + \sum_i \left\{\frac{1}{2}\frac{d}{dt}\left(\frac{\partial m_i}{\partial \dot{q}_j}\mathbf{v_i}^2\right) - \frac{1}{2}\frac{\partial m_i}{\partial q_j}\mathbf{v_i}^2\right\} \end{aligned} \tag{19}$$

[6]Recall that $v_i = v_i(q_j; \dot{q}_j; t); j = 1, \ldots, M$

Moreover, if the active forces $\mathbf{f_i}$ are split into

$$\mathbf{f_i} = \mathbf{f_i^c} + \mathbf{f_i^{nc}} \tag{20}$$

being $\mathbf{f_i^c}$ conservative and $\mathbf{f_i^{nc}}$ non-conservative parcels, respectively, and such that

$$\sum_i \mathbf{f_i^c}.\delta P_i = -\sum_j \frac{\partial V}{\partial q_j}\delta q_j \tag{21}$$

being V the potential energy function, equation (19) can be written in the convenient form,

$$\begin{aligned}
&\frac{d}{dt}\frac{\partial L}{\partial \dot{q}_j} - \frac{\partial L}{\partial q_j} = \widehat{Q}_j^{nc}; \quad j = 1, \ldots, n \\
&\widehat{Q}_j^{nc} = \sum_i (\mathbf{f_i^{nc}} + \dot{m}_i \mathbf{w_i}).\frac{\partial P_i}{\partial q_j} + \sum_i \left\{ \frac{1}{2}\frac{d}{dt}\left(\frac{\partial m_i}{\partial \dot{q}_j}\mathbf{v_i}^2\right) - \frac{1}{2}\frac{\partial m_i}{\partial q_j}\mathbf{v_i}^2 \right\}
\end{aligned} \tag{22}$$

where $L = T - V$ is the Lagrangean function. As a matter of fact, problems in classical mechanics where mass is an explicit function of velocities are not so common, to say the least. Therefore, if just time and position dependence is considered, one obtains,

$$\begin{aligned}
&\frac{d}{dt}\frac{\partial L}{\partial \dot{q}_j} - \frac{\partial L}{\partial q_j} = \widehat{Q}_j^{nc}; \quad j = 1, \ldots, n \\
&\widehat{Q}_j^{nc} = \sum_i (\mathbf{f_i^{nc}} + \dot{m}_i \mathbf{w_i}).\frac{\partial P_i}{\partial q_j} - \sum_i \left\{ \frac{1}{2}\frac{\partial m_i}{\partial q_j}\mathbf{v_i}^2 \right\}
\end{aligned} \tag{23}$$

Equation (23) can be verified to agree with the derivation by Cvetićanin (1993), for the practical case where mass is solely dependent on time and on generalized coordinates (not in velocities). It must be observed that the first term appearing in Cvetićanin's equation (8) is exactly our term given by equation (16). In the present derivation, the Meshcherski's reactive force has been split in the form $\mathbf{\Phi_i} = \dot{m}_i(\mathbf{w_i} - \mathbf{v_i}) = \mathbf{h_i} - \dot{m}_i\mathbf{v_i}$, such that the term given by equation (16) ends up to be cancelled out, as observed before, turning the final equation (23) somewhat simpler in form.

The Hamiltonian approach[7]

As pointed out by Cvetićanin (1993), depending on the form of the variable mass reactive force, two types of systems may be considered: (i) those completely describable by means of a Lagrangean, despite the mass variation, pertaining to the Levi-Civita case; (ii) those which are not, being purely nonconservative in nature. In the second type, *strictly speaking*, Hamilton's action integral does not generally exits what turns the problem much more complex. Cvetićanin (1993, 1998) takes D'Alembert's differential variational principle as being a basic physical invariant and obtains conservation laws for this kind of systems.

On the other hand, following Pars (1965) words, "the beautiful theorem known as Hamilton's principle is, *en essence*, only an integrated form of the fundamental equation, being the deduction of Lagrange's equation from this principle essentially the same process as that used in the deduction from the fundamental equation itself". In this respect, the extended form of Hamilton's principle for dissipative systems, primarily conceived for constant mass systems, is, see, e.g., Meirovitch (2003), page 68,

$$\int_{t_1}^{t_2} (\delta\overline{T} + \sum_j Q_j \delta q_j) dt = 0 \tag{24}$$

where $\delta(.)$ is the usual variational derivative operator; see, e.g., Gelfand & Fomin (2000). The particular notation, $\delta\overline{T}$, present in 24, refers to the first variation of the kinetic energy of an invariant mass system. In the scenario of discrete systems of variable mass, this form would represent the first variation of the kinetic energy as if the particles of the system had, instantaneously, invariant mass - it is here referred to as the *"instantaneous"*[8] variational derivative; see Appendix. As shown below, in the most general case, $m_i(q_k, \dot{q_k}, t)$, the extended Hamilton's principle also follows from the integration of the fundamental equation.

Firstly, notice that, Eq. (19) assumes what is sometimes called a *fourth fundamental* form,

[7] From another perspective, a comprehensive account of Lagrange's equation, from the point of view of open systems, or non material volumes, may be found in Irschik & Holl, 2004. Accordingly, a recent new derivation, from a Hamiltonian approach applied to non material volumes, may be found in Casetta & Pesce(2013a).

[8] This term, though concise, may sound somewhat ambiguous, if virtual dislocations are concerned.

$$\sum_j \left(\frac{d}{dt}\frac{\partial T}{\partial \dot{q}_j} - \frac{\partial T}{\partial q_j} - \widehat{Q}_j \right) \delta q_j = 0; \quad j = 1, \ldots, n$$
$$\text{with} \tag{25}$$
$$\widehat{Q}_j = \sum_i (\mathbf{f_i} + \dot{m}_i \mathbf{w_i}).\frac{\partial P_i}{\partial q_j} + \sum_i \left\{ \frac{1}{2}\frac{d}{dt}\left(\frac{\partial m_i}{\partial \dot{q}_j}\mathbf{v_i}^2 \right) - \frac{1}{2}\frac{\partial m_i}{\partial q_j}\mathbf{v_i}^2 \right\}$$

On the other hand, as shown in the Appendix, the *"instantaneous"* variational derivative of the kinetic energy of the variable mass system turns to be

$$\delta\overline{T} = \delta T - \sum_i \frac{1}{2}\delta m_i \mathbf{v_i}^2 =$$
$$= \sum_j \left(-\frac{d}{dt}\frac{\partial T}{\partial \dot{q}_j} + \frac{\partial T}{\partial q_j} + \sum_i \left(\frac{d}{dt}\left(\frac{1}{2}\frac{\partial m_i}{\partial \dot{q}_j}\mathbf{v_i}^2 \right) - \frac{1}{2}\frac{\partial m_i}{\partial q_j}\mathbf{v_i}^2 \right) \right) \delta q_j + \tag{26}$$
$$+ \sum_j \frac{d}{dt}\left(\frac{\partial T}{\partial \dot{q}_j}\delta q_j \right) - \sum_j \sum_i \frac{d}{dt}\left(\frac{1}{2}\frac{\partial m_i}{\partial \dot{q}_j}\mathbf{v_i}^2 \delta q_j \right)$$

which, for the simpler time-dependent case, $m_i = m_i(t)$, simplifies as

$$\delta\overline{T} = \sum_j \left\{ -\frac{d}{dt}\frac{\partial T}{\partial \dot{q}_j} + \frac{\partial T}{\partial q_j} \right\} \delta q_j + \frac{d}{dt}\sum_j \left(\frac{\partial T}{\partial \dot{q}_j}\delta q_j \right) = \delta T \tag{27}$$

By using the fourth fundamental form, (25), the variational derivative, (26), reads

$$\delta\overline{T} + \sum_j Q_j \delta q_j - \sum_j \frac{d}{dt}\left(\frac{\partial T}{\partial \dot{q}_j}\delta q_j \right) + \sum_j \sum_i \frac{d}{dt}\left(\frac{1}{2}\frac{\partial m_i}{\partial \dot{q}_j}\mathbf{v_i}^2 \delta q_j \right) = 0$$
$$\text{with} \tag{28}$$
$$Q_j = \sum_i (\mathbf{f_i} + \dot{m}_i \mathbf{w_i}).\frac{\partial P_i}{\partial q_j}$$

Equation (28) may be further integrated from instant t_1 to t_2, leading to the same integral form given by (24). In fact, notice that the third and fourth terms in the left-hand side of Eq. (28) integrates to zero, under the usual assumption of prescribed configurations for the system at the instants of time t_1 and t_2; i.e., $\delta q_j|_{t_1,t_2} = 0$. Conversely, for independent variations δq_j, and considering (26) - or the simpler form (27), valid for the case $m_i = m_i(t)$ - equation (24) leads to the extented Lagrange's equation, (19) (or to the corresponding simpler form), as expected[9].

Applications

Attention is now turned on to some applications. Illustrative examples are addressed in two engineering fields, *offshore* and *civil* engineering. In the first category are included: (i) the reel laying operation of marine cables; (ii) the dynamics of a water column inside a free surface piercing open pipe (and the analogous moon pool problem) and (iii) the hydrodynamic impact of a solid body against a free surface of water. In the second category, the governing equation of motion of vertically collapsing towers is properly derived. At last, considerations on the classical falling chain problem are made.

Ocean Engineering problems

The Ocean Engineering field is an extraordinary source of interesting dynamical problems. In this section three idealized problems are selected to illustrate how the extended Lagrange Equations can help modeling tasks and interpretations. They are ordered in an increasing degree of physical complexity. The text below is essentially based on Pesce (2003), Pesce et al (2006) and Casetta et al (2011)[10].

The reeling cable deployed from a laying ocean barge

A common task in ocean and offshore engineering is the deployment of cables on the sea bottom. Power supply, umbilical and telecommunication cables are just few examples to be mentioned. Usually, the cable is deployed from a reel, installed on the deck of a laying-vessel, sometimes through a moon-pool, as schematically illustrated in Figure 1. The cable is supposed to be acted on by the vessel, ocean current, sea waves, buoyancy and gravity. Initially, and for simplicity, consider only buoyancy and gravity actions. Also for simplicity, the suspended part of the cable is considered fully im-

[9] Generalized canonical equations of Hamilton can be also worked out, by adapting Kozlov's technique Kozlov (1998); see, e.g., Casetta & Pesce (2012) for the $m_i(t)$ case.

[10] Respectively, under kind permission of: ASME - the American Society of Mechanical Engineers; ABCM - the Brazilian Society of Mechanical Sciences and Engineering; SOBENA - the Brazilian Society of Naval Architects.

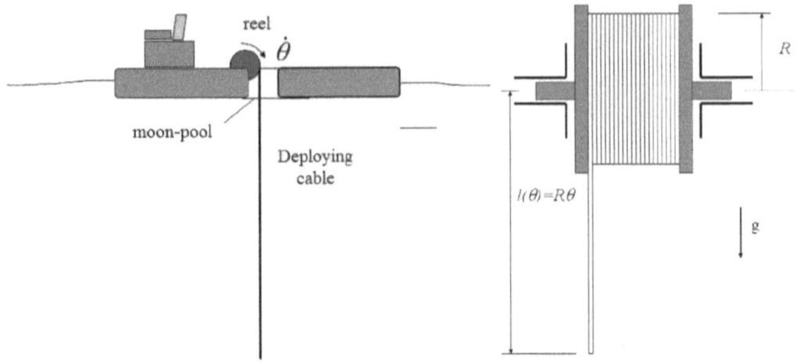

(a) (b)

Figure 1. Figure 1. (a) Cable being deployed from a barge, through a moon-pool. (b) The laying reel. From Pesce et al (2006).

mersed into the water. The reel radius is R and its moment of inertia around the axis of rotation is I. Let μ be the mass per unit of length of the cable, supposed non-extensible and infinitely flexible. Let θ be the generalized coordinate, measured, without loss of generality, from horizontal, such that at a given instant t the suspended length is $l(\theta) = R\theta$. Let also L be the total length of the cable such that $m = \mu L$ is the total mass of the cable. For simplicity, take the cable diameter very small compared to the radius of the reel such that the winding pitch is small enough for all the turns be considered accommodated into a single winding layer. Let also $m_s(\theta) = \mu l(\theta) = \mu R\theta$ and $m_R(\theta) = m - m_s(\theta) = \mu(L - \theta R)$ be, respectively, the suspended and the wound masses of the cable.

Obviously, for this particular problem, the best and shortest way to directly apply the Lagrange equation would be to consider the whole (invariant mass) system. In this case, kinetic energy is simply

$$T = \frac{1}{2}(I + mR^2)\dot{\theta}^2 \tag{29}$$

Accordingly, potential energy is given by,

$$\begin{aligned} V = -g\int_0^{R\theta}(\mu - \rho A)\zeta d\zeta = -\frac{1}{2}(\mu - \rho A)R^2\theta^2 = \\ -\frac{1}{2}((m_s(\theta) - \rho A R\theta)gR\theta) = -\frac{1}{2}(1 - \beta)\mu g R^2\theta^2 \end{aligned} \tag{30}$$

where ρ is the density of water, A the area of the cross section of the cable and $\beta = \rho A/\mu$ is the mass density ratio. An extra non-conservative force has to be considered to model the hydrodynamic friction force acting along the cable during the free-falling deployment. Otherwise, no limit speed would be achieved, and the rotation speed of the reel would increase indefinitely. This force may be written in the form

$$F_f(\theta,\dot{\theta}) = -\frac{1}{2}C_d\rho D(R\dot{\theta})^2 l(\theta) = -\frac{1}{2}C_f\rho D R^3\theta\dot{\theta}^2 \tag{31}$$

with the viscous friction force nondimensional coefficient $C_f = \mathcal{O}(10^{-3})$. The direct application of the usual Lagrange equation to this *invariant mass* system, in the form,

$$\frac{d}{dt}\left(\frac{\partial T}{\partial\dot{\theta}}\right) - \frac{\partial T}{\partial\theta} + \frac{\partial V}{\partial\theta} = Q_\theta = F_f R \tag{32}$$

leads to the consistent equation of motion

$$(I + mR^2)\ddot{\theta} - (1-\beta)\mu R^2\theta = F_f(\theta,\dot{\theta})R \tag{33}$$

Suppose now that, for some practical reason, the analyst decides to take a sub-system composed by the reel and by the wound part of the cable, considering the suspended part of the cable as a second sub-system. Note that the suspended part of the cable can be considered as a material point gaining mass at rate $\dot{m}_s(\theta)$, with velocity $w = R\dot{\theta}$. The resultant of the active forces applied to the suspended part is, therefore,

$$f(\theta) = (m_s(\theta) - \rho A R\theta)g - \tau(\theta) \tag{34}$$

being $\tau(\theta)$ the traction force at the upper section. Applying the extended Levi-Civita form of Newton's law to the suspended part, it is easily obtained

$$\frac{d}{dt}(m_s(\theta)R\dot{\theta}) = (1-\beta)\mu R\theta g - \tau(\theta) + \dot{m}_s(\theta)R\dot{\theta} + F_f(\theta,\dot{\theta}) \tag{35}$$

Hence, the traction applied by the wound part to the suspended part of the cable is simply

$$\tau(\theta) = \mu R\theta((1-\beta)g - R\ddot{\theta}) + F_f(\theta,\dot{\theta}) \tag{36}$$

Let, now, $J = I + M_R(\theta)R^2 = I + \mu R^2(L - R\theta)$ be the moment of inertia of the first sub-system (reel + wound cable), such that the corresponding kinetic energy is given by $T_1 = 1/2J\dot{\theta}^2$. Note that mass exits the wound part with velocity $w = R\dot{\theta}$, at a rate $\dot{m}_R(\theta) = -\mu R\dot{\theta}$. If, *erroneously*, the usual Lagrange equation is applied to the first sub-system in the form,

$$\begin{aligned}&\frac{d}{dt}\left(\frac{\partial T_1}{\partial \dot{\theta}}\right)-\frac{\partial T_1}{\partial \theta}=\widehat{Q}_\theta;\\&with\\&\widehat{Q}_\theta=R(\tau(\theta)+\dot{m}_R(\theta)R\dot{\theta})\end{aligned} \tag{37}$$

the following and obviously incorrect equation of motion is obtained (compare it with (33)),

$$(I+mR^2)\ddot{\theta}+\frac{1}{2}\mu R^3\dot{\theta}^2-(1-\beta)\mu g R^2\theta=F_f(\theta,\dot{\theta})R \tag{38}$$

Note the presence of an erroneous quadratic term in velocity, namely, $1/2\mu R^3\dot{\theta}^2$. This term is quadratic in the angular velocity of the reel. Therefore, apart the conceptual error, it could lead to significant discrepancies in the calculated traction, if the rotation speed were large enough. On the other hand, if the correct form, the extended Lagrange equation, given by (19) or (23), is applied to this variable mass sub-system, i.e.,

$$\begin{aligned}&\frac{d}{dt}\left(\frac{\partial T_1}{\partial \dot{\theta}}\right)-\frac{\partial T_1}{\partial \theta}=\widehat{Q}_\theta\\&with\\&\widehat{Q}_\theta=(\tau(\theta)+\dot{m}_R(\theta)R\dot{\theta})R-\frac{1}{2}\frac{dm_R}{d\theta}R^2\dot{\theta}^2\end{aligned} \tag{39}$$

the *consistent* equation of motion, Eq. (33), previously derived when the whole system was considered, is readily recovered.

As an example, the case of a multi-functional electric cable being deployed vertically in deep water is taken. The cable has a diameter $D = 100mm$ and a weight per unit length, $\gamma = 0.15kN/m$. The reel has radius R=1.0m and inertia $I = 4t.m^2$. The total length of the cable is $L = 3000m$. Figure 2 shows the simulation of an "immersed-free-fall" deployment. The depth is supposed to be $1500m$ and the simulation is carried out up to the instant the cable touches the soil. The solution, $\dot{\theta}(t)$ and $\tau(t)$, obtained from both equations, the consistent and the erroneous ones, are compared. Initial conditions were chosen as $\dot{\theta}(0) = 0$ and $l(\theta(0)) = 10m$ (the initial suspended length). As can be noticed, there is not a significant difference between both results, as the quadratic term in velocity is not dominant for this operation, rendering, the importance of the present analysis much more theoretical than practical.

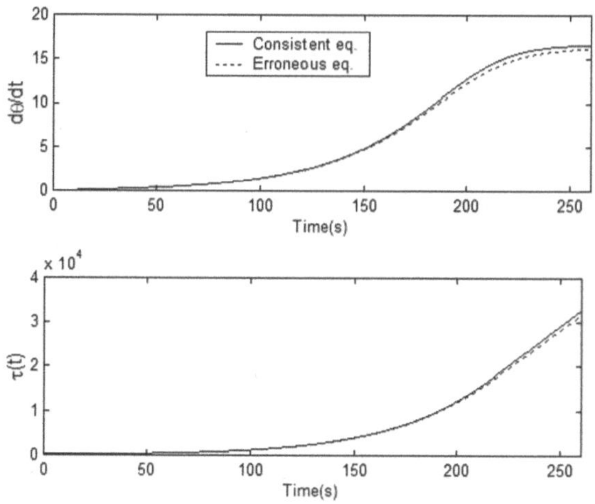

Figure 2. The "free-fall" deployment of a multi-functional electric cable from a reel barge. Wave and ocean current actins not considered. From Pesce et al (2006).

The dynamics of the water column in a free-surface piercing tube and the moon-pool problem

Moon-pools are commonly found in many floating offshore structures as in pipe-laying and work barges. Figure 3.a presents a mono-column oil production platform, with a cylindrical moon-pool. Pipes and cables are suspended through the moon-pool to the sea bottom. The main purpose is to provide safer operational conditions, regarding the action of waves. Nevertheless, the water column inside the moon-pool may resonate due to the wave action and to the motions of the floating platform. Resonance in this case should be avoided. Another interesting analogous problem is the dynamics of free surface piercing pipes used as elements of hydro-electrical power devices driven by the action of waves. In this latter case, however, resonance is the key to a good performance.

Either case, the nonlinear dynamics of the water column must be modeled properly. For the purpose of the present text, just the simplest case of a free-surface piercing pipes opened to the atmosphere is considered. Only the unforced problem is addressed. The forced problem, due to the action of ocean waves, might then be readily assessed. A full account of the mono-

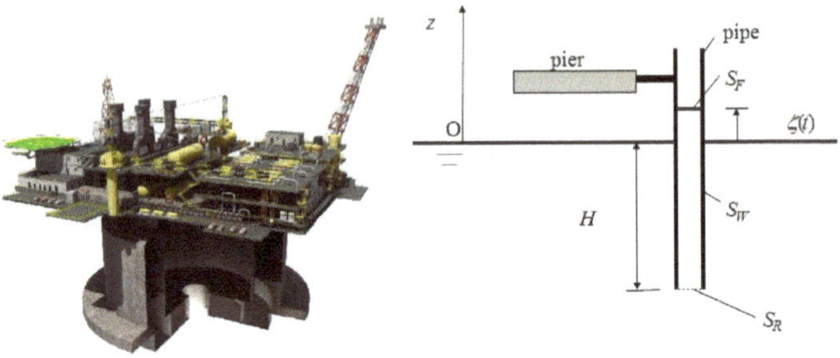

(a) (b)

Figure 3. (a) A mono-column, floating oil production platform. The risers and umbilical cables that connect the production plant to the well heads are suspended from the platform through the moon-pool. (b) The free surface piercing, open pipe problem. Unit normal vectors are positive outwards the surfaces which enclose the mass of water inside the pipe. From Pesce et al (2006).

column platform moon-pool problem may be found in the master thesis by Torres (2007).

Consider an open vertical circular pipe of internal radius R piercing a quiescent external free surface of an incompressible and inviscid liquid. Let H be the draft of the pipe. Let g be the acceleration of gravity. For simplicity, let $\zeta(t)$ describe the position of the free surface of the column of liquid in the interior of the pipe. Clearly, a simplified model with just one degree of freedom (one generalized coordinate) can be used, $\zeta(t)$. Other free surface vibration modes are not considered in this simplified model.

Before the Lagrangean approach is applied, the equation of motion is derived from the point of view of potential theory in hydrodynamics. This equation will serve as a basis of comparison.

The classical hydrodynamic approach

Take the material sub-system as composed solely by the liquid inside the pipe. That is, the liquid that in a given instant fills the volume bounded by $\partial\Omega = S_F \cup S_R \cup S_W$. S_F is the (material and non-permeable) free surface, $z = \zeta(t)$. S_W is the material, fixed and non-permeable surface, corresponding to the interior wetted surface of the pipe and S_R the non-material (permeable) fixed control surface at the lower end of the pipe, given by $z_R = -H$. An exchanging flux of mass clearly exists between the sub-system and the external fluid. Note that the vertical components of the outwardly positive normal unit vector are $n_z = 1$ on S_F and $n_z = -1$ on S_R. Let the flow be non-rotational and $\phi(z)$ the potential velocity function. The kinematic (Neumann) boundary condition on S_F is

$$\frac{\partial\phi}{\partial z} = \frac{\partial\zeta}{\partial t} = \dot{\zeta} \tag{40}$$

The velocity potential, inside the pipe, can then be written

$$\phi(x, y, z, t) = z\dot{\zeta} \tag{41}$$

Note that $\phi_t = z\ddot{\zeta}$. Let the fluid be unbounded in the far field. The dynamic pressure on S_R is given by

$$p_D(x, y)\big|_{S_R} = -\frac{1}{2}\rho\dot{\zeta}^2 \tag{42}$$

Pressure on S_F is taken as null, as usual. Therefore, from momentum considerations, the dynamic equation for $\zeta(t)$ is readily derived. In fact, let p_z be the linear momentum of the fluid inside the pipe. Then, from classical potential hydrodynamics, see, e.g. Newman (1978),

$$\frac{dp_z}{dt} = \rho\frac{d}{dt}\int_S \phi n_z dS = F_H + F_D - \psi \tag{43}$$

where

$$\begin{aligned} F_H &= -\rho g A\zeta \\ F_D &= -\int_{S_R} p_D(x, y, z, t) n_z dS = -\rho A\frac{1}{2}\dot{\zeta}^2 \end{aligned} \tag{44}$$

are respectively the forces due to the differential hydrostatic pressure and to the hydrodynamic pressure applied to the water column, on S_R, and

$$\psi = \rho\int_{S_R} \frac{\partial\phi}{\partial z}\left(\frac{\partial\phi}{\partial n} - U_n\right) dS = -\rho\int_{S_R}\left(\frac{\partial\phi}{\partial z}\right)^2 dS = -\rho A\dot{\zeta}^2 \tag{45}$$

is the flux of linear momentum across the fluid boundary, S_R. The mass of fluid inside the pipe at a given instant is an explicit function of position, $M = \rho A(\zeta + H)$. Therefore, the time rate of linear momentum inside the pipe can be directly calculated,

$$\frac{dp_z}{dt} = \frac{d}{dt}(\rho A(\zeta + H)\dot{\zeta}) = \rho A(\zeta + H)\ddot{\zeta} + \rho A\dot{\zeta}^2 \tag{46}$$

Note that this result could also be obtained by recalling that the derivative and integral signs are interchangeable on the fixed control surface S_R. Therefore,

$$\begin{aligned}
\frac{dp_z}{dt} &= \rho\frac{d}{dt}\int_S \phi n_z dS = \\
&= \rho\frac{d}{dt}\int_{S_F} \phi n_z dS + \rho\frac{d}{dt}\int_{S_R} \phi n_z dS = \\
&= \rho\frac{d}{dt}\int_{S_F} \phi dS - \rho\frac{d}{dt}\int_{S_R} \phi dS = \rho\frac{d}{dt}\int_{S_F} \phi dS - \rho\int_{S_R} \frac{\partial\phi}{\partial t} dS = \\
&= \rho\frac{d}{dt}\int_{S_F} \zeta\dot{\zeta} dS + \rho\int_{S_R} H\ddot{\zeta} dS = \\
&= \rho A(\zeta\ddot{\zeta} + \dot{\zeta}^2 + H\ddot{\zeta}) = \rho A(\zeta + H)\ddot{\zeta} + \rho A\dot{\zeta}^2
\end{aligned} \tag{47}$$

Collecting terms, from (43) - (46), one obtains

$$\rho A(\zeta + H)\ddot{\zeta} + \rho A\dot{\zeta}^2 = -\rho A g\zeta - \frac{1}{2}\rho A\dot{\zeta}^2 + \rho A\dot{\zeta}^2 \tag{48}$$

This reduces to the following nonlinear homogeneous equation

$$\ddot{\zeta} + \frac{1}{2}\frac{\dot{\zeta}^2}{(\zeta + H)} + g\frac{\zeta}{(\zeta + H)} = 0 \tag{49}$$

Let $\eta = \zeta/H$ be the dimensionless free surface position. Scaling nondimensional time with $t' = \omega t$, with $\omega = \sqrt{g/H}$, Eq. (49) may be written in dimensionless form as,

$$\ddot{\eta} + \frac{1}{2}\frac{\dot{\eta}^2}{\eta + 1} + \frac{\eta}{\eta + 1} = 0 \tag{50}$$

The constant ω can be readily recognized as the dimensional natural frequency of the corresponding linear oscillator $\ddot{\eta} + \eta = 0$, obtained from Eq. (50) in the case of small displacements and small velocities. Note also that the term that is quadratic in velocity is, in fact, conservative. Notice also that Eq. (50) is valid for $\eta > -1$. A singular behavior, leading to infinity acceleration, arises when $\eta = -1$, i.e. $\zeta = -H$. Physically, this corresponds to the water-column surface reaching the bottom of the pipe, the mass of the system becoming zero. Beyond this point, a cavity would form, and a proper modeling should consider this other highly nonlinear phenomenon.

The Lagrange equation approach

From another point of view, the dynamics of the fluid inside the pipe may be modeled as a single degree of freedom (hydro-) mechanical system. The kinetic energy is then given by

$$T = \frac{1}{2}\rho A(\zeta + H)\dot{\zeta}^2 \tag{51}$$

In this case, where effluxes of mass and kinetic energy do exist from the domain under analysis (the fluid inside the pipe), one must use the extended Lagrange equation (19). One obtains

$$\begin{aligned} \frac{d}{dt}\frac{\partial T}{\partial \dot{\zeta}} &= \rho A(\zeta + H)\ddot{\zeta} + \rho A\dot{\zeta}^2 \\ \frac{\partial T}{\partial \zeta} &= \frac{1}{2}\rho A\dot{\zeta}^2 \end{aligned} \tag{52}$$

Note that, if the system were defined starting from the kinetic energy, the mass dependence on position could not be promptly recognized. As can be clearly seen, the quantity

$$\frac{d}{dt}\frac{\partial T}{\partial \dot{\zeta}} - \frac{\partial T}{\partial \zeta} = \rho A\left((\zeta + H)\ddot{\zeta} + \frac{1}{2}\rho A\dot{\zeta}^2\right) \tag{53}$$

that arises when the usual Lagrange equation is applied, is not the time rate of change of linear momentum inside the pipe, which is given by Eq. (46). To this quantity it should be added $\frac{1}{2}\rho A\dot{\zeta}^2$, that is exactly the quantity one would obtain from the additional term $\frac{1}{2}\sum_i \frac{\partial m_i}{\partial \zeta}\mathbf{v_i}^2$, that appears on the right hand side of equation (19). In fact,

$$\begin{aligned} &\frac{1}{2}\sum_i \frac{\partial m_i}{\partial \zeta}\mathbf{v_i}^2 = \frac{1}{2}\sum_i \frac{\partial m_i}{\partial \zeta}\dot{\zeta}^2 = \frac{1}{2}\frac{\partial}{\partial \zeta}\left(\sum_i m_i\right)\dot{\zeta}^2 = \\ &= \frac{1}{2}\rho A\frac{\partial}{\partial \zeta}\left(\int_{-H}^{\zeta} dz\right)\dot{\zeta}^2 = \frac{1}{2}\rho A\frac{\partial}{\partial \zeta}(\zeta + H)\,\dot{\zeta}^2 = \frac{1}{2}\rho A\dot{\zeta}^2 \end{aligned} \tag{54}$$

To consistently apply the extended Lagrange Equation, one must consider the equivalent non-conservative generalized force, according to Eq. (19b), that in this case reads,

$$\begin{aligned}\widehat{F}_z = f + \dot{m}w - \frac{1}{2}\sum_i \frac{\partial m_i}{\partial \zeta}\mathbf{v_i}^2 = (F_H + F_D) + \dot{m}w - \frac{1}{2}\frac{\partial m}{\partial \zeta}\dot{\zeta}^2 = \\ \left(-\rho A g \zeta - \frac{1}{2}\rho A \dot{\zeta}^2\right) + (\rho A \dot{\zeta}^2) - \left(\frac{1}{2}\rho A \dot{\zeta}^2\right) = -\rho A g \zeta\end{aligned} \tag{55}$$

Note that, in this case, the term given by Eq. (44.b) is, quantitatively, half the momentum flux and exactly the same as that corresponding to the dynamic pressure. Note also that, curiously, only the (conservative) hydrostatic term is left. Collecting results, (53) - (55), *recovers the consistent dynamic equation, given by Eqs.* (49) or (50). Otherwise, *disregarding the term given by Eq. (54) would lead to the erroneous equation of motion,*

$$\ddot{\eta} + \frac{\dot{\eta}^2}{\eta + 1} + \frac{\eta}{\eta + 1} = 0 \tag{56}$$

Apart the conceptual correctness, from the point of view of practical application, significant differences between Eq. (50) and Eq. (56) arise only if the motion is large enough. Figure 4 presents a comparison between results obtained by using Eq. (50) and the erroneous Eq. (56). The phase trajectories are closed curves, since no dissipation was considered. The quadratic terms in velocity are conservative, as already anticipated. For all initial displacements, the acceleration attains a maximum when the water column level reaches its minimum value (mass inside the pipe is minimum), as already mentioned. This is qualitatively observed in reality.

The vertical hydrodynamic impact problem

Consider a body impacting a quiescent free surface of a liquid. In the offshore engineering context, important examples that could be mentioned are the deployment of lifeboats from platforms, ship slamming and wave impacts against structures. Von Kármán (1929) first addressed the simplest problem (of an impacting rigid body), in order to estimate the loading on seaplane floaters during "landing".

The duration of the impact is so short that inertia forces dominate other ones. This makes consistent to treat the problem within potential flow theory. As well known, it is usual practice to treat potential hydrodynamic problems involving motion of solid bodies within the frame of system dynamics. This is done whenever a finite number of generalized coordinates can be used as a proper representation for the motion of the whole fluid. Terming this approach as 'hydro mechanical' the impact force acting upon

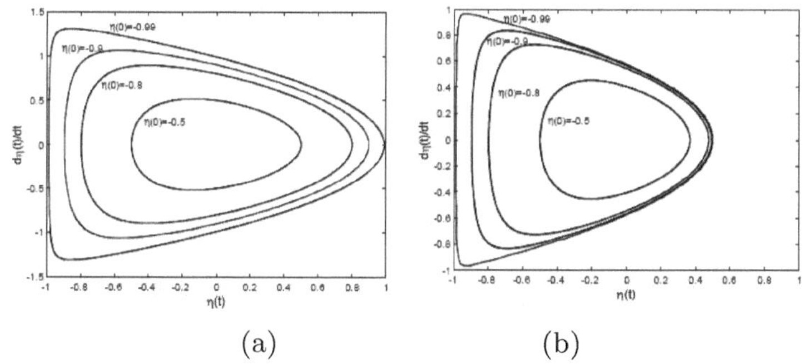

Figure 4. (a) Phase portraits of the water column dynamics. Comparison between results from the consistent (left) and the erroneous equations (right). Initial conditions: $\eta(0) = -0.99; \cdots - 0.5; \dot{\eta}(0) = 0$. From Pesce et al (2006).

the body, for a purely vertical impact, is simply written, in the manner of Von Kármán - see, e.g., Faltinsen (1990), chapter 9,

$$F_z = -\frac{d}{dt}(M_{zz}W) \tag{57}$$

being W the vertical velocity, positively oriented downwards, and M_{zz} the corresponding added mass.

Note that, in this case, the added mass may be written as an explicit function of the position of the body and has to be determined at each instant of time, during the impact phenomenon. This is not an easy task, as the hydrodynamic problem is geometrically nonlinear due to the presence of the free surface and the moving body. Usually in hydrodynamics, the added-mass is defined only *in the bulk of fluid*, excluding the jets. In this case, an efflux of kinetic energy does exist from the domain under analysis (the bulk of fluid) to the jets. In other words, there is an effective loss of energy gauged by a “loss in the added mass” through the jets. The extended Lagrange equation is the one that should be used. Otherwise, if the added mass is defined considering the whole liquid, including the bulk and the jets, there is no loss of kinetic energy, or equivalently, “no loss of added-mass”, the usual Lagrange equation should be used instead.

The formulation of the impact problem under the Lagrangean formalism, should recall the explicit added mass dependence on the position of the body. However, restraining the analysis to the bulk of the fluid, an

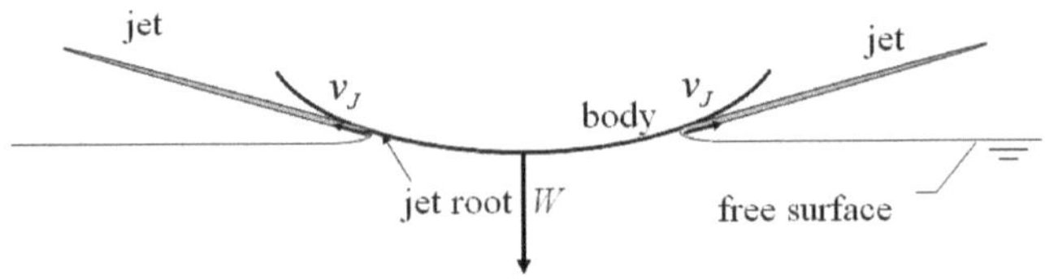

Figure 5. A convex and axisymmetric rigid body vertically impacting a quiescent free surface of a liquid. From Pesce et al (2006).

erroneous equation of motion would be obtained if the Lagrange equation were not properly applied, namely, the extended form given by Eq. (19) or (23). This subtle aspect has been the source of controversies and discussions on false paradoxes by many authors in the specialized literature; see, e.g., Cointe et al (2004), Molin & Cointe (1996), Wu (1998), with contributions by the present authors in Casetta & Pesce (2005, 2006, 2007) and Pesce (2003). Those and other theoretical issues are discussed in Casetta et al (2011), where Analytical Mechanics arguments, based on both discrete and continuum representations - with the application of the Reynolds transport theorem - enabled a deeper analysis, accounting for energy balance and confirming the proper formulation of the problem.

Take, for simplicity, the purely vertical impact case of an axisymmetric rigid body against a free surface and let ζ be defined as the (positive downward) vertical displacement of the body into the water, measured from the quiescent free surface. Let $W(t)$ be the downward vertical velocity. The kinetic energy in the bulk of the liquid may be written as

$$\begin{aligned} T &= \frac{1}{2} M_{zz} W^2 \\ M_{zz} &= M_{zz}(\zeta) \\ \zeta &= \int_{0+}^{t} W dt \end{aligned} \tag{58}$$

The added mass, consistently defined *in the bulk of the liquid*, at each instant of time, takes into account the so-called wetted correction, due to the marching of the jet root. In this case, as already observed, the correct Lagrange equation approach is to use Eq. (19), restricted to position dependence of mass only, such that the total vertical force applied by the body (and the jets) on the bulk of the fluid is given by

$$-F_z^B = -\frac{d}{dt}\left(\frac{\partial T}{\partial W}\right) + \frac{\partial T}{\partial \zeta} - \frac{1}{2}\frac{dM_{zz}}{d\zeta}W^2 \tag{59}$$

Notice that the added mass is supposed to be accreted at null velocity. The force applied by the bulk of fluid on the body is then, simply,

$$F_z = -F_z^B = -\frac{d}{dt}\left(\frac{\partial T}{\partial W}\right) + \frac{\partial T}{\partial \zeta} - \frac{1}{2}\frac{dM_{zz}}{d\zeta}W^2 \tag{60}$$

Equation (60) transforms, as expected, into

$$F_z = -\frac{d}{dz}(M_{zz}W) + \frac{1}{2}W^2\frac{dM_{zz}}{d\zeta} - \frac{1}{2}\frac{dM_{zz}}{d\zeta}W^2 = -\frac{d}{dt}(M_{zz}W) \tag{61}$$

The third term appearing on the right hand side of Eq. (60), if not considered, would lead to an *erroneous* assertive, according to which,

$$F_z = -\frac{1}{2}\frac{M_{zz}}{dt}W - M_{zz}\frac{dW}{dt} \tag{62}$$

As mentioned, Eq. (60) recovers the simple expected result. Note that in the present case the changing in the added mass is due to an actual changing of size and shape of the body in contact with the liquid.

Equation (62) would be correct in form, however, if the analysis had considered the whole fluid domain, including not only the bulk but also the jets. In that case the added mass $= M_{zz} = M_{zz}^{wfd}$ should be interpreted as a measure of kinetic energy of the whole fluid domain; see Casetta & Pesce (2006) and Casetta et al (2011). Obviously, in that case, there would be no efflux of kinetic energy - neither an "efflux of added mass". The extended Lagrange equation, for systems with mass explicitly dependent on position would be no longer applicable. One should then apply the usual form of the Lagrange equation, as in Lamb (1932), art. 137. In this latter case the computation of the added mass corresponding to the whole fluid domain, M_{zz}^{wfd}, would be even more difficult than that corresponding to the bulk of the fluid.

A Rayleigh-like dissipation function

Note that, generally, for a single degree of freedom model, with mass explicitly dependent on position, in this case, $M_{zz}(\zeta) = M_b(\zeta)$, where b denotes "bulk of liquid", a Rayleigh-like function could be defined as

$$R(\zeta, \dot{\zeta}) = \frac{1}{2}C(\zeta, \dot{\zeta})\dot{\zeta}^2 \tag{63}$$

with

$$C(\zeta,\dot{\zeta}) = \frac{1}{3}\frac{\partial M_b}{\partial \zeta}\dot{\zeta} = \frac{1}{3}\dot{M}_b \tag{64}$$

i.e.;

$$R(\zeta,\dot{\zeta}) = \frac{1}{6}\frac{\partial M_b}{\partial \zeta}\dot{\zeta}^3 \tag{65}$$

such that its partial derivative with respect to the generalized velocity would then be given by

$$\frac{\partial R}{\partial \dot{\zeta}} = \frac{1}{2}\frac{\partial M_b}{\partial \zeta}\dot{\zeta}^2 \tag{66}$$

Clearly, $R(\zeta,\dot{\zeta})$ is a non conservative function, draining energy from the system or feeding energy to it, depending on if $\dot{M}_b = \left(\frac{\partial M_b}{\partial \zeta}\dot{\zeta}\right)$ is positive (added mass accreted to the system) or negative (added mass "expelled" from the system). In the present case $\dot{M}_b(t) > 0$, always. The energy is drained to the jets. The extended Lagrange equation for the impact problem might then be written in a usual Rayleighian form,

$$F_z = -\left(\frac{d}{dt}\frac{\partial T}{\partial \dot{\zeta}} - \frac{\partial T}{\partial \zeta} + \frac{\partial R}{\partial \dot{\zeta}}\right) \tag{67}$$

It should be remarked however that the Rayleigh-like function has not been introduced in an *ad-hoc* manner. It has been defined after the extended dissipative generalized force term, $-\frac{1}{2}\frac{\partial M_b}{\partial \zeta}\dot{\zeta}^2$, already present in Eq. (60), was derived from fundamental variational principles. The introduction of such a Rayleigh-like function should here be taken just as an additional physical interpretation to the problem, from the point of view of energy dissipation: the energy drained from the bulk of the liquid to the jets.

The impacting sphere

To finalize the present analysis, an analytical result will be shown, applying a still very useful approximate approach due to Wagner (1931)[11]. In this approach the added mass is defined in the bulk of the fluid only and the flux of kinetic energy to the jets must be properly considered.

[11] A comprehensive classification on analytical methods to deal with the hydrodynamic impact problem is given in Korobkin (2004).

The impacting surface of the body is taken as the equivalent surface of a "time-varying floating plate". In other words, the interaction problem is treated as the "continuous impact of a floating plate" whose area changes in time. The usual free-surface condition is replaced by an equipotential boundary condition, $\phi = 0$, that corresponds to the limit of infinity frequency in the sense of the wave radiation problem.

At the very start stage, the condition $\frac{\partial \phi}{\partial z} = 0$ is valid on an equipotential control surface that replaces the actual free surface, except at the surface-body intersection, where jets are formed. Actually, to impose such a condition everywhere, even at the body intersection, is equivalent to disregard the flux of kinetic energy through the jets. A more detailed analysis is presented in Casetta et al (2011) and Pesce (2005).

As a simple example, we take the case of a sphere of radius R and mass m, reaching the free surface with initial velocity W_0. Let the dimensionless time be defined as

$$\mathrm{t} = W_0 t/R \tag{68}$$

such that the dimensionless position, velocity and acceleration are given by

$$\eta = \zeta/R; \quad \dot{\eta} = \frac{d\eta}{d\mathrm{t}} = \frac{1}{W_0}\frac{d\zeta}{dt}; \quad \ddot{\eta} = \frac{d^2\eta}{d\mathrm{t}^2} = \frac{R}{W_0^2}\frac{d^2\zeta}{dt^2} \tag{69}$$

The dimensional hydrodynamic impact force is then written,

$$F_z = -m_D \frac{W_0^2}{R}\frac{d}{dt'}(\mu_b(\eta)\dot{\eta}) \tag{70}$$

where $\mu_b(\eta) = M_b(\zeta)/m_D$ is the "specific added mass", i.e., a nondimensional form of the added mass with respect to $m_D = 4\pi R^3/3$, the mass of liquid displaced by a totally immersed sphere. Since no external force is considered impressed to the body other than the hydrodynamic impact force itself, the equation of motion may be easily derived in nondimensional form as,

$$\ddot{\eta} + \frac{1}{\beta + \mu_b(\eta)}\frac{d\mu_b}{d\eta}\dot{\eta}^2 = 0 \tag{71}$$

where $\beta = m/m_D$ is the specific mass of the sphere.

If the erroneous Eq. (62) were supposed to hold, the nondimensional form for the equation of motion would read, instead,

$$\ddot{\eta} + \frac{1}{2}\frac{1}{\beta + \mu_b(\eta)}\frac{d\mu_b}{d\eta}\dot{\eta}^2 = 0 \tag{72}$$

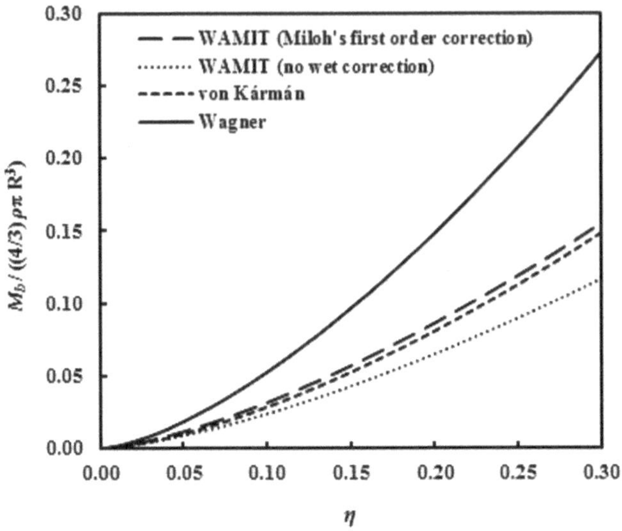

Figure 6. Added mass for an impacting sphere: Wagner's and von Kármán asymptotic approximation compared to "exact body shape approach" numerical solutions, *via* $WAMIT^{®}$. From Casetta et al (2011).

The dimensionless added mass function $\mu_b(\eta)$ can be obtained, for example, through a numerical procedure. For instance, the added mass for the penetrating sphere shown in Figure 6 was determined by taking the infinity frequency limit provided by $WAMIT^{®}$[12]. Notice that, the free surface boundary condition $\phi = 0$ is the corresponding infinite frequency asymptotic limit in the usual oscillating floating body problem (see, e.g., Newman (1978)).

For bodies of regular shape, as edges, cylinders and spheres, asymptotic techniques and singular perturbation methods can be applied successfully; see, e.g., Korobkin (2004). Wagner's approach leads to the well-known approximate expression for the added mass of the impacting sphere, see, e.g, Faltinsen & Zhao (1997),

$$\mu_b(\eta) \approx \frac{3\sqrt{3}}{\pi}\eta^{3/2} \tag{73}$$

[12] $WAMIT^{®}$: 'Wave Analysis Massachusetts Institute of Technology'.

On the other hand, von Kármán approximation leads to, see, e.g., Casetta (2004)

$$\mu_b(\eta) \approx \frac{2\sqrt{2}}{\pi}\eta^{3/2} \tag{74}$$

Wagner's approach, together with the generally valid Eq. (71), leads then to the following consistent dimensionless equation of motion,

$$\ddot{\eta} + \frac{\frac{9\sqrt{3}}{2\pi}\eta^{1/2}\dot{\eta}^2}{\beta + \frac{3\sqrt{3}}{\pi}\eta^{3/2}} = 0 \tag{75}$$

Observe that the specific mass of the sphere, β, is the only parameter left in Eq. (75). However, if Eq. (72) were supposed to hold, the equation of motion would read,

$$\ddot{\eta} + \frac{1}{2}\frac{\frac{9\sqrt{3}}{2\pi}\eta^{1/2}\dot{\eta}^2}{\beta + \frac{3\sqrt{3}}{\pi}\eta^{3/2}} = 0 \tag{76}$$

Additionally, notice that the magnitude of the dimensional impacting force may be written in terms of the body weight as,

$$F_z(t) = \left(F_R^2 \frac{d^2\eta}{dt\prime^2}\right) mg \tag{77}$$

where

$$F_R = \frac{W_0}{\sqrt{gR}} \tag{78}$$

may be defined as the "impact Froude number".

Equation (71) was asymptotically derived assuming small submergence, say $\eta < 0.1$. In this stage the impacting force reaches its maximum value. Moreover, the impacting force usually dominates the buoyancy force and that is the reason why buoyancy has not been considered. As can be easily inspected from Eq. (75), the impacting force peak decreases the mass ratio and increases with the square of the Froude number. In fact, the impacting force peak is of order $F_I = \mathcal{O}(\beta^{-1}F_R^2 mg) = \mathcal{O}(F_R^2 m_D g)$. On the other hand, the maximum buoyancy force (totally immersed sphere) is given by $F_B = m_D g = \beta^{-1}mg$. Therefore, $F_I/F_B \approx \mathcal{O}(F_R^2) >> 1$, for high-speed impacts. As a figure, if the sphere is dropped (in vacuum) to the free surface, from a height H, then $F_R^2 = 2H/R$.

Equation (75) is to be integrated under initial conditions $\eta(0) = 0$ and $\dot{\eta}(0) = 1$. Figure 7 exemplifies the large discrepancies existing between the

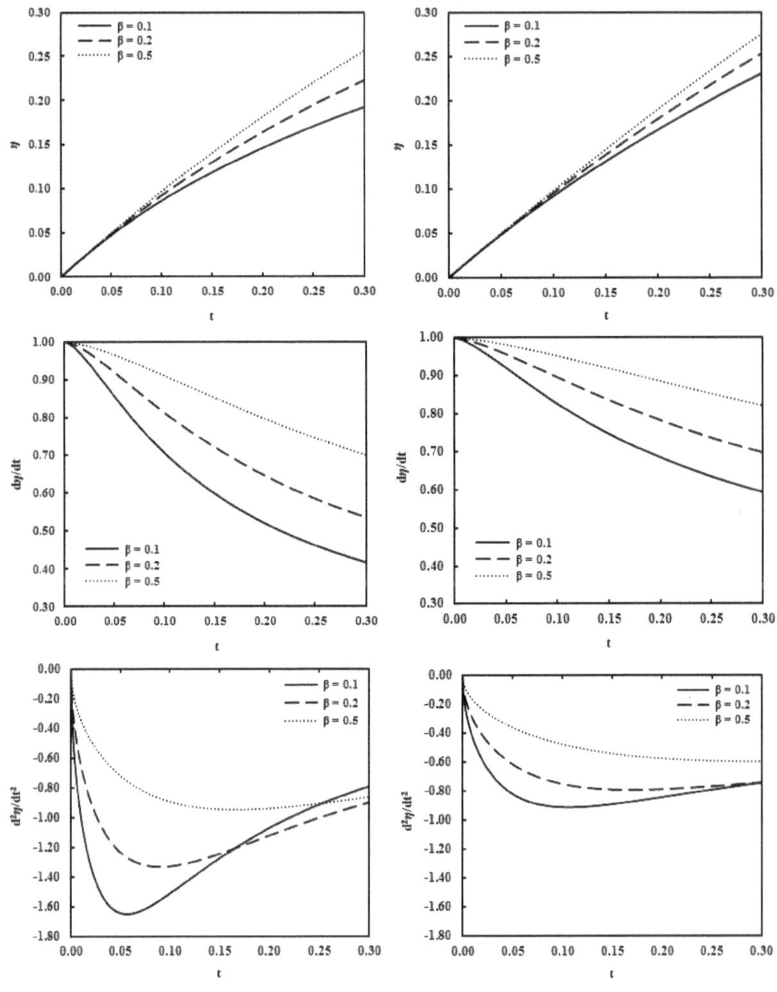

Figure 7. Dimensionless penetration, velocity and acceleration of an impacting sphere versus dimensionless time. Left: Consistent equation (70). Right: non-consistent equation (71). Added mass determined through Wagner's approximation. From Casetta et al (2011).

results obtained from both equations: the consistent equation, Eq. (75) and the erroneous one, Eq. (76). In fact, if the erroneous equation is used, the acceleration peak is approximately 50% smaller, what implies a rather significant underestimation for the magnitude of the maximum impact force. Nevertheless, penetration is not affected as much. Note also that, for usual offshore and naval engineering applications, practical relevance exists for mass ratio values smaller than 1.

A Civil Engineering problem: the vertically collapsing tower

This section brings a summary of the discussion contained in Pesce et al (2012), from where reasoning, writing, illustrations, figures and tables are extracted[13]. The reader is directed to that reference for details and further discussions. As a matter of fact, the initial motivation came from two recent papers by Bažant & Verdure (2007), and by Seffen (2008). Within a simple - though quite representative - single degree of motion modeling, an interesting conceptual point was raised: what should be the proper form of the equation of motion which governs the dynamics of vertically collapsing towers? Apart some differences on the consideration of a parameter known as the compaction factor, both authors ended up with two distinct equations of motion, derived by applying either the usual or the extended form of the Lagrange equation, Eq. (19), this latter by Pesce (2003). The two distinct equations essentially differ from each other by a term of form $\frac{1}{2}\frac{\dot{y}}{y}$, similarly to the previous examples treated in the present text. Such a discussion, opened and addressed in both papers, was let inconclusive, though. In fact, a exciting discussion followed, as testimony a series of papers, published from 2008 on; see, e.g., Bažant et al (2008), Bažant & Le (2008), Beck (2007, 2008) and Le & Bažant (2010, 2011).

A simple single degree of freedom model

The following model refers to the motion phase preceding the stacking caused by the accumulation of material collapsed on the ground. This first phase is named "crush-down". The second phase, i.e. the stacking of collapsed material, is called "crush-up"[14] and is not treated in the present

[13] Under kind permission of: ASCE - the American Society of Civil Engineers.

[14] A starting combined crush mode was also investigated in Bažant et al (2008) and in Bažant & Le (2008). It is shown that, in this combined mode, "crush-down" and "crush-up" modes may concomitantly exist for a brief period of time. However, the "crush-up" is arrested before having penetrated upwards a full story height, whereas the "crush-down" mode persist till the avalanche front reaches the soil.

text. Similarly to current models, as those by Bažant & Verdure (2007), by Seffen (2008) or by Beck (2007, 2008), the present model admits that such breakdown occurs due to gravity, starting from a particular floor whose structural strength was sufficiently damaged, with an "avalanche front", or an instability front, propagating downwards[15].

In an arbitrary instant of time, the tower is divided into two main regions, i.e., the moving and the non-moving ones; see Figure 8. The moving region is located above the avalanche front and the one at rest lies beneath it. As the avalanche front propagates, the mass of the region at rest is transferred to the moving region and compressed behind the "avalanche front", the moving part "swallowing" the immobile one, through a moving surface: the avalanche front, a non material, control surface. The region at rest, uncompressed, corresponds to a region of (the same) constant density, $\sigma_{nc} = \sigma_0$, but variable volume, so with variable mass. The moving region can, in turn, be divided in two parts: (i) the compacted one, with density per unit length σ_c and variable height, so with variable mass; and (ii) the non-compacted one, with constant density, $\sigma_{nc} = \sigma_0$, and non-variable height, so non variable mass. The non compacted part of the moving region (ii) translates downwards as a rigid body.

The density of the compacted region is also assumed constant. Such a very strong hypothesis implies the existence of a density jump across the avalanche front. This, in turn, implies that a velocity jump is assumed across the control surface, a plausible approximation, despite being an unusual condition. In other words, the constant density hypothesis ends up modeling the compacted part of the moving region as a translating and "growing" rigid body.

Variable mass and kinetic energy

Following Beck (2007), firstly, the mass of the moving region, m_{mov} may be conveniently expressed as,

$$m_{mov} = \int_{y_T}^{y_A^-} \sigma_{nc} dy + \int_{y_A}^{y_B^-} \sigma_c dy = \sigma_{nc}(y_A - y_T) + \sigma_c(y_B - y_A) \qquad (79)$$

With H as the initial height of the tower, the mass of the non-moving region, m_{rest}, may be written

[15]A thorough discussion on the modeling of such a collapse mechanism may be found in Bažant et al (2008), where supporting arguments are consistently proved; see also Bažant & Le (2008) and Le & Bažant (2010, 2011).

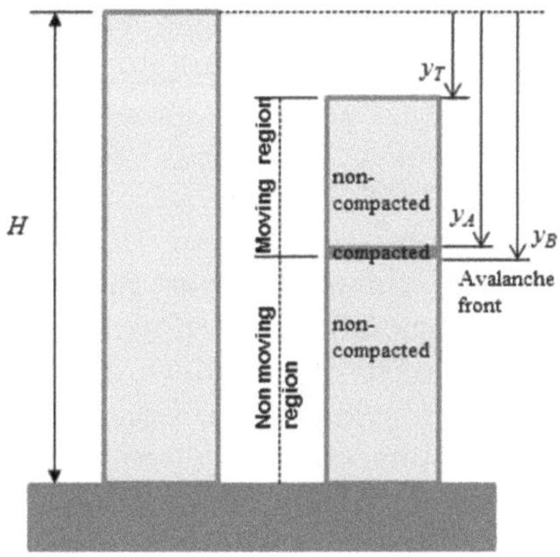

Figure 8. Schematics of the vertical collapse of buildings. From Pesce et al (2012).

$$m_{rest} = \int_{y_B}^{H} \sigma_{nc} dy = \sigma_{nc}(H - y_B) \tag{80}$$

Conservation of mass for the whole tower is written,

$$M = \sigma_{nc} H = m_{mov} + m_{rest} \tag{81}$$

i.e.,

$$\sigma_{nc}(y_A - y_T - y_B) + \sigma_c(y_B - y_A) = 0 \tag{82}$$

Taking $h = (y_A - y_T)$ as the constant height of the non-compacted part of the moving region, such that, $\dot{y_T} = \dot{y}_A$ and differentiating (82) with respect to time leads to the kinematic constraint,

$$\dot{y}_A = (1 - K)\dot{y}_B \tag{83}$$

where K is the "compacting factor", or "compaction coefficient", defined as $K = \sigma_{nc}/\sigma_c$. Notice from (82) that $y_A = (1 - K)y_B + Kh$ and from (83) that $\ddot{y}_A = (1 - K)\ddot{y}_B$. Also notice that, at $t = 0^+$, $y_A = y_B = h$, irrespective the value of K. As $0 < K < 1$, then $\dot{y}_A < \dot{y}_B$. In words, the avalanche front moves faster than the moving region, making the compacted region to grow, "swallowing" mass initially at rest. The "swallowed" mass suddenly starts falling as if it were a rigid body. The moving part can then be thought of as if it were a translating single particle, accreting mass, as an explicit function of position. Indeed, from (80) and (81),

$$m_{mov} = \sigma_{nc} y_B = \sigma_{nc} \left(\frac{y_A - Kh}{1 - K} \right) \tag{84}$$

In other words, the mass of the moving part may be seen as an explicit linear function of either the generalized coordinates, y_A or y_B. The latter gives the position of the avalanche front, which defines the boundary between the moving and the non moving region. Consistently, during the mass transfer process through the avalanche front, from the region at rest to the moving region, the captured particles abruptly change their velocity from zero to $\dot{y}_A$, in a dissipative and continuous inelastic "impact" approach.

The kinetic energy, T_{mov}, of the moving part may be then simply expressed as,

$$T_{mov} = \frac{1}{2} m_{mov} \dot{y}_A^2 = \frac{1}{2} [\sigma_{nc} yb] \dot{y}_A^2 = \frac{1}{2} \sigma_{nc} \left(\frac{y_A - Kh}{1 - K} \right) \dot{y}_A^2 \tag{85}$$

Obviously, the kinetic energy of the moving part is actually identical to the kinetic energy of the whole building. From the point of view of the mechanical system restricted to the moving part, the dependence of the kinetic energy on position is given in an explicit form, through its variable mass. This is, in fact, the kind of subtlety already discussed. Alternatively, using the kinematic relation (83) the kinetic energy of y_B,

$$T_{mov} = \frac{1}{2} \sigma_{nc} (1 - K)^2 y_B \dot{y}_B^2 \tag{86}$$

The proper and non proper equations of motion

To begin with the equation of motion derivation, let y_A be chosen as the generalized coordinate. We note that y_A represents the position of an equivalent accreting mass particle that translates vertically with velocity $\dot{y}_A$ and acceleration $\ddot{y}_A$. Obviously, the velocity of the (initially at rest)[16] accreted mass is $\mathbf{w}=\mathbf{0}$. The extended Lagrange equation (19) is then written as,

[16] Notice that the velocity jump would not be mandatory, if the compacted density were not taken constant along the compacted part of the moving region, what, in turn, would require an improvement on this simplistic single degree of freedom model.

$$\begin{aligned} &\frac{d}{dt}\left(\frac{\partial T_{mov}}{\partial \dot{y}_A}\right) - \frac{\partial T_{mov}}{\partial y_A} = \widehat{Q}_A \\ &\widehat{Q}_A = m_{mov} g - F - \frac{1}{2}\frac{\partial m_{mov}}{\partial y_A}\dot{y}_A^2 \end{aligned} \tag{87}$$

In Eq. (87), F models the resistive force applied by the intact region at rest to the moving region that is represented by the accreting mass particle. Notice that the dissipative virtual work done by the resistive force[17] is given by $\delta W_R = -F\delta y_A = -(1-K)F\delta y_B < 0$.

From Eq. (85), and recalling that $\frac{\partial y_B}{\partial y_A} = \frac{\partial \dot{y}_B}{\partial \dot{y}_A} = (1-K)^{-1}$, it follows

$$\frac{\partial T_{mov}}{\partial \dot{y}_A} = \sigma_{nc} y_B \dot{y}_A = (1-K)\sigma_{nc} y_B \dot{y}_B \tag{88}$$

such that

$$\frac{d}{dt}\left(\frac{\partial T_{mov}}{\partial \dot{y}_A}\right) = \sigma_{nc}\frac{d}{dt}\left(y_B \dot{y}_A\right) = (1-K)\sigma_{nc}\frac{d}{dt}\left(y_B \dot{y}_B\right) \tag{89}$$

Also, from Eq. (85), the second term on the left hand side of the extended Lagrange Equation, Eq. (87), is simply

$$-\frac{\partial T_{mov}}{\partial y_A} = -\frac{1}{2}\sigma_{nc}\frac{\dot{y}_A^2}{1-K} = -\frac{1}{2}(1-K)\sigma_{nc}\dot{y}_B^2 \tag{90}$$

which cancels out, exactly, as expected, the term of the extended generalized force, given by:

$$-\frac{1}{2}\frac{\partial m_{mov}}{\partial y_A}\dot{y}_A^2 = -\frac{1}{2}\frac{\sigma_{nc}}{1-K}\dot{y}_A^2 = -\frac{1}{2}(1-K)\sigma_{nc}\dot{y}_B^2 \tag{91}$$

Using (89)-(91), the extended Lagrange equation (87) simplifies then to,

$$(1-K)\sigma_{nc}\frac{d}{dt}\left(y_B \dot{y}_B\right) = \sigma_{nc} g y_B - F \tag{92}$$

with $F > 0$, which can be written in the final form,

$$\ddot{y}_B = \frac{g}{1-K} - \frac{\dot{y}_B^2}{y_B} - \frac{1}{1-K}\frac{F}{\sigma_{nc} y_B} \tag{93}$$

[17] For a thorough discussion on the resistive force see Bažant et al (2008), Bažant et al (2008) and Le & Bažant (2010).

Equation (93) is the proper equation and can be straightforwardly derived from Newton's law; see a straightforward derivation in Pesce et al (2012). If, instead, y_B were chosen, at first, as the generalized coordinate, recalling once more that $\frac{\partial y_B}{\partial y_A} = \frac{\dot{y}_B}{\partial \dot{y}_A} = (1-K)^{-1}$, the extended Lagrange equation would read

$$\begin{aligned} &\frac{d}{dt}\left(\frac{\partial T_{mov}}{\partial \dot{y}_B}\right) - \frac{\partial T_{mov}}{\partial y_B} = \widehat{Q}_B \\ &\widehat{Q}_B = (1-K)\sigma_{nc} y_B g - (1-K)F - \frac{1}{2}(1-K)^2 \frac{\partial m_{mov}}{\partial y_B}\dot{y}_B^2 \end{aligned} \tag{94}$$

Note the appearance of the multiplicative factors in the generalized force, $(1-K)$, in the first and second terms and $(1-K)^2$, in the third term. Proceeding, Eq. (86) gives

$$\frac{\partial T_{mov}}{\partial \dot{y}_B} = \sigma_{nc}(1-K)^2 y_B \dot{y}_B \tag{95}$$

such that

$$\frac{d}{dt}\left(\frac{\partial T_{mov}}{\partial \dot{y}_B}\right) = \sigma_{nc}(1-K)^2 \frac{d}{dt}(y_B \dot{y}_B) \tag{96}$$

The second term on the left hand side of the Lagrange Equation, Eq. (94), is simply

$$-\frac{\partial T_{mov}}{\partial y_B} = -\frac{1}{2}\sigma_{nc}(1-K)^2 \dot{y}_B^2 \tag{97}$$

which, as before, cancels out exactly, the second term of the extended generalized force, given by $-\frac{1}{2}(1-K)^2 \frac{\partial m_{mov}}{\partial y_B}\dot{y}_B^2$. Eq. 94 simplifies then to,

$$\sigma_{nc}(1-K)^2 \frac{d}{dt}(y_B \dot{y}_B) = (1-K)\sigma_{nc} y_B g - (1-K)F \tag{98}$$

Eq. (98) recovers, as it should, the proper equation of motion, Eq. (93). If the usual form of the Lagrange Equation were erroneously applied, not taking into account the explicit dependence of the mass of the moving part on the position, the term $-\frac{\partial T_{mov}}{\partial y_A}$ in Eq. (87) (or, equivalently, $-\frac{\partial T_{mov}}{\partial y_B}$ in Eq. (94)) would not be canceled out and the following non-proper equation of motion[18] would be obtained,

[18] The application of the usual Lagrange equation to the whole tower as the system under analysis leads, as well, to the non proper equation of motion, unless energy loss due to compaction is introduced. This can be worked out with the introduction of a proper Rayleigh-like dissipative function; see next section and Pesce et al (2012).

$$\ddot{y}_B = \frac{g}{1-K} - \frac{1}{2}\frac{\dot{y}_B^2}{y_B} - \frac{1}{1-K}\frac{F}{\sigma_{nc} y_B} \tag{99}$$

We promptly see that the decelerating term $-\frac{1}{2}\frac{\partial \dot{y}_B^2}{y_B}$, present in the non-proper form (99), is exactly half the corresponding term shown in the proper form (93), what would give an augment in the acceleration, consequently leading to an under estimate value for the "crushing-down time". Interesting is that, irrespective the equation, (93) or (99), the decelerating term, which is quadratic in velocity, is not multiplied by any term dependent on the compaction factor parameter. It is inherent to both equations and *given solely in terms of the kinematic state* $(y_B, \dot{y}_B)$ *of the system.*

Going back and taking y_A as the generalized coordinate, the proper equation, (93), and the non proper one, (99), may be written,

$$\ddot{y}_A = g - \frac{\dot{y}_A^2}{y_A - Kh} - \frac{(1-K)F}{\sigma_{nc}(y_A - Kh)} \tag{100}$$

and

$$\ddot{y}_A = g - \frac{1}{2}\frac{\dot{y}_A^2}{y_A - Kh} - \frac{(1-K)F}{\sigma_{nc}(y_A - Kh)} \tag{101}$$

At $t = 0^+$, as $y_A = h$ and $\dot{y}_A = 0$, both equations give the same initial acceleration,

$$\ddot{y}_A(0^+) = g\left(1 - \frac{F}{\sigma_{nc} h g}\right) \tag{102}$$

an expected result, clearly showing that, $\ddot{y}_A < g$ at $t = 0^+$. Notice also that $\ddot{y}_A(0^+) = g(1-\epsilon), 0 < \epsilon << 1, \epsilon = F(\sigma_{nc} g h)^{-1}$, if the magnitude of the resistive force is very small compared to the initial weight of the moving region. Obviously, the initial acceleration of the avalanche front is larger than $\ddot{y}_A(0)^+)$ and is given by $\ddot{y}_B(0^+) = (1-K)^{-1}\ddot{y}_A(0^+)$.

A Rayleighian approach

As observed in the last section, a dissipative Rayleigh-like function could be introduced in the usual Lagrange equation formulation, to deal with the energy loss due to compaction. Let a Rayleigh-like function be defined as

$$
\begin{aligned}
R(y_A, \dot{y}_A) &= \frac{1}{2} f(y_A, \dot{y}_A) \dot{y}_A^2 \\
f(y_A, \dot{y}_A) &= \frac{1}{3} \dot{m}_{mov} = \frac{1}{3} \frac{\partial m_{mov}}{\partial y_A} \dot{y}_A
\end{aligned}
\tag{103}
$$

or, equivalently, as

$$
\begin{aligned}
& R(y_B, \dot{y}_B) = \frac{1}{2} g(y_B, \dot{y}_B) \dot{y}_B^2 \\
& \textit{with} \\
& g(y_B, \dot{y}_B) = \frac{1}{3}(1-K)^2 \dot{m}_{mov} = \frac{1}{3}(1-K)^2 \frac{\partial m_{mov}}{\partial y_B} \dot{y}_B
\end{aligned}
\tag{104}
$$

Equation (94) could then be promptly re-written in the form,

$$
\begin{aligned}
& \frac{d}{dt}\left(\frac{\partial T_{mov}}{\partial \dot{y}_B}\right) - \frac{\partial T_{mov}}{\partial y_B} = \widehat{Q}_B \\
& \widehat{Q}_B = (1-K)(\sigma_{nc} y_B g - F) - \frac{\partial R}{\partial \dot{y}_B}
\end{aligned}
\tag{105}
$$

In fact, the *Rayleighian-like* dissipative generalized force,

$$
-\frac{\partial R}{\partial \dot{y}_B} = -\frac{1}{2}(1-K)^2 \dot{m}_{mov} \dot{y}_B = -\frac{1}{2}(1-K)^2 \frac{\partial m_{mov}}{\partial y_B} \dot{y}_B^2 = -\frac{1}{2} \sigma_{nc} \dot{y}_B^2 (1-K)^2
\tag{106}
$$

cancels out, precisely, the term given by Eq. (97), recovering once more the proper equation (93). Notice that (104) (or (103)) is absolutely similar to the Rayleigh-like function introduced in the hydrodynamic impact problem, equations (63-64). Both Rayleighians represent energy dissipation: in the impact problem, the energy drained to the jets; in the vertical collapsing problem, the energy drained trough compaction. And the extended form of the Lagrange equation inherently takes into account such energy loss, through the term of mass depending explicitly on position. In fact, this term naturally appears in the derivation of the extended form of the Lagrange equation, irrespective how smooth the velocity transition might be and may be interpreted as derived from the Rayleigh-like function. This is a rather subtle aspect, indeed, also present in the classic falling chain problems.

Illustrative example - the WTC towers

Let $y^* = y_B/H$, $t^* = t\sqrt{g/H}$ and $\Phi = F/(\sigma_{nc}gH) = F/P$ be, respectively, the dimensionless variables giving, the position of the avalanche front, time and the resistive force to collapse[19]. The resistive force naturally appears normalized by the intact building weight, $P = Mg = \sigma_{nc}Hg$. The proper and the non-proper equations of motion, respectively equations (93) and (99), can be written in dimensionless form:

$$\ddot{y}^* = \frac{1}{1-K}\left(1 - \frac{\Phi}{y^*}\right) - \frac{\dot{y}^{*2}}{y^*} \tag{107}$$

and

$$\ddot{y}^* = \frac{1}{1-K}\left(1 - \frac{\Phi}{y^*}\right) - \frac{1}{2}\frac{\dot{y}^{*2}}{y^*} \tag{108}$$

According to Seffen (2008), the compacting factor in the case of the WTC towers may be approximately evaluated as $K \approx 0.2$. Both papers, by Bažant & Verdure (2007) as well as by Seffen (2008), present parametric analyses, varying the magnitude of the resistive force to collapse between zero, $\Phi = 0$, (free-fall hypothesis) and a representative limit value Φ_L, which would completely refrain the collapse initiation. As the initial conditions are, always, $y^*(0) = h^*$; $\dot{y}^*(0) = 0$, then, $\Phi_L = h^*$. Moreover, the initial acceleration $\ddot{y}^*(0) = a^*(0)$ will, therefore, decrease linearly with Φ, with rate $(1-K)^{-1}$, being equal to $(1-K)^{-1}$ if "free fall" is assumed ($\Phi = 0$) and equal to zero, if $\Phi_L = h^*$. Notice that due to the initial condition $\dot{y}^*(0) = 0$, the initial acceleration is the same for both equations (107) or (108). This is illustrated in Figure 9.

Taking the following data: $H = 407m$, $\sigma_{nc} = 770 \times 10^3 t/m$ and $P = 3.073GN$, Bažant & Verdure (2007) point typical values of the resistive force in the interval $0 < \Phi < 0.21$. Those authors suggest, as a good first estimate, an average value $\Phi \approx 0.044$ (4.4% of the tower weight). Figure 10 illustrates the downward propagation of the avalanche front, during the crushing-down phase, computed with both equations: the proper (107) and

[19] Same nomenclature, but not to be confused with Meshcherski's force.

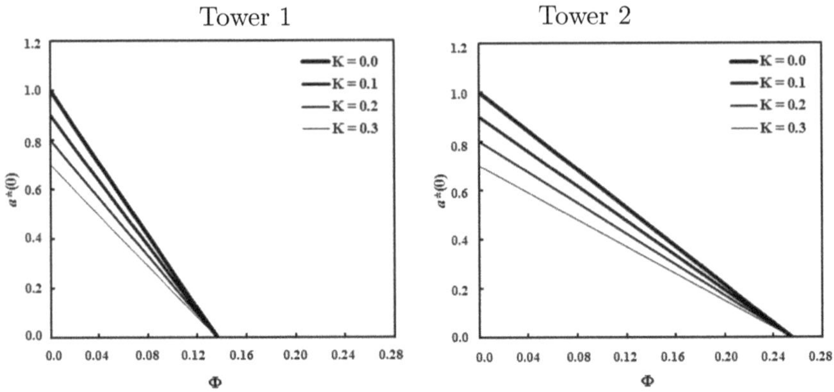

Figure 9. Initial acceleration of the avalanche front, $a^*(0)$, normalized with respect to gravity, as function of the resistive force, normalized with respect to the tower weight. K is the compaction factor. WTC: Tower 1: $y^*(0) = h^* = 0.1364$; Tower 2: $y^*(0) = h^* = 0.2545$. From Pesce et al (2012).

the non proper one (108). Position, velocity and acceleration are shown as function of the non dimensional time.

Both towers had 110 stores. The graphs on the left refer to tower number 1. According to Seffen (2008), the collapse initiated at the floor 95°, i.e., $y^*(0) = h^* = (110 - 15)/110 = 0.1364$. The graphs on the right refer to tower number 2, for which the collapse initiated at the floor 82°, i.e., $y^*(0) = h^* = (110 - 28)/110 = 0.2545$. The collapse front reaches the soil when $y^* = 1$. Both references, Bažant & Verdure (2007) and Seffen (2008), report collapse times of order 11 seconds for tower 1. As shown in Table 1, Eq. (107) - the proper one - predicts this event at $t^* = t_C^* \approx 1.75$, for the tower 1 and $t^* = t_C^* \approx 1.45$, for the tower 2. Such values correspond to crush-down times of $t_C \approx 11.3s$ and $t_C \approx 9.3s$, respectively. On the other hand, Eq. (108) - the erroneous one - predicts, respectively, $t^* = t_C^* \approx 1.55$ and $t^* = t_C^* \approx 1.32$, i.e., $t_C^* \approx 10.0s$ and $t_C \approx 8.5s$, values substantially smaller than those predicted by Eq. (107).

It is clear that parameter uncertainties, regarding not only the resistive collapse load but also the compacting factor, may induce a relatively large range of variation for the crushing-down time. This is illustrated in Figure 12, where the non dimensional "crush-down" time is plotted as a function of Φ, considering four distinct values of the compaction factor, including a

Table 1. "Crush-down" time. WTC: towers 1 and 2. Comparing the results from equations (107) and (108), the proper and non proper ones. Adapted from Pesce et al (2012).

		$K = 0.2$	$\Phi = 0.044$	$K = 0.2$	$\Phi = 0$
Tower	Equation	t_C^*	$t_C(s)$	t_C^*	$t_C(s)$
1	107 - proper	1.75	11.3	1.59	10.2
1	108 - non-proper	1.55	10.0	1.39	9.0
2	107 - proper	1.45	9.3	1.36	8.8
2	108 - non-proper	1.32	8.5	1.23	7.9

hypothetical lower limit, $K = 0$. Results from both equations, (107) and (108), are compared.

For reference sake, simulations of the "free-fall case", $\Phi = 0$, with the same compaction factor ($K = 0.2$), are shown in Figure 11.

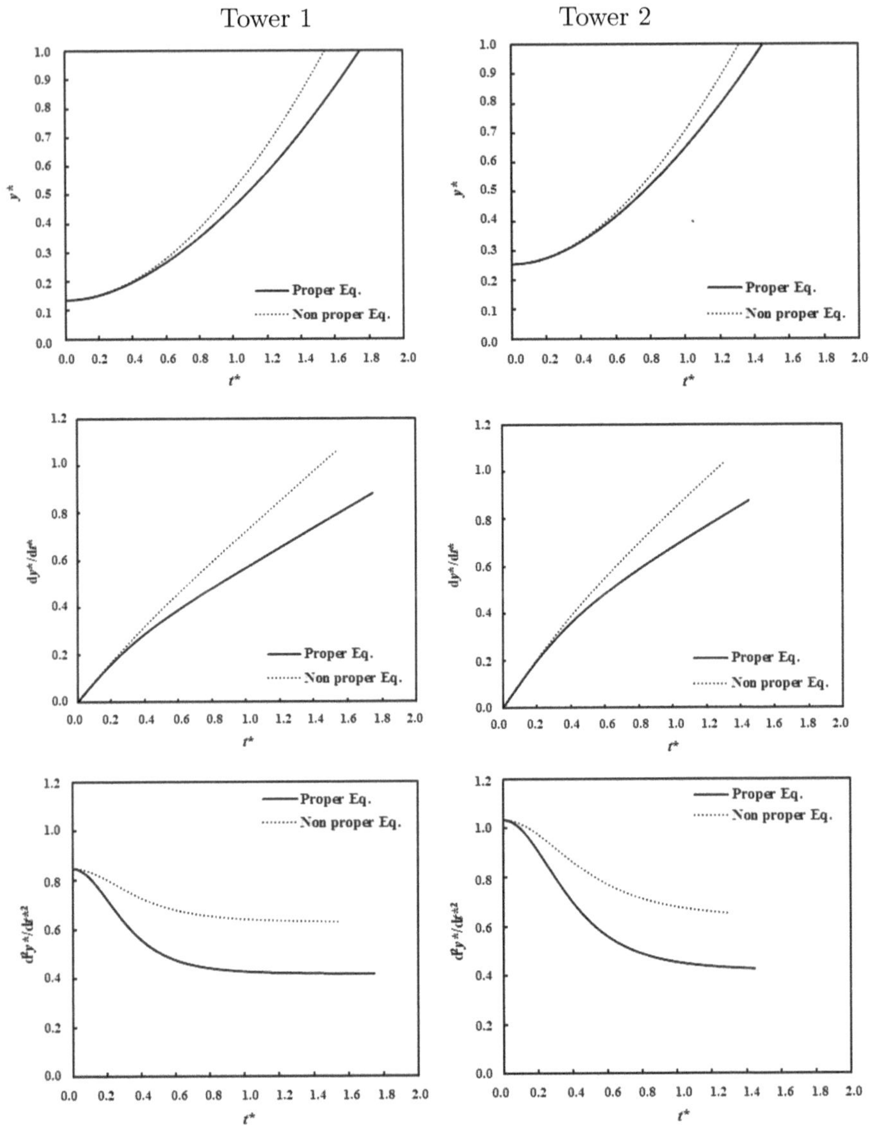

Figure 10. Time simulation of the avalanche front propagation. Downward (dimensionless) position, velocity and acceleration. Comparing solutions of the proper (107) and the non proper (108) equations. Typical representative parameters values: $K = 0.2$, $\Phi = 0.044$. WTC: Tower 1: $y^*(0) = h^* = 0.1364$; Tower 2: $y^*(0) = h^* = 0.2545$. From Pesce et al (2012).

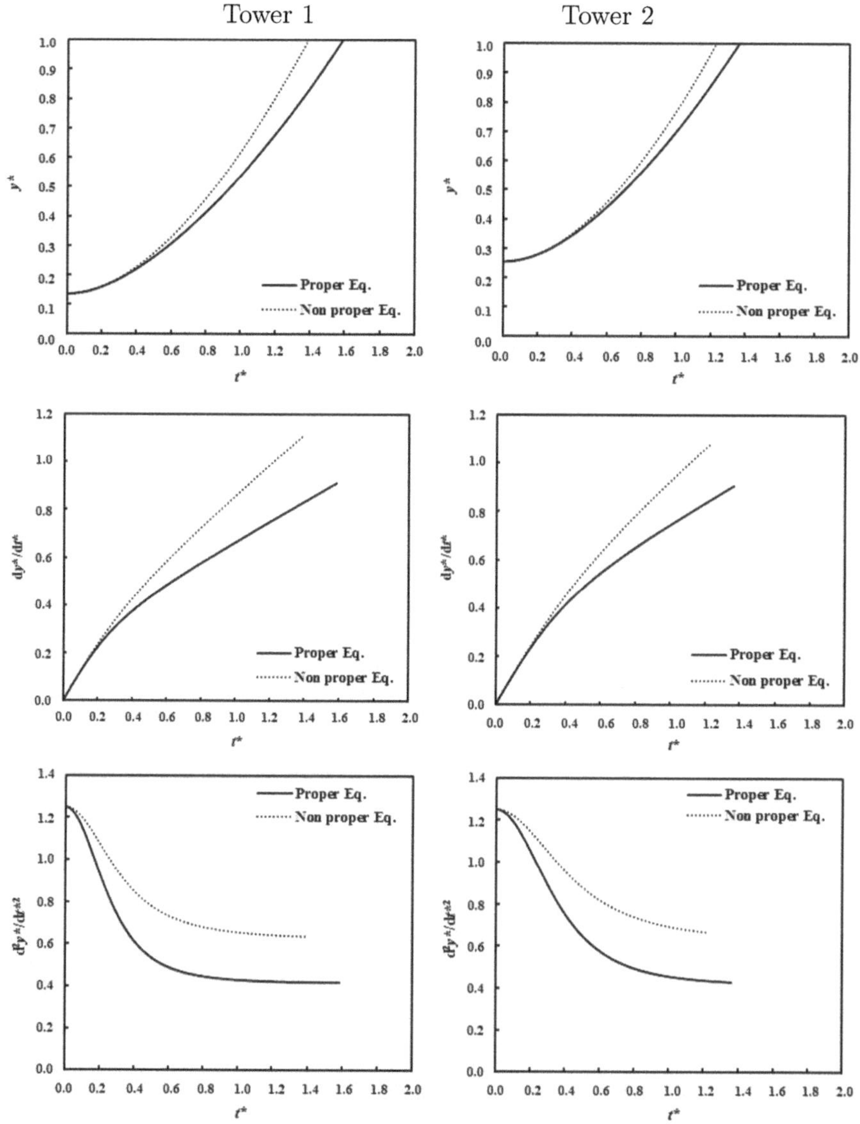

Figure 11. Time simulation of the avalanche front propagation. Hypothetical lower bound: $\Phi = 0$. Downward (dimensionless) position, velocity and acceleration. Comparing solutions of the proper (107) and the non proper (108) equations. Typical representative parameters values: $K = 0.2$, $\Phi = 0.044$. WTC: Tower 1: $y^*(0) = h^* = 0.1364$; Tower 2: $y^*(0) = h^* = 0.2545$. From Pesce et al (2012).

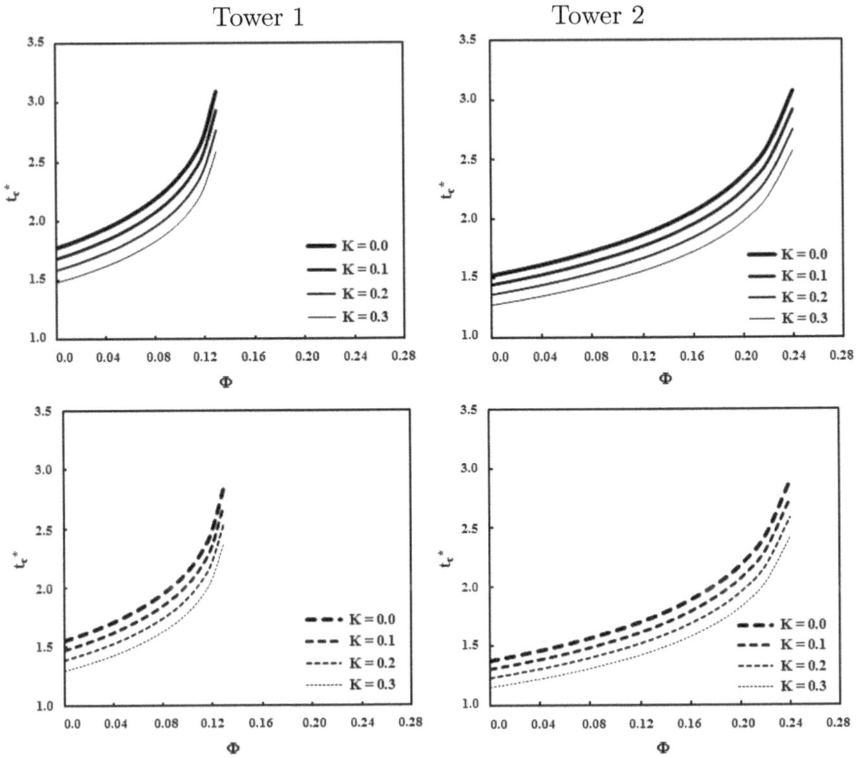

Figure 12. Non dimensional "crush-down" time, t_C^*, as function of the non dimensional resistive force, Φ. Comparing results from simulations with the proper Eq. (107) (upper graphs) and the non proper Eq. (108) (bottom graphs); compaction factor K as parameter, including a hypothetical lower limit, $K = 0$. WTC: Tower 1: $y^*(0) = h^* = 0.1364$; Tower 2: $y^*(0) = h^* = 0.2545$. From Pesce et al (2012).

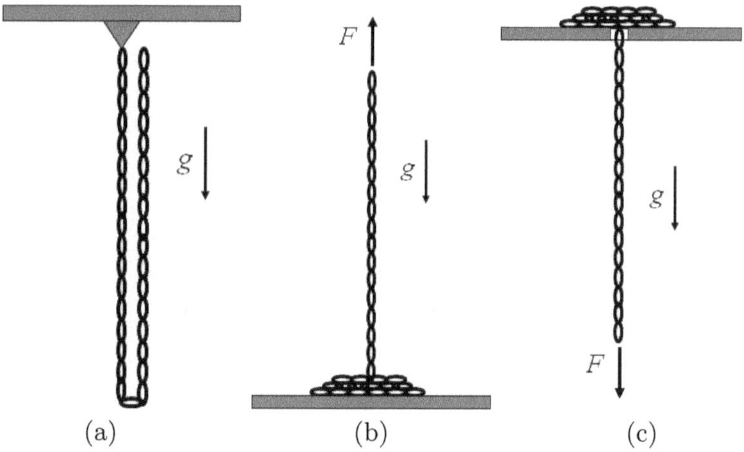

Figure 13. The classic "falling chains". (a)The U-chain; (b) the bottom-pile chain (Buquoy's) and (c) the top-pile chain (Cayley's).

Similarity with the classic falling chain problem

The technical literature on this classic problem is vast. Extensive discussion has been done since the $18th$ century, with a renewed interest after the American scholar debate, in the $1960's$, and even in the latest twenty years. Apparent paradoxes are discussed, back and forth, arguing on energy conservation, rejecting old arguments and sometimes defending and introducing erroneous ones.

One of the last accounts on the subject is due to Grewal et al (2011), motivated by intriguing questionings from whom they called 'the persistent student'. In that paper, motivated by didactic aspects, the authors make a detailed description on the three classic cases: (a) the U-chain, (b) the bottom-pile chain and(c) the falling top-pile chain and . The first and second cases are divided in two sub-cases, according if the tip is lifted or lowered. After questioning common assumptions, they dedicated a part of the analysis to the falling-bottom pile chain, discussing internal reactions involving the colliding link. The authors end up with the design of special falling chains, whose links collide against the soil obliquely, accelerating downwards due to linear momentum fed by the angular momentum variation involved in the collision.

It is not the aim of the present discussion to present a historic perspective, neither to promote a full discussion on the subject, but only to point out the existent similarity with the problem under analysis: the vertically collapsing towers.

To be remarked, apart the compaction phenomenon, exactly the same kind of 'apparent paradox' involving two distinct equations of motion has been the focus of recent discussion on the classical and well known "falling chain problem"; see, e.g., Šima & Podolsky (2005),Wong & Yasui (2006) and Wong et al (2007).

Of particular interest of the present text - the similarity with the vertically collapsing tower problem - if, both, K and F are taken to be null, the proper equation (93) takes exactly the same form of Cayley's equation of motion,

$$\ddot{y} = g - \frac{\dot{y}^2}{y} \tag{109}$$

being $y(t)$ the position of the lower tip of the chain, falling from a table. On the other hand, taking $z(t)$ as the vertical position of the upper tip of a coiled chain being pulled from or falling on a table, positively oriented contrary to gravity, equation (93) takes exactly the form of Buquoy's equation,

$$\ddot{z} = -g - \frac{\dot{z}^2}{z} \tag{110}$$

As a matter of fact, Šima & Podolsky (2005), as well as Wong & Yasui (2006), based on the same erroneous approach, questioned the classical solution by Cayley to the falling chain problem. By applying the usual Lagrange equation, they obtained the non proper equation that is also promptly obtained from Eq. (99) with both, K and F, nulls:

$$\ddot{y} = g - \frac{1}{2}\frac{\dot{y}^2}{y} \tag{111}$$

Not surprisingly, just one year later, Wong et al (2007), presented an experimental evidence of the correctness of the classical Cayley solution, contradicting their own previous claim made in 2006, in a meritorious scientific attitude. They measured the asymptotic value of the acceleration as $\ddot{y_{lim}} = (0.3204 \pm 0.0010)g$. The value predicted by Cayley's solution is $\ddot{y_{lim}} = g/3$! However, the value predicted by their own equation, which coincides with that of Šima & Podolsky (2005), is $\ddot{y_{lim}} = g/2$. It appears that those authors were not aware of the work by Cvetićanin (1993)- or of that by Pesce (2003)- and, as a consequence, could not be conclusive on the (subtle) reasons hidden behind their own mistake. A full discussion of the general falling chain problem is left to further publications.

Concluding remarks

This chapter aimed at presenting a didactic perspective on the subtleties hidden behind the application of the Lagragian formalism to mechanical systems with explicit mass dependence on position. Rather than introducing new material, the authors preferred to present a step-by-step derivation of the extended Lagrange equation, recollecting, from their own previous publications, some one-degree-of-freedom examples inspired in practical engineering applications.

Many formal points still deserve attention, as those related to open systems, see, Irschik & Holl (2004), or to classical procedures, Casetta & Pesce (2012). Those are left for further analysis and publications and as motivating issues for the interested readers.

Acknowledgements

The authors acknowledge (i) FAPESP - the State of São Paulo Research Funding Agency - for the research grants no 12/10253–0 and and 2013/02997–2 and for a previous PhD scholarship no 04/04611–5; (ii) CNPq - the Brazilian National Research Council - for the research grant no 303838/2008 – 6 and for the post-doctoral grant no150731/2011 – 6. Our sincere gratitude to Professors H. Irschik, G.K. Mikhailov, V. Šima, T. Miloh and A. A. Korobkin, for references, discussions and scientific incentives. The authors also thank Prof. Eduardo A. Tannuri and the Dr. Flavia M. Santos for co-authoring and collaborating with results here collected, respectively from Pesce et al (2006) and from Pesce et al (2012), Casetta et al (2011). Especial thanks to ASME - the American Society of Mechanical Engineers; ABCM - the Brazilian Society of Mechanical Sciences and Engineering; SOBENA - the Brazilian Society of Naval Architects; ASCE - the American Society of Civil Engineers, for kindly permitting the use of material previously published in the references Pesce (2003), Pesce et al (2006), Casetta et al (2011), Pesce et al (2012). Finally, our deepest gratitude to Dr. Guilherme R. Franzini, for the volunteer and careful manuscript editing.

This text is dedicated to Professors Luís N. F. França and Giorgio E.O. Giacaglia, who, more than 35 years ago, gave the first author the opportunity of entering the beauty of Mechanics.

Bibliography

C. Agostinelli. Sui Sistemi Dinamici di Masse Variabili. In *Atti della Reale Accademia delle Scienze di Torino*, Classe di Scienze Fisiche, Matemat. Naturali, Vol.71, I, pp. 254–272, 1936.

A.B. Arons, A.M. Bork. Newton's law of motion and the 17th century laws of impact. *American Journal of Physics*, Vol.32, n.4, pp. 313–317, 1964.

Z.P. Bažant, M. Verdure. Mechanics of progressive collapse: learning from world trade center and building demolitions. *Journal of Engineering Mechanics*, Vol. 133, n.3, pp. 308–19, 2007.

Z.P. Bažant, J.-L. Le, F.R. Greening, D.B. Benson. What did and did not cause collapse of World Trade Center twin towers in New York? *Journal of Engineering Mechanics*, Vol. 134, n.10, pp. 892–906, 2008.

Z.P. Bažant, J.-L. Le. Closure to J.R. Gourley and G. Suladzinski's discussions of Zdenek P. Baant and Mathieu Verdure's paper "Mechanics of Progressive Collapse: Learning from World Trade Center and Building Demolitions. *Journal of Engineering Mechanics*, Vol. 134, n.10, pp. 917–924, 2008.

C.M. Beck. Mathematical models of progressive collapse and the question on how did the World Trade Centers perish, 2007.(http://arxiv.org; see articlephysics/0609105).

C.M. Beck. Role of Compaction in the mathematical model of progressive collapse, 2008.(http://arxiv.org; see article 0806.4792).

A.M. Bork, A.B. Arons. Newtons law and variable mass. *American Journal of Physics*, Vol.32, n.8, pp. 646, 1964.

M.G. Calkin, R.H. March. The dynamics of a falling chain: I. *American Journal of Physics*, Vol.57, n.2, pp. 154–7, 1989a.

M.G. Calkin, R.H. March. The dynamics of a falling chain: II. *American Journal of Physics*, Vol.57, n.2, pp. 157–9, 1989b.

L. Casetta. Theoretical Investigation on the Hydrodynamic Impact Problem (in Portuguese). São Paulo, Escola Politécnica, University of São Paulo. Master Dissertation in Mechanical Engineering, 2004, 151pp.

L. Casetta. Contributions to the Mechanics of Variable Mass Systems (in Portuguese). São Paulo. Escola Politécnica, University of São Paulo. Doctoral Thesis in Mechanical Engineering, 2008, 185pp.

L. Casetta, C.P. Pesce. The generalized Hamilton's principle for a non-material volume. *Acta Mechanica*, Vol. 224, n. 4, p. 919-924, 2013.

L. Casetta, C.P. Pesce. The inverse problem of Lagrangian mechanics for Meshchersky's equation. *Acta Mechanica*, d.o.i. 10.1007/s00707-013-1004-1, 2013.

L. Casetta, C.P. Pesce. On the generalized canonical equations of Hamilton for a time-dependent mass-particle. *Acta Mechanica*, Vol. 223, n. 12, pp. 2723-2726, 2012.

L. Casetta, C.P. Pesce. Hamilton's principle for dissipative systems and Wagner's problem. In *22nd International Workshop on Water Waves and Floating Bodies*, 2007, Plitvice, Croatia.

L. Casetta, C.P. Pesce. The proper definition of the added mass for the water entry problem. In *21st International Workshop on Water Waves and Floating Bodies*, 2006, Loughborough, United Kingdom.

L. Casetta, C.P. Pesce. A noticeable question of the water entry problem: the split of kinetic energy during the initial stage.In *COBEM 2005, 18th International Congress of Mechanical Engineering*, 2005, Ouro Preto, Brazil.

L. Casetta, C.P. Pesce., F.M. Santos. On the hydrodynamic vertical impact problem: an analytical mechanics approach. *Marine Systems & Ocean Technology*, Vol.6, pp .47–57, 2011.

A. Cayley. On a class of dynamical problems. *Proceedings of the Royal Society of London*, Vol.8, pp. 506-11, 1857.

R. Chicon. Comment on The falling chain and energy loss, by David Keiffer. *American Journal of Physics*, Vol.71, n.9, pp. 951, 2003.

R. Cointe, E. Fontaine, B. Molin, Y-.M. Scolan. On energy arguments applied to the hydrodynamic impact force. *J Eng Mathematics*, Vol.48, pp. 305–319, 2004.

J. Copeland. Work-energy theorem for variable mass systems. *American Journal of Physics*, Vol.50, n.7, pp. 599–601, 1982.

E.B. Crellin., F.L. Janssens, D. Poelaert, W. Steiner, H. Troger. On balance and variational formulations of the equation of a motion of a body deploying along a cable. *J of Applied Mechanics*, Vol.64, pp. 369–374, 1997.

L. Cvetićanin. Conservation laws in systems with variable mass. *Journal of Applied Mechanics*, Vol.60, pp. 954–958, 1993.

L. Cvetićanin. Dynamics of machines with variable mass. Gordon and Breach Science, London, 1998, 235 p.

L. Cvetićanin. Dynamics of body separation analytical procedure. *Nonlinear Dynamics*, Vol.55, n.3, pp. 269–278, 2009.

L. Cveticanin, I. Kovacic. On the dynamics of bodies with continual mass variation. *J. Appl. Mech.* Vol.74, pp. 810–815 (2007)

R. Dugas. A History of Mechanics. New York, Dover Publications, 1988.

F.O. Eke, T.C. Mao. On the dynamics of variable mass systems. *International Journal of Mechanical Engineering Education*, Vol.30, n.2, pp. 123–37, 2002.

O.M. Faltinsen. Sea Loads on Ships and Offshore Structures. Cambridge Ocean Technology Series, Cambridge University Press, 1990, 328 pp.

O.M. Faltinsen, R. Zhao. Water Entry of Ship Sections and Axisymmetric Bodies. *Agard Ukraine Inst. on Hydromechanics. Workshop on High Speed Body Motion in Water*, 1997.

C.P. Gadsden. Laws of motion for variable mass systems. *American Journal of Physics*, Vol.34, n.10, pp. 987–8, 1966.

C.P. Gadsden. Newtons law with variable mass. *American Journal of Physics*, Vol.32, n.1, pp. 61, 1964.

Z.-M. Ge. The Equations of Motion of Nonlinear Nonholonomic Variable Mass System with Applications. *J of Applied Mechanics*, Vol.51, pp. 435–437, 1984.

A. Grewal, P. Johnson, A. Ruina. A chain that speeds up, rather than slows down, due to collisions: how compression can cause tension. *American Journal of Physics*, Vol.79, n.7, pp. 723–729, 2011.

C. Grubin. Mechanics of variable mass systems. *Journal of the Franklin Institute*, Vol.276, pp. 305–12, 1963.

I.M. Gelfand, S.V. Fomin. Calculus of variations. Dover, New York, 2000.

H.J. Holl, F. Hammelmuller, H. Irschik. The effect of variable radius onto the oscillations in a coiling process. In *6th European Solid Mechanics Conference*, Budapest, Hungary, 2006.

H.J. Holl, F. Hammelmuller. Analysis of the vibrations due to thermal deflection of the drum in the coiling process. *Proc. Appl. Math. Mech.* Vol.11, pp. 317–318, 2011.

H. Irschick. The Cayley variational principle for continuous-impact problems: a continuum mechanics based version in the presence of a singular surface. *J. Theor. Appl. Mech.* Vol 50, n. 3, pp. 717–727, 2012.

H. Irschik, H.J. Holl. Mechanics of variable-mass systems part 1: balance of mass and linear momentum. *Applied Mechanics Review*, Vol.57, n.2, pp. 145–60, 2004.

H. Irschik, H.J. Holl. The equations of Lagrange written for a non-material volume. *Acta Mechanica*, Vol.153, pp. 231–48, 2002.

F.L. Janssens, D. Poelaert, E.B. Crellin. Deployment and retraction of a continuous tethered satelitte - the equations reviseted. *Proceedings of the Fourth Int Conference on Tethers in Space*, Hampton, VA, pp. 1415–1423, 1995.

D. Jeltsema, A. Doria-Cerezo. Modeling of systems with position-dependent mass revisited: a Port-Hamiltonian ppproach. *Journal of Applied Mechanics*, Vol.78, 2011, doi: 10.115/1.4003910.

D. Jeltsema Memory Elements: A paradigm shift in Lagrangian modeling of electrical circuits. *Math Modeling 2012, contribution 448.*

J.L. Jimenez et al. Newtonian and canonical analysis of the motion of a rope falling from a table. *European Journal of Physics*, Vol.26, pp. 1127–37, 2005.

D. Keifer. The falling chain and the energy loss. *American Journal of Physics*, Vol.68, n.3, pp. 385–6, 2001.

A.A. Korobkin. Analytical models of water impact. *European J. of Applied Mathematics*, vol.15: pp. 821–838, 2004.

V.V. Kozlov. Hydrodynamic theory of a class of finite-dimensional dissipative systems. *Proc. Stek. Inst. Math.* Vol.223, pp. 178–184 (1998)

H. Lamb. Hydrodynamics, Cambridge University Press, Cambridge, 1932.

C. Lanczos. The variational principles of mechanics. Dover, New York (1970).

Le. J.-L.. Closure to A. Bjorkman's discussion of "What did and did not cause collapse of World Trade Center Twin Towers in New York?" by Z.P. Bažant et al. *Journal of Engineering Mechanics*, vol. 136, n.7, 934-935,(2010).

Le. J.-L. Why the observed motion history of World Trade Center towers is smooth. *Journal of Engineering Mechanics*, vol. 137, n.1, 82-84,(2011).

T. Levi-Civita. Sul Moto di un Corpo de Massa Variabile. *Rendiconti delle Sedute della Reale Accademia Nazionale dei Lincei.* Vol.VIII, pp. 329–333, Aggiunta alla nota, pp. 621–622. 1928.

V.A. Lubarda, A. Hoger. On the mechanics of solids with a growing mass. *International Journal of Solids and Structures*, Vol.39, pp. 4627–64, 2002.

D.B. McIver. Hamiltons principle for systems of changing mass. *Journal of Engineering Mathematics*, Vol.7, n.3, pp. 249–61, 1973.

L. Meirovitch. Methods of Analytical Dynamics. Dover Publications Inc., New York, 2003, 524 pp.

J.L. Meriam. Variable-mass dynamics. *Journal of Engineering Education*, Vol.51, n.3, pp. 241–3, 1960.

I.V. Meshcherski. Dinamika tochki Peremnoj Massy, St Petesburg, Akademia Nauk, Peterburskij Universitet, 1897.

I.V. Meshcherski. Equations of motion of a variable mass point in the general case (in Russian), *St. Petersburg Polytechnic University News*, Vol.1, pp. 77–118, 1904.

G.K. Mikhailov. On the history of variable-mass system dynamics. *Mechanics of Solids*, Vol.10, n.5, pp. 32–40, 1975.

B. Molin, R. Cointe, E. Fontaine. On energy arguments applied to the slamming force. In *11st International Workshop on Water Waves and Floating Bodies*, Hamburg, pp. 17–20 March 1996.

D. Mušicki. Extended Lagrangian formalism and the corresponding energy relations. *European Journal of Mechanics A/Solids*, Vol.23, pp. 975–91, 2004.

D. Mušicki. Generalization of a new parametric formulation of mechanics for systems with variable mass. *European Journal of Mechanics A/Solids*, Vol.19, p-. 1059–76, 2000

D. Mušicki. General energy change law for systems with variable mass. *European Journal of Mechanics A/Solids*, Vol.18, pp. 719–30, 1999.

J.N. Newman. Marine Hydrodynamics, The MIT Press, 1978, 402 pp.

J.J. Ong, O.M. O'Reilly. On the equations of motion for rigid bodies with surface growth. *International Journal of Engineering Science*, Vol.42, pp. 2159–74, 2004.

L.A. Pars. A Treatise on Analytical Dynamics. Heinemann, London, 1965, 641pp.

C.P. Pesce. The application of Lagrange equations to mechanical systems with mass explicitly dependent on position. *Journal of Applied Mechanics*, Vol.70, pp. 751–6, 2003.

C.P. Pesce. A note on the classical free surface hydrodynamic impact problem. Advances in Engineering Mechanics, reflections and outlooks, World Scientific Publishing Co., Singapore, 2005.

C.P. Pesce, L. Casetta. Variable mass system dynamics in Engineering Mechanics education. In *COBEM2007, 19th International Congress of Mechanical Engineering*, 2007, Braslia, Brazil.

C.P. Pesce, L. Casetta, F.M. Santos. On the equation of motion governing the dynamics of vertically collapsing buildings. *Journal of Engineering Mechanics*Vol. 138, n. 12, pp. 1420-1431, 2012.

C.P. Pesce, E.A. Tannuri, L. Casetta. The Lagrange equations for systems with mass varying explicitly with position: some applications to offshore engineering. *Journal of the Brazilian Society of Mechanical Sciences and Engineering*, Vol.28, pp .496–504, 2006.

A.R. Plastino, J.C. Muzzio. On the use and abuse of Newtons law for variable mass problems. Celestial Mechanics and Dynamical Astronomy, Vol.53, pp. 227 32, 1992.

S.D. Poisson. Sur le Mouvement d'un Systme de Corps, en supposant le Masses Variables. *Bull. Sci. Soc. Philomat*, Paris, avril, pp. 227–32, 1819.

K.B. Pomeranz. Newton's law for systems with a variable mass. *American Journal of Physics*, Vol.32, n.5, p. 386, 1964.

D. Prato, R.J. Gleiser. Another look at the uniform rope sliding over the edge of a smooth table. *American Journal of Physics*, Vol.50, n.6, pp. 536–539, 1982.

M. Schagerl et al. On the paradox of the free falling folded chain. *Acta Mechanica*, Vol.125, p. 155-68, 1997.

M. Schwarzbart, A. Steindl, W. Steiner, H. Troger, G. Wiedermann. Tethered satellite systems: a challenge for mechanics and applied mathematics. *GAMM-Mitteilungen*, Vol.32, n.1, pp. 105–20, 2009.

K.A. Seffen. Progressive collapse of the World Trade Center: simple analysis. *Journal of Engineering Mechanics*, Vol.134, n.2, pp. 125–32, 2008.

M. Shan-jun, G. Wei-go, H. Pei-tian. The third-order Lagrange equation for mechanical systems of variable mass. *Chinese Physics*, Vol.14, pp. 879–81, 2005.

L. Shao-kai, M. Feng-xiang. The principles of least action of variable mass nonholonomic nonconservative system in noninertial reference frames. *Applied Mathematics and Mechanics*, Vol.13, n.9, pp. 851–9, 1992.

C.A. Souza, V.H. Rodrigues. Mass redistribution in variable mass systems. *European Journal of Physics*, Vol.25, pp. 41–9, 2004.

V.M. Starjinski. Mecânica Teórica, Ed. Mir, Moscou, in Portuguese, translated from the Russian Ed., 1980, 543 pp., 1986.

V. Šima, J. Podolski. Buquoys problem. European Journal of Physics, Vol.26, pp. 1037–1045, 2005.

S. Targ. Theoretical Mechanics, a Short Course, Mir Publishers, Moscow, English translation, from the 2nd Russian Ed., 1974, 525 pp.

P.G. Tait, W.J. Steele. A Treatise on the Dynamics of a Particle, *Cambridge*, 1856, 6th ed. 1889.

J.F. Thorpe. On the momentum theorem for a continuous system of variable mass. *American Journal of Physics*, Vol.30, n.9, pp. 637–40, 1962.

W. Tomaszewski, P. Pieranski, J.-C. Geminard. The motion of a freely falling chain tip. *American Journal of Physics*, Vol.74, n.9, pp. 776–83, 2006.

F.G.S. Torres. Study of Moonpool as a Motion Minimization Device in a Monocolumn type Platform [online]. (in Portuguese); São Paulo. Escola Politécnica, University of São Paulo. *MSc Dissertation in Ocean Engineering*, 2007, 239pp.

C. Truesdell, R.A. Toupin. The classical field theories. *Handbuch der Physik, v. II/1: Prinzipien der klassischen Mechanik und Feldtheorie*, pp. 347, 1960.

J.A. Van den Akker. Newtons law with variable mass. *American Journal of Physics*, Vol.32, n.5, pp. 387–8, 1964.

T. Von Karman. The Impact of Seaplane Floats During Land. National Advisory Committee for Aeronautics (NACA). *Technical Note no. 321*, 1929.

B.D. Vujanovic, S.E. Jones. Variational Methods in Nonconservative Phenomena. Mathematics in Science and Engineering, v. 2, Academic Press Inc, London, 370 pp., 1989.

B.D. Vujanovic, T.M. Atanackovic. An Introduction to Modern Variational Techniques in Mechanics and Engineering. Birkhauser, Boston, 346 pp., 2004.

H. Wagner. Landing of seaplanes. *Tech Mem. 622. NACA/NASA Langley Research Center*, 1931.

C.W. Wong, K. Yasui. Falling chains. *American Journal of Physics*, Vol.6, pp. 490–496, 2006.

C.W. Wong, S.H. Youn, K. Yasui. The falling chain of Hopkins, Tait, Steele and Cayley. *European Journal of Physics*, Vol.28, pp. 385–400, 2007.
G.X. Wu. Hydrodynamic force on a rigid body during impact with liquid. *Journal of Fluids and Structures* Vol.12: pp. 549–559, 1998.

Appendix-Complementary to the Hamiltonian approach

Fundamental principles of Mechanics were primarily conceived for constant mass systems. Thus, within that original scenario, let $\delta\overline{T}$ refer to the first variation of the kinetic energy,

$$\begin{aligned} &\delta\overline{T} = \delta\sum_i \frac{1}{2}\overline{m}_i\mathbf{v}_i^2 = \sum_i \overline{m}_i\mathbf{v}_i\delta\mathbf{v}_i \\ &with \\ &\overline{m}_i = const., \forall i \end{aligned} \tag{112}$$

However, when considering a discrete system of variable mass, such that $m_i = m_i(q_k, \dot{q_k}, t)$, the virtual variation of the kinetic energy should, instead, be written as

$$\delta T = \delta\sum_i \frac{1}{2}m_i\mathbf{v}_i^2 = \sum_i \frac{1}{2}\delta m_i\mathbf{v}_i^2 + \sum_i m_i\mathbf{v}_i.\delta\mathbf{v}_i \tag{113}$$

With no loss of generality, take the *instant* of time τ at which $m_i(q_k(\tau), \dot{q_k}(\tau), \tau) = \overline{m}_i$, $i = 1, \ldots, N$. In this sense, equations (112) and (113) lead to

$$\delta\overline{T} = \delta T - \sum_i \frac{1}{2}\delta m_i{\mathbf{v}_\mathbf{i}}^2 \tag{114}$$

For conciseness sake this is here called the "instantaneous" first variation. The reader may notice that, when dealing with the particular case in which $m_i = m_i(t)$, Eq. (114) is obviously simplified as

$$\delta\overline{T} = \delta T \tag{115}$$

Equation (114) shows that, in the most general case, i.e.,$m_i = m_i(q_k, \dot{q_k}, t)$, the original form of Hamilton's principle turns to be

$$\int_{t_1}^{t_2}(\delta\overline{T} + \sum_j Q_j\delta q_j)dt = \int_{t_1}^{t_2}(\delta T - \sum_i \frac{1}{2}\delta m_i{\mathbf{v}_\mathbf{i}}^2 + \sum_j Q_j\delta q_j)dt = 0 \tag{116}$$

with Q_j as in Eq. (28), therefore already including Mescherski's forces. However, in terms of generalized coordinates,

$$\delta T = \sum_j \left(\frac{\partial T}{\partial q_j} \delta q_j + \frac{\partial T}{\partial \dot{q}_j} \delta \dot{q}_j \right) \tag{117}$$

Moreover, the following identity also holds true,

$$\frac{d}{dt} \sum_j \frac{\partial T}{\partial \dot{q}_j} \delta q_j = \sum_j \left(\frac{d}{dt} \left(\frac{\partial T}{\partial \dot{q}_j} \right) \delta q_j + \frac{\partial T}{\partial \dot{q}_j} \delta \dot{q}_j \right) \tag{118}$$

Analogously,

$$\sum_i \frac{1}{2} \delta m_i \mathbf{v}_i^2 = \sum_j \sum_i \left(\left(\frac{1}{2} \frac{\partial m_i}{\partial q_j} \mathbf{v}_i^2 \right) \delta q_j + \left(\frac{1}{2} \frac{\partial m_i}{\partial \dot{q}_j} \mathbf{v}_i^2 \right) \delta \dot{q}_j \right) \tag{119}$$

and

$$\frac{d}{dt} \sum_j \sum_i \left(\frac{1}{2} \frac{\partial m_i}{\partial \dot{q}_j} \mathbf{v}_i^2 \delta q_j \right) = \sum_j \sum_i \left(\frac{d}{dt} \left(\frac{1}{2} \frac{\partial m_i}{\partial \dot{q}_j} \mathbf{v}_i^2 \right) \delta q_j + \frac{1}{2} \frac{\partial m_i}{\partial \dot{q}_j} \mathbf{v}_i^2 \delta \dot{q}_j \right) \tag{120}$$

Substituing (117)-(120)

$$\begin{aligned} \delta \overline{T} = \delta T - \sum_i \frac{1}{2} \delta m_i \mathbf{v}_i^2 &= \sum_j \left(-\frac{d}{dt} \frac{\partial T}{\partial \dot{q}_j} + \frac{\partial T}{\partial q_j} \right) \delta q_j + \frac{d}{dt} \sum_j \left(\frac{\partial T}{\partial \dot{q}_j} \delta q_j \right) + \\ -\sum_j \sum_i \left(-\frac{d}{dt} \left(\frac{1}{2} \frac{\partial m_i}{\partial \dot{q}_j} \mathbf{v}_i^2 \right) \right) \delta q_j &- \frac{d}{dt} \sum_j \sum_i \left(\frac{1}{2} \frac{\partial m_i}{\partial \dot{q}_j} \mathbf{v}_i^2 \delta q_j \right) \end{aligned} \tag{121}$$

Or, rearranging terms,

$$\begin{aligned} &\delta \overline{T} = \delta T - \sum_i \frac{1}{2} \delta m_i \mathbf{v}_i^2 = \\ &\sum_j \left(-\frac{d}{dt} \frac{\partial T}{\partial \dot{q}_j} + \frac{\partial T}{\partial q_j} + \sum_i \left(\frac{d}{dt} \left(\frac{1}{2} \frac{\partial m_i}{\partial \dot{q}_j} \mathbf{v}_i^2 \right) - \frac{1}{2} \frac{\partial m_i}{\partial q_j} \mathbf{v}_i^2 \right) \right) \delta q_j \\ &+ \sum_j \frac{d}{dt} \left(\frac{\partial T}{\partial \dot{q}_j} \delta q_j \right) - \sum_j \sum_i \frac{d}{dt} \left(\frac{1}{2} \frac{\partial m_i}{\partial \dot{q}_j} \mathbf{v}_i^2 \delta q_j \right) \end{aligned} \tag{122}$$

By usual procedures, substituting Eq. (122) into (116) leads to the extended form of the Hamilton principle, Eq. (24) and, therefore, to the corresponding extended Lagrange's equation, (25), or (19).

Dynamics of the Mass Variable Body

Livija Cveticanin*

* Faculty of Technical Sciences, 21000 Novi Sad, Serbia

Abstract In this Chapter the dynamics of the body with time variable mass and time variable moment of inertia is treated. The both cases of mass and moment of inertia variation are considered: the discontinual and the continual. The basic lows in dynamics are extended to the case when the mass is varying in time. The principle of momentum and of angular momentum are applied to obtain the velocity and angular velocity of the body after discontinual mass variation. The same results are applied analytically by introducing the procedures of analytical mechanics. The dynamics of mass addition is treated as the plastic impact, and of the separation as the inverse process to plastic impact. For the case of the continual variation of the mass and of the moment of inertia in time, beside the reactive force, the reactive torque is introduced. The case of the free motion of the mass variable body is investigated. The Lagrange's equations of motion are derived. As the special motion of the mass variable body, the vibration is considered. The main attention is given to approximate solving of the strong non-linear differential equations of motion. The influence of the reactive force on the vibration properties of the body is analyzed.

1 Introduction

The problem of the motion of mass variable systems is evident since the 17th century. Galileo discovered the anomaly in the Moon motion which he believed is the function of the system mass variation. Lately, Laplace theoretically explained the phenomena of the secular acceleration of the Moon. Dufour, 1886, explained that the mass of the earth varies continuously due to the falling shooting-stars and also due to combustion or spending in the atmosphere. He found that the dust of shooting-stars which fall on the surface of France in one year can cover a volume of 0.1 m^3. Oppalzer, 1884, was the first to analyze the reason for secular acceleration of the Moon as the result of Earth and Moon mass increase. Namely, during a hundred year a 2.8 mm dust layer is formed on the Earth. Gylden, 1884, extended the

H. Irschik, A. K. Belyaev (Eds.), *Dynamics of Mechanical Systems with Variable Mass*,
CISM International Centre for Mechanical Sciences DOI 10.1007/ 978-3-7091-1809-2_3

previous investigations in celestial mechanics by analyzing of the relative motion of two variable mass systems under influence of the Newton force. Meshchersky, 1893, continued the investigation and found that the body with variable mass would move along a spiral, tending toward zero, or it would increase the distance to the other mass variable system.

Cayley in his works (Cayley, 1857; 1858) was the first to consider the influence of the continual mass variation on the motion of the body. The class of the dynamic problem he studied was the 'continuous reactive problem', i.e., the problem when continuously the infinitesimal small mass is added to a system which causes the velocity of the system continuously to be changed for a definite value. The two most widely discussed examples were (see Cayley, 1859): one, the chain is on the table and is dropping vertically down from the table, and the second, the chain is moving straightforward on a horizontal plane without friction under the influence of a mass M which is fixed at the end of a chain which is rolling around a drum and changing the length during motion.

At the end of the nineteenth century and at the beginning of the twentieth Meshchersky (1897) laid the foundations of the modern dynamics of a particle with variable mass. After that publication numerous investigations have been done and the dynamics of variable mass systems is developed. In the 'variable mass systems' particles are expelled and /or captured during motion.

Two kinds of systems are identified:

- 'continuously' particle-ejecting systems, where the mass variation is a continual function of the time, position (see Grudtsyn, 1972) or velocity of the particle, and
- 'discretely' particle-ejecting systems, where the mass variation is a discontinual function (for example, automatic weapons that fire rounds, one at a time).

For the case of discontinual mass variation Meshchersky (1952) calculated velocity of the particle after mass variation. The finite discontinual mass variation in a very short time was not of special interest for a long time and was not intensively discussed. Meshchersky was the first to consider the velocity change of a translatory moving body during step-like mass variation. Mass which is separated or added has a finite amount and the mass variation of the body is discontinual. The theory is mostly applied for solving of the Keplerian two-body problem (see Luk'yanov, 2005), and also the three and four-body problems (Cveticanin, 2007).

The motion of the continuously mass variable systems is much more investigated due to its application in rocket theory (Meirovitch, 1970; Cormelisse *et al*, 1979; Tran & Eke), astronomy (Kayuk & Denisenko, 2004), for charged particle motion in a magnetic field with decreasing mass and charge (Howard, 2007), in robotics (McPhee & Djerassi, 1991; Djerassi, 1998) in machinery (Cveticanin, 1984; 1988; 1989; 1991; 1993_1; 1993_2; 1995; 1998_1; 2001) etc. The motion is described with differential equations with variable parameters (Kayuk & Tivalov, 1987; Wang & Eke, 1995; Eke & Mao, 2002; Pesce, 2003). The effect of expulsion and /or capture of particles on the motion of the continuously mass variable system is evident as changes in the integration variables of the governing dynamic equations. In spite of that the variable mass systems may be conservative (Leubner & Krumm, 1990; Cveticanin, 1994) and with non-holonomic properties (Ge & Cheng, 1982; Ge, 1984). For the case when the mass is continually varying in time, the influence of the reactive force on the motion (see Apykhtin & Jakovlev, 1980; Azizov, 1986; Cveticanin, 1992; 1993_3; 2004) and also on the stability (Ignat'yev, 1991; Cveticanin, 1996_1; 1996_2) were investigated. The reactive force is mathematically the product of the mass variation function and the relative velocity of mass separated or added to the particle. Usually, two special cases were considered: first, the relative velocity is zero and second, the absolute velocity of separated or added mass is zero. If the relative velocity is zero, i.e., the absolute velocity of the separated or added particle is equal to the velocity of the basic particle, then the reactive force is also zero. Levi-Civita, 1928, investigated the motion of the particle for the case when the absolute velocity of the separated or added mass is zero and the reactive force exists. The most comprehensive consideration of the dynamics of the body with variable mass is given in the books of Bessonov, 1967, Conelisse *et al.*, 1979, and Cveticanin, 1998_2.

Based on the Dynamics of the particle with time variable mass and the basic laws of dynamics, in this Chapter the theoretical consideration of the dynamics of the body with time variable mass is given. The Chapter is divided into following Sections:

After the Introduction, the linear momentum and the angular momentum of the body with variable mass and moment of inertia is considered. The linear and angular momentum of the body before and after mass modification due to adding or separating of the mass is determined.

The obtained linear and angular momentum relations are applied for calculation of the velocity and angular velocity of the body when the body separation or augmentation is discontinual. The principles of momentum and angular momentum are used as the basic ones. The special case of the in-plane separation is considered. Depending on the type of motion of the separated body, the kinematic properties of the remainder body are discussed.

The dynamics of the discontinual mass variation is also treated by analytical approach using the principles of analytical mechanics. An analytical procedure for the velocity and angular velocity determination of a body in the process of mass variation is developed. The main attention is given to the case when no external forces and torques act. The process of body addition is treated as the plastic impact and the separation as an inverse process of the plastic impact of bodies.

In the next section, we express the free motion of the body with continual time variation of the mass and moment of inertia. Due to mass and also moment of inertia variation, beside the reactive force, the reactive torque acts. In this section the main attention is directed toward investigation of the influence of these two physical actions. As a special type of motion the in-plane motion of the body is considered. The obtained theory is applied for analyzing of the plane motion of the rotor on which the band is winding up.

The Lagrange's equations of the motion for the body with continual variation of mass and moment of inertia is derived. The generalized forces due to the reactive force and the reactive torque are defined. The obtained Lagrange's equations represent the analytical description of the free motion of the body with variable mass.

As a special type of the motion, the vibrations of the body with variable mass are investigated. Based on the general equations of motion given in the previous section, the mathematical model for the oscillatory motion is formed. The main attention is directed to approximate solution procedures for solving the strong nonlinear differential equations with slow-time variable parameters describing the vibrations of various kinds of oscillators.

The Chapter ends with the Reference list.

2 Linear and angular momentums for the mass variable body

Let us consider the discontinual mass variation caused by the body separation or augmentation. The initial body has the mass M and the moment of inertia $\mathbf{I}_S$ with respect to the mass centre S whose position vector due to the fixed point O is $\mathbf{r}_S$ (see Fig.1).

The linear velocity of the mass centre is v_S and the angular velocity of the initial body around the mass centre S is $\boldsymbol{\Omega}$. The position vector of the mass centre S to the fixed point O is $\mathbf{r}_S$. The separating or adding body has mass m and the mass centre S_2 . Moment of inertia of that body is $\mathbf{I}_{S2}$ with respectto the mass centre S_2.The absolute velocity of the mass center S_2 is $\mathbf{v}_{S2}$ while the angular velocity of the body with respect to point of rotation S_2 is $\boldsymbol{\Omega}_2$. The position vector of the mass centre S_2 according to the fixed

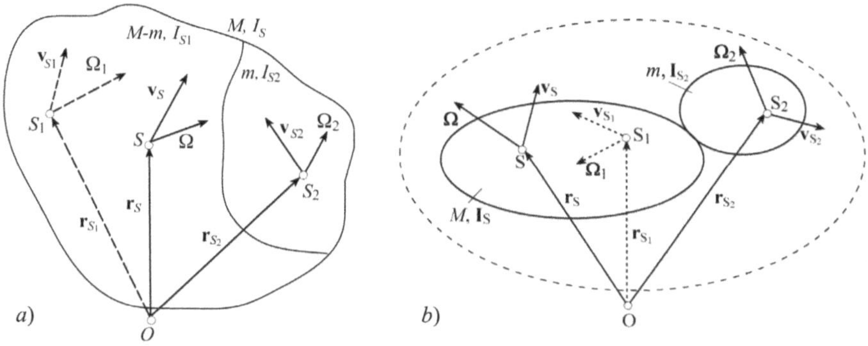

Figure 1. Position vectors, velocities and angular velocities of the system: a) body separation, b) body augmentation.

point O at the moment of adding or separating is $\mathbf{r}_{S2}$. If the separation of the body occurs, the remainder body has the mass $(M - m)$ and if the augmentation occurs the mass is $(M + m)$. The moment of inertia of the final body is $\mathbf{I}_{S1}$ due to the mass centre S_1 whose position to the fixed point O is given with the vector $\mathbf{r}_{S1}$. The unknown linear and angular velocity of the body after mass variation are $\mathbf{v}_{S1}$ and $\boldsymbol{\Omega}_1$. The Nomenclature is as follows:

	Initial body	Final body	Separated/ Added body
Index	-	1	2
Mass center	S	S_1	S_2
Position of mass center		$\mathbf{SS}_1$	$\mathbf{SS}_2$
Position vector of mass center	$\mathbf{r}_S$	$\mathbf{r}_{S1}$	$\mathbf{r}_{S2}$
Mass	M	$M \mp m$	$\mp m$
Absolute velocity of mass center	$\mathbf{v}_S$	$\mathbf{v}_{S1}$	$\mathbf{v}_{S2}$
Relative velocity of mass center		$\mathbf{v}_{r1}$	$\mathbf{v}_r$
Absolute angular velocity	$\boldsymbol{\Omega}$	$\boldsymbol{\Omega}_1$	$\boldsymbol{\Omega}_2$
Relative angular velocity		$\boldsymbol{\Omega}_{r1}$	$\boldsymbol{\Omega}_r$
Moment of inertia tensor	$\mathbf{I}_S$	$\mathbf{I}_{S1}$	$\mathbf{I}_{S2}$
Linear momentum	$\mathbf{K}$	$\mathbf{K}_1$	$\mathbf{K}_2$
Angular momentum relating to O	$\mathbf{L}_O$	$\mathbf{L}_{O1}$	$\mathbf{L}_{O2}$
Angular momentum relating to mass center	$\mathbf{L}_S$	$\mathbf{L}_{S1}$	$\mathbf{L}_{S2}$

The linear momentum of the initial body with mass M and velocity of mass center $\mathbf{v}_S$ is

$$\mathbf{K} = M\mathbf{v}_S. \tag{1}$$

If the separated or added mass m has the velocity $\mathbf{v}_{S2}$, its linear momentum is

$$\mathbf{K}_2 = m\mathbf{v}_{S2}. \tag{2}$$

After process of separation or addition the final mass is $M \mp m$, where the minus sign is for mass separation and plus sign for mass addition.

Remark 2.1. In general in this Chapter in any relation, the minus sign is for change of some quantity caused by mass separatiom and the plus sign is for mass addition.

The unknown velocity of the mass centre S_1 is $\mathbf{v}_{S1}$ and the corresponding linear momentum is

$$\mathbf{K}_1 = (M \mp m)\mathbf{v}_{S1}. \tag{3}$$

Introducing the assumption that the bodies during the mass separation or addition belong to an unique system (see Fig.1), we obtain the linear momentums $\mathbf{K}_b$ before and $\mathbf{K}_a$ after mass variation:

for mass separation

$$\mathbf{K}_b = \mathbf{K} = M\mathbf{v}_S, \qquad \mathbf{K}_a = \mathbf{K}_1 + \mathbf{K}_2 = m\mathbf{v}_{S2} + (M - m)\mathbf{v}_{S1}, \tag{4}$$

for mass augmentation

$$\mathbf{K}_b = \mathbf{K} + \mathbf{K}_2 = M\mathbf{v}_S + m\mathbf{v}_{S2}, \qquad \mathbf{K}_a = \mathbf{K}_1 = (M + m)\mathbf{v}_{S1}, \tag{5}$$

The difference between the linear momentums before and after mass variation is

$$\Delta\mathbf{K} = M(\mathbf{v}_{S1} - \mathbf{v}_S) \pm m(\mathbf{v}_{S2} - \mathbf{v}_{S1}). \tag{6}$$

The angular momentum of the initial body before mass variation with respect to the fixed point O (Fig.1) is

$$\mathbf{L}_O = \mathbf{r}_S \times M\mathbf{v}_S + \mathbf{L}_S. \tag{7}$$

After mass variation the angular of the final body relating to the fixed point O is

$$\mathbf{L}_{O1} = \mathbf{r}_{S1} \times (M \mp m)\mathbf{v}_{S1} + \mathbf{L}_{S1}. \tag{8}$$

The angular momentum of the separated or added body is

$$\mathbf{L}_{O2} = \mathbf{r}_{S2} \times m\mathbf{v}_{S2} + \mathbf{L}_{S2}. \tag{9}$$

Dependently on the type of mass variation the angular momentums before $\mathbf{L}_{Ob}$ and after $\mathbf{L}_{Oa}$ mass variation are:

for mass separation

$$\begin{aligned} \mathbf{L}_{Ob} &= \mathbf{L}_O = \mathbf{r}_S \times M\mathbf{v}_S + \mathbf{L}_S, \\ \mathbf{L}_{Oa} &= \mathbf{L}_{O1} + \mathbf{L}_{O2} = \mathbf{r}_{S1} \times (M-m)\mathbf{v}_{S1} + \mathbf{L}_{S1} + \mathbf{r}_{S2} \times m\mathbf{v}_{S2} + \mathbf{L}_{S2}, \end{aligned} \tag{10}$$

for mass addition

$$\begin{aligned} \mathbf{L}_{Ob} &= \mathbf{L}_O + \mathbf{L}_{O2} = \mathbf{r}_S \times M\mathbf{v}_S + \mathbf{L}_S + \mathbf{r}_{S2} \times m\mathbf{v}_{S2} + \mathbf{L}_{S2}, \\ \mathbf{L}_{Oa} &= \mathbf{L}_{O1} = \mathbf{r}_{S1} \times (M+m)\mathbf{v}_{S1} + \mathbf{L}_{S1}. \end{aligned} \tag{11}$$

Based on (10) and (11), the difference between the angular momentums is

$$\Delta\mathbf{L}_O = \mathbf{L}_{S1} \pm \mathbf{L}_{S2} - \mathbf{L}_S + \mathbf{r}_{S1} \times (M \mp m)\mathbf{v}_{S1} \pm \mathbf{r}_{S2} \times m\mathbf{v}_{S2} - \mathbf{r}_S \times M\mathbf{v}_S. \tag{12}$$

The relation (6) and (12) are the basic ones for dynamic analysis of the mass variation problems.

For the position of the system mass center S

$$\mathbf{r}_S = \frac{M \mp m}{M}\mathbf{r}_{S1} \pm \frac{m}{M}\mathbf{r}_{S2}, \tag{13}$$

and position vectors $\mathbf{r}_{S1}$ and $\mathbf{r}_{S2}$ (Fig.1)

$$\mathbf{r}_{S1} = \mathbf{r}_S + \mathbf{SS}_1, \qquad \mathbf{r}_{S2} = \mathbf{r}_S + \mathbf{SS}_2, \tag{14}$$

we obtain

$$(M \mp m)\mathbf{SS}_1 = \mp m\mathbf{SS}_2, \tag{15}$$

Substituting (14) into (12) we obtain

$$\Delta L_O = \mathbf{L}_{S1} \pm \mathbf{L}_{S2} - \mathbf{L}_S + \mathbf{SS}_1 \times (M \mp m)\mathbf{v}_{S1} \pm \mathbf{SS}_2 \times m\mathbf{v}_{S2}. \tag{16}$$

Introducing the relation (6) into (16) it is

$$\Delta\mathbf{L}_O = \mathbf{L}_{S1} \pm \mathbf{L}_{S2} - \mathbf{L}_S + \mathbf{r}_S \times \Delta K \mp \mathbf{SS}_2 \times m(\mathbf{v}_{S1} - \mathbf{v}_{S2}). \tag{17}$$

Due to (15) the relation (16) transforms into

$$\Delta L_O = \mathbf{L}_{S1} - \mathbf{L}_S \pm \mathbf{L}_{S2} \mp \mathbf{SS}_2 \times (\mathbf{v}_{S1} - \mathbf{v}_{S2})m, \tag{18}$$

i.e.,

$$\Delta L_O = \mathbf{I}_{S1}\boldsymbol{\Omega}_1 - \mathbf{I}_S\boldsymbol{\Omega} \pm \mathbf{I}_{S2}\boldsymbol{\Omega}_2 \mp \mathbf{SS}_2 \times (\mathbf{v}_{S1} - \mathbf{v}_{S2})m, \tag{19}$$

where $\mathbf{L}_S = \mathbf{I}_S\boldsymbol{\Omega}$ is the known angular momentum of the initial body before mass variation, $\mathbf{L}_{S1} = \mathbf{I}_{S1}\boldsymbol{\Omega}_1$ depends on the angular velocity $\boldsymbol{\Omega}_1$ of the final body with respect to S_1 and is proportional to the moment of inertia $\mathbf{I}_{S1}$ for S_1 and $\mathbf{L}_{S2} = \mathbf{I}_{S2}\boldsymbol{\Omega}_2$ is the known angular momentum of the separated or added body with the angular velocity $\boldsymbol{\Omega}_2$ with respect to S_2 and moment of inertia $\mathbf{I}_{S2}$ for S_2.

3 Dynamics of the body with discontinual mass variation

Dynamics of the discontinual mass variation requires some assumptions during the process of body separation or addition and are as follows (Cveticanin & Djukic, 2008):

1. The separated and the final body, and also the initial body and the added body, form a unique system during mass variation (see Fig. 1);

2. The separated body leaves the system after the process of separation. The added body gets into the system before mass augmentation;

3. Separation or adding of the body is done in a very short time interval τ;

4. The considered bodies are rigid during mass separation;

5. Due to the assumption 1) we can regard the two parts of the body as a complex system, where the reaction forces and torques between these parts are internal within the system;

6. During the process of mass variation the external forces $\mathbf{F}_i$ and torques $\mathfrak{M}_j$, which act on the system, produce the impulses.

Let $\mathbf{F}_r$ be the resultant force of all external activ forces and constraint reactions, which are acting on the bodies. According to the principle of the momentum, the variation of the linear momentum (6) for the time interval from $\tau = \Delta t$ is equal to the impulse $\mathbf{I}^{Fr}$ of the resultant force $\mathbf{F}_r$

$$\Delta\mathbf{K} = \mathbf{F}_r\Delta t \equiv \mathbf{I}^{Fr}. \tag{20}$$

Substituting (6) into (20) it is

$$M(\mathbf{v}_{S1} - \mathbf{v}_S) \pm m(\mathbf{v}_{S2} - \mathbf{v}_{S1}) = \mathbf{I}^{Fr}. \tag{21}$$

According to the principle of the angular momentum, the variation of the angular momentum (19) in the time interval Δt is equal to the impulse $\mathbf{I}^M$ which is the sum of the impulse of the moment of resultant force for the point O, $\mathbf{M}_0^{Fr}$, and of the impulse of the resultant torque $\mathfrak{M}$, caused by activ torque and reaction torque, i.e.,

$$\Delta\mathbf{L}_O = (\mathbf{M}_0^{Fr} + \mathfrak{M})\Delta t = \mathbf{I}^M. \tag{22}$$

Substituting (16) into (22) we have

$$\mathbf{L}_{S1} \pm \mathbf{L}_{S2} - \mathbf{L}_S + \mathbf{SS}_1 \times (M \mp m)\mathbf{v}_{S1} \pm \mathbf{SS}_2 \times m\mathbf{v}_{S2} = \mathbf{I}^M. \tag{23}$$

Usually, the impulses of the external forces and torques are quite small due to the short time τ. It is the reason, that the system is usually assumed to be without action of the external forces and torques. Then, the linear momentum of the system before and after mass variation remains invariable. The same conclusion is valid also for the angular momentum.

3.1 Velocity of the body after mass variation

According to (21) the velocity of the body after mass variation is

$$\mathbf{v}_{S1} = \frac{1}{(M \mp m)}(M\mathbf{v}_S \mp m\mathbf{v}_{S2} + \mathbf{I}^{Fr}). \tag{24}$$

Using the fact that the absolute velocity of the mass center S_2 of the separated or added body is the sum of the dragging velocity of S_2 and the relative velocity $\mathbf{v}_r$ of the point S_2 with respect to the point S. The dragging velocity of S_2 has two components: translatory $\mathbf{v}_S$ and velocity of rotation $\boldsymbol{\Omega} \times \mathbf{SS}_2$ of the point S_2 with respect to point S. The absolute velocity of the point S_2 for mass varaiation is

$$\mathbf{v}_{S2} = \mathbf{v}_S + \boldsymbol{\Omega} \times \mathbf{SS}_2 + \mathbf{v}_r. \tag{25}$$

The absolute velocity of the mass center S_1 of the final body after mass variation has the form

$$\mathbf{v}_{S1} = \mathbf{v}_S + \boldsymbol{\Omega} \times \mathbf{SS}_1 + \mathbf{v}_{r1}, \tag{26}$$

where $\mathbf{v}_S$ and velocity of rotation $\boldsymbol{\Omega} \times \mathbf{SS}_1$ are the translatory and rotational components of velocity of mass center S_1 and $\mathbf{v}_{r1}$ is the relative velocity of the point S_1. Substituting (25) and (26) into (24), the following relation is

$$(M \mp m)\mathbf{v}_{r1} \pm m\mathbf{v}_r = \mp m(\boldsymbol{\Omega} \times \mathbf{SS}_2) - \boldsymbol{\Omega} \times (M \mp m)\mathbf{SS}_1) + \mathbf{I}^{Fr}. \tag{27}$$

The relations (27) and (15) give the relative velocity of the mass center S_1 of the final body

$$(M \mp m)\mathbf{v}_{r1} = \mathbf{I}^{Fr} \mp m\mathbf{v}_r, \tag{28}$$

and the corresponding absolute velocity

$$\mathbf{v}_{S1} = \mathbf{v}_S + \boldsymbol{\Omega} \times \mathbf{SS}_1 + \frac{\mathbf{I}^{Fr} \mp m\mathbf{v}_r}{(M \mp m)}. \tag{29}$$

Let us assume the system without external forces. In that case the linear momentum of the system is same before and after mass variation, hence $\Delta\mathbf{K} = 0$. According to the relation (29), we determine the velocity of mass center S_1 of the body after mass variation as

$$\mathbf{v}_{S1} = \mathbf{v}_S + \boldsymbol{\Omega} \times \mathbf{SS}_1 \mp \frac{m\mathbf{v}_r}{(M \mp m)}. \tag{30}$$

For the case when the separation of the body occurs, the velocity of the body after mass transformation is due to (30)

$$\mathbf{v}_{S1} = \mathbf{v}_S + \boldsymbol{\Omega} \times \mathbf{SS}_1 - \frac{m}{M - m}\mathbf{v}_r. \tag{31}$$

3.2 Angular velocity of the body after mass variation

Using the relations (22) and (19), it follows

$$\mathbf{I}_{S1}\boldsymbol{\Omega}_1 = \mathbf{I}^M + \mathbf{I}_S\boldsymbol{\Omega} \mp \mathbf{I}_{S2}\boldsymbol{\Omega}_2 \pm \mathbf{SS}_2 \times (\mathbf{v}_{S1} - \mathbf{v}_{S2})m. \tag{32}$$

The relation is suitable for calculation of the angular velocity $\boldsymbol{\Omega}_1$ of the final body after mass variation.

Using the assumption 1) that two bodies form one unique system and no external forces and torques act, it is stated that the angular momentums of the body before and after mass variation are invariable, i.e., $\Delta\mathbf{L}_O = 0$, and due to (32) the angular velocity $\boldsymbol{\Omega}_1$ is

$$\mathbf{I}_{S1}\boldsymbol{\Omega}_1 = \mathbf{I}_S\boldsymbol{\Omega} \mp \mathbf{I}_{S2}\boldsymbol{\Omega}_2 \pm \mathbf{SS}_2 \times (\mathbf{v}_{S1} - \mathbf{v}_{S2})m. \tag{33}$$

For the case when mass separation occurs, the angular velocity of the final body is

$$\mathbf{I}_{S1}\boldsymbol{\Omega}_1 = \mathbf{I}_S\boldsymbol{\Omega} - \mathbf{I}_{S2}\boldsymbol{\Omega}_2 + \mathbf{SS}_2 \times (\mathbf{v}_{S1} - \mathbf{v}_{S2})m, \tag{34}$$

where $\mathbf{v}_{S1}$ satisfies the Eq. (31).

The value of the angular velocity $\boldsymbol{\Omega}_1$ and the velocity of mass center after mass variation $\mathbf{v}_{S1}$ represent the initial values for the motion of the final body.

Remarks Due to the assumptions and the relations (31) and (34), it can be concluded:

Remark 3.1. During the process of mass variation, which lasts for the infinitesimal time interval, $t \in [t_1, t_1 + \tau]$, the interaction of the separated and final body or initial and added bodies results in a finite change of the linear and the angular velocity of the body parts. The linear momentum and the velocity of the final body and also the angular momentum and the angular velocity of the final body receive finite increments during the infinitesimal time, that is, these quantities change in a jump-like manner.

Remark 3.2. According to the assumption 3), during the process of mass variation the position change of the bodies is negligible, i.e., the position vectors of mass centers and the angle position of bodies are not varying during mass variation.

Remark 3.3. Due to the aforementioned Remarks it is concluded that the body additon corresponds to the perfect plastic impact where the relative velocity of the adding body is zero.

Remark 3.4. Accordint to here obtained results it is obvious that the body separation is the inverse process to the perfectly plastic impact where the relative velocity of the separated body is zero. As for the plastic impact the restitution coefficient is zero, the same is evident for the body separation. It means that the motion does not depend on the geometric and dynamical properties of the separation surface.

Remark 3.5. In this Chapter more attention is given to the mass separation as the dynamics of mass augmentation can be treated as the plastic impact.

3.3 In-plane separation of the body

Consider the body which moves in-plane before and after body separation. The absolute velocity of the initial body, remainder (final) and separated body are defined by (24) and (28), with projections

$$\mathbf{v}_S = v_{Sx}\mathbf{i} + v_{Sy}\mathbf{j}, \qquad \mathbf{v}_{S1} = v_{S1x}\mathbf{i} + v_{S1y}\mathbf{j}, \qquad \mathbf{v}_{S2} = v_{S2x}\mathbf{i} + v_{S2y}\mathbf{j}, \quad (35)$$

where $\mathbf{i}$ and $\mathbf{j}$ are unit vectors in the plane of motion (Fig.2).

For the in-plane motion the angular moments of the initial body, separated body and remainder body are

$$\mathbf{L}_{S1} = I_{S1}\Omega_1\mathbf{k}, \qquad \mathbf{L}_{Sb} = I_S\Omega\mathbf{k}, \qquad \mathbf{L}_{S2} = I_{S2}\Omega_2\mathbf{k}, \quad (36)$$

where $\mathbf{k}$ is the unit vector orthogonal to the plane of motion. Substituting (36) into (34) we obtain the angular velocity of the remainder body as a function of the angular velocity of the separated body

$$I_{S1}\Omega_1\mathbf{k} = I_S\Omega\mathbf{k} - I_{S2}\Omega_2\mathbf{k} + \mathbf{SS}_2 \times (\mathbf{v}_{S1} - \mathbf{v}_{S2})m, \quad (37)$$

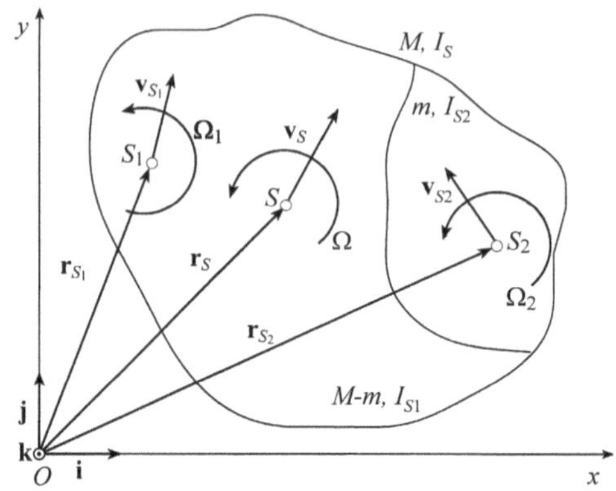

Figure 2. Position vectors of the plane body centers S, S_1 and S_2 with respect to a fixed point O.

where

$$\mathbf{SS}_2 = SS_{2x}\mathbf{i} + SS_{2y}\mathbf{j}, \tag{38}$$

In relation (37) the absolute angular velocity of the remainder body is the function of the absolute angular velocity of the separated body.

Introducing (25) and (26) and also (36) into (32) leads to

$$\begin{aligned} I_S\Omega\mathbf{k} \quad = \quad & I_{S1}\Omega_1\mathbf{k} + \mathbf{SS}_1 \times (M-m)(\mathbf{v}_S + \boldsymbol{\Omega} \times \mathbf{SS}_1 + \mathbf{v}_{r1}) \\ & + I_{S2}\Omega_2\mathbf{k} + \mathbf{SS}_2 \times m(\mathbf{v}_S + \boldsymbol{\Omega} \times \mathbf{SS}_2 + \mathbf{v}_r). \end{aligned} \tag{39}$$

Using the relation (15), the Stainer formulas for the moment of inertia for the axis in S parallel to the axis in S_1 and S_2

$$I_{S1} = I_1 + (SS_1)^2(M-m), \qquad I_{S2} = I_2 + (SS_2)^2 m, \tag{40}$$

and the relative angular velocities

$$\Omega_1 = \Omega + \Omega_{r1}, \qquad \Omega_2 = \Omega + \Omega_r, \tag{41}$$

the relation (39) yields

$$I_{S1}\Omega_{r1}\mathbf{k} = -\, I_{S2}\Omega_r\mathbf{k} - \mathbf{SS}_2 \times m(\mathbf{v}_r - \mathbf{v}_{r1}). \tag{42}$$

For the difference of the relative velocities

$$\mathbf{v}_r - \mathbf{v}_{r1} = (v_{rx} - v_{r1x})\mathbf{i} + (v_{ry} - v_{r1y})\mathbf{j}, \tag{43}$$

and (42), the relative angular velocity of the remainder body is obtained

$$\Omega_{r1}=\frac{m}{I_{S1}}[SS_{2y}(v_{rx}-v_{r1x})-SS_{2x}(v_{ry}-v_{r1y})]-\frac{I_{S2}}{I_{S1}}\Omega_r. \tag{44}$$

The relations (28) and (44) define the relative velocity and angular velocity of the remainder body during in-plane body separation.

Some special cases Depending on the velocity and the angular velocity of the separated body, the various cases of body motion are possible. In the following Tables some special cases of body separation are shown, together with the corresponding properties of the remainder body, calculated on the bases of (28) and (42).

TRANSLATION OF THE WHOLE BODY
$\mathbf{v}_S \neq 0, \quad \Omega = 0.$

No.	Separated body	Remainder body
1	$\mathbf{v}_{S2} \neq 0,\ \Omega_2 \neq 0$	$\mathbf{v}_{r1} = -\frac{m}{M-m}\mathbf{v}_r,\ I_{S1}\Omega_1\mathbf{k} = \mathbf{SS}_2 \times (\mathbf{v}_{S1} - \mathbf{v}_{S2})m - I_{S2}\Omega_2\mathbf{k}$
2	$\mathbf{v}_{S2} \neq 0,\ \Omega_2 = 0$	$\mathbf{v}_{r1} = -\frac{m}{M-m}\mathbf{v}_r,\ I_{S1}\Omega_1\mathbf{k} = \mathbf{SS}_2 \times (\mathbf{v}_{S1} - \mathbf{v}_{S2})m$
3	$\mathbf{v}_{S2} = 0,\ \Omega_2 \neq 0$	$\mathbf{v}_{S1} = \frac{M}{M-m}\mathbf{v}_S,\ I_{S1}\Omega_1\mathbf{k} = \frac{Mm}{M-m}\mathbf{SS}_2 \times \mathbf{v}_S - I_{S2}\Omega_2\mathbf{k}$
4	$\mathbf{v}_{S2} = 0,\ \Omega_2 = 0$	$\mathbf{v}_{S1} = \frac{M}{M-m}\mathbf{v}_S,$ $I_{S1}\Omega_1\mathbf{k} = \frac{Mm}{M-m}\mathbf{SS}_2 \times \mathbf{v}_S$
5	$\mathbf{v}_{S2} \neq 0,$ $\Omega_2\mathbf{k} = -\frac{mM}{I_{S2}(M-m)}\mathbf{SS}_2 \times \mathbf{v}_r$	$\mathbf{v}_{S1} = \mathbf{v}_S - \frac{m}{M-m}\mathbf{v}_r,$ $\Omega_1 = 0$
6	$\mathbf{v}_{S2} = \frac{M}{m}\mathbf{v}_S,\ \Omega_2 \neq 0$	$\mathbf{v}_{S1} = 0,\ I_{S1}\Omega_1\mathbf{k} = -M(\mathbf{SS}_2 \times \mathbf{v}_S) - I_{S2}\Omega_2\mathbf{k}$
7	$\mathbf{v}_{S2} = \frac{M}{m}\mathbf{v}_S,\ \Omega_2\mathbf{k} = \frac{M(\mathbf{v}_S \times \mathbf{SS}_2)}{I_{S2}}$	$\mathbf{v}_{S1} = 0,\ \Omega_1 = 0$
8	$\mathbf{v}_{S2} = \mathbf{v}_S, (\mathbf{v}_r = 0),\ \Omega_2 \neq 0$	$\mathbf{v}_{S1} = \mathbf{v}_S,\ I_{S1}\Omega_{r1}\mathbf{k} = -I_{S2}\Omega_r\mathbf{k}$

Table 1.

Table 1. provides properties of the remainder body for the case when the initial body is in translation. Subcases for various absolute velocity $\mathbf{v}_{S2}$ of mass center and absolute angular velocity Ω_2 of the separated body are considered.

In Table 2. the velocity and the angular velocity of the remainder body for the case when the initial body is rotating are shown. Subcases are

ROTATING OF THE WHOLE BODY

$\mathrm{v}_S = 0, \qquad \Omega \neq 0.$

No.	Separated body	Remainder body
1	$\mathbf{v}_r \neq 0, \qquad \Omega_r \neq 0$	$\mathbf{v}_{S1} = \boldsymbol{\Omega} \times \mathbf{SS}_1 - \frac{m}{M-m}\mathbf{v}_r$, $\Omega_{r1}\mathbf{k} = \frac{M(\mathbf{SS}_1 \times \mathbf{v}_r)}{I_{S1}} - \frac{I_{S2}}{I_{S1}}\Omega_r\mathbf{k}$
2	$\mathbf{v}_r = 0, \qquad \Omega_r \neq 0$	$\mathbf{v}_{S1} = \boldsymbol{\Omega} \times \mathbf{SS}_1$, $\Omega_{r1}\mathbf{k} = -\frac{I_{S2}}{I_{S1}}\Omega_r\mathbf{k}$
3	$\mathbf{v}_r = 0, \qquad \Omega_r = 0$	$\mathbf{v}_{S1} = \boldsymbol{\Omega} \times \mathbf{SS}_1$, $\Omega_{r1} = 0, \qquad \Omega_1 = \Omega$
4	$\mathbf{v}_r \parallel \mathbf{SS}_1, \qquad \Omega_r \neq 0$	$\mathbf{v}_{S1} = \boldsymbol{\Omega} \times \mathbf{SS}_1 - \frac{m}{M-m}\mathbf{v}_r$, $\Omega_{r1}\mathbf{k} = -\frac{I_{S2}}{I_{S1}}\Omega_r\mathbf{k}$
5	$\mathbf{v}_{S2} = \boldsymbol{\Omega} \times \mathbf{SS}_2 + \mathbf{v}_r$, $\boldsymbol{\Omega}_r = -\boldsymbol{\Omega}$, $(\Omega_2 = 0)$	$\mathbf{v}_{S1} = -\frac{m}{M-m}\mathbf{v}_{S2}$, $I_{S1}\Omega_1\mathbf{k} = I_S\Omega\mathbf{k} - \frac{Mm(\mathbf{SS}_2 \times \mathbf{v}_{S2})}{M-m}$
6	$\mathbf{v}_r = \frac{M-m}{m}(\boldsymbol{\Omega} \times \mathbf{SS}_1)$, $\Omega_r \neq 0$	$\mathbf{v}_{S1} = 0$, $\Omega_{r1} = \frac{M}{I_{S1}}\frac{M-m}{m}\Omega(SS_1)^2 + \frac{I_{S2}}{I_{S1}}\Omega_r$
7	$\mathbf{v}_r \neq 0, \quad I_{S2}\Omega_r\mathbf{k} = M(\mathbf{SS}_1 \times \mathbf{v}_r) + I_{S1}\Omega\mathbf{k}$	$\mathbf{v}_{S1} = \boldsymbol{\Omega} \times \mathbf{SS}_1 - \frac{m\mathbf{v}_r}{M-m}$, $\Omega_{r1} = 0, \qquad \Omega_1 = \Omega$
8	$\mathbf{v}_r = \frac{M-m}{m}(\boldsymbol{\Omega} \times \mathbf{SS}_1)$, $\Omega_r = \left(\frac{I_{S1}}{I_{S2}} - \frac{M(M-m)}{m}\frac{SS_1^2}{I_{S2}}\right)\Omega$	$\mathbf{v}_{S1} = 0$, $\Omega_{r1} = 0, \qquad \Omega_1 = \Omega$

Table 2.

considered with respect to various values of the relative velocity v_r of mass center and the relative angular velocity Ω_r of the separated body.

Table 3. provides properties of the remainder body for the case when the whole body has the plane motion.

Analyzing the results in the Tables 1-3 the following is concluded:

1. If the relative velocity and the relative angular velocity of body separation are zero, the relative velocity and the relative angular velocity of the remainder body are also zero, independently of the type of motion of the initial body (see Table 1 case 4, Table 2 case 3 and Table 3 case 5). The absolute velocity of mass center of the remainder body is equal to the dragging velocity of S_1 before body separation. The angular velocity of the remainder body is equal to the angular velocity of the initial body before separation.

2. If the motion of the initial body and of the separated body is translatory with velocity $\mathbf{v}_S$ (the relative velocity $\mathbf{v}_r$ is zero), the velocity of mass center of the remainder body is also $\mathbf{v}_S$. This result was previously obtained by I.V. Meshchersky, 1896, for the continual mass variation of the transla-

PLANE MOTION OF THE WHOLE BODY

$\mathbf{v}_S \neq 0, \qquad \Omega \neq 0.$

No.	Separated body	Remainder body
1	$\mathbf{v}_r = \mathbf{SS}_2 \times \boldsymbol{\Omega}$, $(\mathbf{v}_{S2} = \mathbf{v}_S)$, $\Omega_2 \neq 0$	$\mathbf{v}_{r1} = \mathbf{SS}_1 \times \boldsymbol{\Omega}$, $(\mathbf{v}_{S1} = \mathbf{v}_S)$, $\Omega_1 = \frac{I_S}{I_{S1}}\Omega - \frac{I_{S2}}{I_{S1}}\Omega_2.$
2	$\mathbf{v}_r = \mathbf{SS}_2 \times \boldsymbol{\Omega}, \quad \Omega_2 = (I_S/I_{S2})\Omega$	$\mathbf{v}_{S1} = \mathbf{v}_S, \quad \Omega_1 = 0.$
3	$\mathbf{v}_r = \mathbf{SS}_2 \times \boldsymbol{\Omega}, \quad \Omega_2 = 0$	$\mathbf{v}_{S1} = \mathbf{v}_S$, $\Omega_1 = (I_S/I_{S1})\Omega.$
4	$\mathbf{v}_r = \mathbf{SS}_2 \times \boldsymbol{\Omega}, \quad \Omega_r = 0$	$\mathbf{v}_{S1} = \mathbf{v}_S$, $\Omega_1 = \Omega, \quad \Omega_{r1} = 0.$
5	$\mathbf{v}_r = 0, \quad \Omega_r = 0$	$\mathbf{v}_{r1} = 0, \quad \Omega_{r1} = 0.$

Table 3.

tory moving particle. Namely, due to the fact that the relative velocity of mass separation is zero, the reactive force is also zero and the equation of motion is the same as for the body without mass change.

3. If the motion of the initial body is translatory with the velocity $\mathbf{v}_S$ and the absolute velocity and the absolute angular velocity of separation of the body are zero, the absolute velocity of mass center of the remainder body differs from the velocity of the initial body. The velocity depends on the mass which is separated: for the higher value of separated mass m i.e., for smaller value of the remainder mass, the velocity is higher. The solution of the Levi-Civita equation (Levi Civita, 1928)

$$\mathbf{v} = \frac{\mathbf{Q}}{M}, \tag{45}$$

shows that if mass decreases the velocity of motion increases. $\mathbf{Q}$ is a constant which depends on the velocity properties of the system.

4. If the motion of the initial body is translatory, the relative angular velocity and the absolute angular velocity of the remainder body are equal (see Table 1).

3.4 Conclusion

Based on the general laws of dynamics, the procedure for obtaining the velocity and the angular velocity of the final body for discontinual mass variation is developed. It is concluded that the position and the angular position of the body during separation is approximately unchanged. The velocity and the angular velocity of the body has a jump-like variation during the process of body variation which is caused by mass and geometry variation of the body. The determined velocity of the mass center and the angular velocity of the final body represent the initial values for its further motion. If the relative velocity and the relative angular velocity of the separated or added body are zero, the relative velocity and the relative angular velocity of the final body are also zero, independently of the type of motion of the initial body.

4 Analytical procedures applied in dynamics of the body with discontinual mass variation

In this Section, applying the procedures of analytical dynamics the results obtained previously are rederived.

Let us start with the Lagrange-D'Alambert principle. Here, $\delta\mathbf{r}_S$ is virtual displacement of position of the mass center S and $.\delta\phi$ is virtual change of the angle position ϕ of the initial body

Multiplying the Eq. (21) with a virtual displacement $\delta\mathbf{r}_S$ and using the relations (14), we obtain

$$\delta\mathbf{r}_{S1} = \delta\mathbf{r}_S + \delta\phi \times \mathbf{SS}_1, \qquad \delta\mathbf{r}_{S2} = \delta\mathbf{r}_S + \delta\phi \times \mathbf{SS}_2, \tag{46}$$

and

$$\begin{aligned}\mathbf{I}^{Fr}\delta\mathbf{r}_S = & \pm m\mathbf{v}_{S2}\delta\mathbf{r}_{S2} + (M \mp m)\mathbf{v}_{S1}\delta\mathbf{r}_{S1} - (M \mp m)\mathbf{v}_{S1}(\delta\phi \times \mathbf{SS}_1) \\ & \mp m\mathbf{v}_{S2}(\delta\phi \times \mathbf{SS}_2) - M\mathbf{v}_S\delta\mathbf{r}_S.\end{aligned} \tag{47}$$

Multiplying the Eq. (23) with $\delta\phi$ and using the relations $\mathbf{L}_{S1} = \mathbf{I}_{S1}\mathbf{\Omega}_1$, $\mathbf{L}_{S2} = \mathbf{I}_{S2}\mathbf{\Omega}_2$ and $\mathbf{L}_S = \mathbf{I}_S\mathbf{\Omega}$, we have

$$\begin{aligned}&\mathbf{I}_{S1}\mathbf{\Omega}_1\delta\phi_1 \pm \mathbf{I}_{S2}\mathbf{\Omega}_2\delta\phi_2 + \mathbf{SS}_1 \times (M \mp m)\mathbf{v}_{S1}\delta\phi \pm \\ &\qquad \mathbf{SS}_2 \times m\mathbf{v}_{S2}\delta\phi - \mathbf{I}_S\mathbf{\Omega}\delta\phi = \mathbf{I}^M\delta\phi,\end{aligned} \tag{48}$$

where $\phi_1 = \phi_2 = \phi$ and the angle variations are $\delta\phi_1 = \delta\phi_2 = \delta\phi$. Adding the equations (47) and (48), we obtain

$$\begin{aligned}\mathbf{I}^{Fr}\delta\mathbf{r}_S + \mathbf{I}^M\delta\phi &= \pm m\mathbf{v}_{S2}\delta\mathbf{r}_{S2} \pm \mathbf{I}_{S2}\mathbf{\Omega}_2\delta\phi_2 + (M \mp m)\mathbf{v}_{S1}\delta\mathbf{r}_{S1} \\ &\quad + \mathbf{I}_{S1}\mathbf{\Omega}_1\delta\phi_1 - (M\mathbf{v}_S\delta\mathbf{r}_S - \mathbf{I}_S\mathbf{\Omega}\delta\phi.\end{aligned} \tag{49}$$

Introducing the generalized coordinates q_i where $i = 1, 2, ..., N$, and assuming that all quantities $\mathbf{r}_S$, $\mathbf{r}_{S1}, \mathbf{r}_{S2}$, $\phi_,$, ϕ_1, ϕ_2 are functions of the generalized coordinates we have

$$\delta\mathbf{r}_S=\sum_{i=1}^N\frac{\partial\mathbf{r}_S}{\partial q_i}\delta q_i,\qquad \delta\mathbf{r}_{S1}=\sum_{i=1}^N\frac{\partial\mathbf{r}_{S1}}{\partial q_i}\delta q_i,\qquad \delta\mathbf{r}_{S2}=\sum_{i=1}^N\frac{\partial\mathbf{r}_{S2}}{\partial q_i}\delta q_i,\tag{50}$$

$$\delta\phi=\sum_{i=1}^N\frac{\partial\phi}{\partial q_i}\delta q_i,\qquad \delta\phi_1=\sum_{i=1}^N\frac{\partial\phi_1}{\partial q_i}\delta q_i,\qquad \delta\phi_2=\sum_{i=1}^N\frac{\partial\phi_2}{\partial q_i}\delta q_i.\tag{51}$$

Using the relations (50), (51) and the equalities

$$\frac{\partial\mathbf{r}_S}{\partial q_i}=\frac{\partial\dot{\mathbf{r}}_S}{\partial\dot{q}_i}=\frac{\partial\mathbf{v}_S}{\partial\dot{q}_i},\quad \frac{\partial\mathbf{r}_{S1}}{\partial q_i}=\frac{\partial\dot{\mathbf{r}}_{S1}}{\partial\dot{q}_i}=\frac{\partial\mathbf{v}_{S1}}{\partial\dot{q}_i},\quad \frac{\partial\mathbf{r}_{S2}}{\partial q_i}=\frac{\partial\dot{\mathbf{r}}_{S2}}{\partial\dot{q}_i}=\frac{\partial\mathbf{v}_{S2}}{\partial\dot{q}_i},\tag{52}$$

$$\frac{\partial\phi}{\partial q_i}=\frac{\partial\dot{\phi}}{\partial\dot{q}_i}=\frac{\partial\mathbf{\Omega}}{\partial\dot{q}_i},\quad \frac{\partial\phi_1}{\partial q_i}=\frac{\partial\dot{\phi}_1}{\partial\dot{q}_i}=\frac{\partial\mathbf{\Omega}_1}{\partial\dot{q}_i},\quad \frac{\partial\phi_2}{\partial q_i}=\frac{\partial\dot{\phi}_2}{\partial\dot{q}_i}=\frac{\partial\mathbf{\Omega}_2}{\partial\dot{q}_i},\tag{53}$$

the Eq. (49) becomes

$$\begin{aligned}\sum_{i=1}^N(M\mathbf{v}_S\frac{\partial\mathbf{v}_S}{\partial\dot{q}_i}+\mathbf{I}_S\mathbf{\Omega}\frac{\partial\mathbf{\Omega}}{\partial\dot{q}_i})\delta q_i-\sum_{i=1}^N[\pm(m\mathbf{v}_{S2}\frac{\partial\mathbf{v}_{S2}}{\partial\dot{q}_i}+\mathbf{I}_{S2}\mathbf{\Omega}_2\frac{\partial\mathbf{\Omega}_2}{\partial\dot{q}_i})\\ +(M\mp m)\mathbf{v}_{S1}\frac{\partial\mathbf{v}_{S1}}{\partial\dot{q}_i}+\mathbf{I}_{S1}\mathbf{\Omega}_1\frac{\partial\mathbf{\Omega}_1}{\partial\dot{q}_i}]\delta q_i\\ =-\sum_{i=1}^N\{\frac{\partial\mathbf{r}_S}{\partial q_i}(\mathbf{F}_r\Delta t)+[(\mathbf{M}_0^{Fr}+\mathfrak{M})\Delta t]\frac{\partial\phi}{\partial q_i}\}\delta q_i,\end{aligned}$$

i.e.,

$$\begin{aligned}\sum_{i=1}^N\frac{\partial}{\partial\dot{q}_i}[\pm(\frac{1}{2}m\mathbf{v}_{S2}\mathbf{v}_{S2}+\frac{1}{2}\mathbf{I}_{S2}\mathbf{\Omega}_2\mathbf{\Omega}_2)+(\frac{1}{2}(M\mp m)\mathbf{v}_{S1}\mathbf{v}_{S1}+\frac{1}{2}\mathbf{I}_{S1}\mathbf{\Omega}_1\mathbf{\Omega}_1)]\delta q_i\\ -\sum_{i=1}^N\frac{\partial}{\partial\dot{q}_i}(\frac{1}{2}M\mathbf{v}_S\mathbf{v}_S+\frac{1}{2}\mathbf{I}_S\mathbf{\Omega}\mathbf{\Omega})\delta q_i=\sum_{i=1}^N(\mathbf{I}^{Fr}\frac{\partial\mathbf{r}_S}{\partial q_i}+\mathbf{I}^M\frac{\partial\phi}{\partial q_i})\delta q_i.\end{aligned}\tag{54}$$

The first group of terms on the left side of the equation represent the kinetic energy of the system after, whereas the second group of terms is equal to the kinetic energy before mass variation

$$T_1=\frac{M\mathbf{v}_S\mathbf{v}_S}{2}+\frac{\mathbf{I}_S\mathbf{\Omega}\mathbf{\Omega}}{2},\tag{55}$$

$$T_2=\pm(\frac{m\mathbf{v}_{S2}\mathbf{v}_{S2}}{2}+\frac{\mathbf{I}_{S2}\mathbf{\Omega}_2\mathbf{\Omega}_2}{2})+\frac{(M\mp m)\mathbf{v}_{S1}\mathbf{v}_{S1}}{2}+\frac{\mathbf{I}_{S1}\mathbf{\Omega}_1\mathbf{\Omega}_1}{2},\tag{56}$$

while the terms on the right side of (54) give the generalized impulse

$$Q_i^I = \mathbf{I}^{Fr} \frac{\partial \mathbf{r}_S}{\partial q_i} + \mathbf{I}^M \frac{\partial \phi}{\partial q_i}. \tag{57}$$

Substituting the notations for the kinetic energy (55) and (56) and also of the generalized impulse (57) into (54) and separating the equations with the same variation of the generalized coordinates, the following system of equations is obtained

$$\frac{\partial (T_2 - T_1)}{\partial \dot{q}_i} = Q_i^I, \qquad i = 1, 2, ..., N.$$

If the body mass variation is without external impulses, the system of equations is modified to (see Cveticanin, 2009_1)

$$\frac{\partial (T_1 - T_2)}{\partial \dot{q}_i} = 0, \qquad i = 1, 2, ..., N. \tag{58}$$

Using these equations, one can calculate the velocity and angular velocity of the system after mass variation if the velocities and angular velocities before mass variation are known. In practical applications of these equations the kinetic energy functions before and after separation have to be differentiable and continual functions.

Let us consider the case of mass separation when the velocity and angular velocity of the body before mass separation and also the velocity and angular velocity of the separated mass are given. For the Eq. (58) it follows: The partial derivative in generalized velocity of the difference of the kinetic energy of the body before and the sum of kinetic energies of the separated and remainder bodies after separation is equal to zero.

Let us apply this results to analyze the mass separation for the case of the free motion of the body. The motion has six degrees of freedom.

The kinetic energy of the initial body with fee motion in the space is

$$T_1 = \frac{M}{2}(\dot{x}_S^2 + \dot{y}_S^2 + \dot{z}_S^2) + \frac{1}{2}(I_{xx}\Omega_x^2 + I_{yy}\Omega_y^2 + I_{zz}\Omega_z^2 + \\ 2I_{xy}\Omega_x\Omega_y + 2I_{xz}\Omega_x\Omega_z + 2I_{zy}\Omega_z\Omega_y), \tag{59}$$

where I_{xx}, I_{yy}, I_{zz} are axial moments of inertia; I_{xy}, I_{xz}, I_{zy} centrifugal moments of inertia; $\dot{x}_S$, $\dot{y}_S$ and $\dot{z}_S$ projections of the velocity $\mathbf{v}_S$ of mass center; Ω_x, Ω_y and Ω_z projections of the angular velocity $\mathbf{\Omega}$.

For the velocity of the mass center of the separated body

$$\mathbf{v}_{S2} = \mathbf{v}_S + \mathbf{\Omega} \times \mathbf{SS}_2 + \mathbf{u}, \tag{60}$$

and the angular velocity

$$\boldsymbol{\Omega}_2 = \boldsymbol{\Omega} + \boldsymbol{\Omega}^*, \tag{61}$$

the kinetic energy of the free motion of separated body is

$$\begin{aligned} T_{S2} = \frac{1}{2}m[&\dot{x}_S^2 + \dot{y}_S^2 + \dot{z}_S^2 + u_x^2 + u_y^2 + u_z^2 + 2\dot{x}_S u_x + 2\dot{y}_S u_y + 2\dot{z}_S u_z \\ &+ (\Omega_y z_{SS2} - \Omega_z y_{SS2})^2 + (\Omega_z x_{SS2} - \Omega_x z_{SS2})^2 \\ &+ (\Omega_x y_{SS2} - \Omega_y x_{SS2})^2 \\ &+ 2(\Omega_y z_{SS2} - \Omega_z y_{SS2})u_x + 2(\Omega_z x_{SS2} - \Omega_x z_{SS2})u_y \\ &+ 2(\Omega_x y_{SS2} - \Omega_y x_{SS2})u_z + 2(\Omega_y z_{SS2} - \Omega_z y_{SS2})\dot{x}_S \\ &+ 2(\Omega_z x_{SS2} - \Omega_x z_{SS2})\dot{y}_S + 2(\Omega_x y_{SS2} - \Omega_y x_{SS2})\dot{z}_S] \\ &+ \frac{1}{2}[I_{xx2}(\Omega_x + \Omega_x^*)^2 + I_{yy2}(\Omega_y + \Omega_y^*)^2 + I_{zz2}(\Omega_z + \Omega_z^*)^2 \\ &+ 2I_{xy2}(\Omega_x + \Omega_x^*)(\Omega_y + \Omega_y^*) \\ &+ 2I_{xz2}(\Omega_x + \Omega_x^*)(\Omega_z + \Omega_z^*) + 2I_{zy2}(\Omega_z + \Omega_z^*)(\Omega_y + \Omega_y^*)], \end{aligned} \tag{62}$$

where $\mathbf{u}$ is the relative velocity of the separation with projections u_x, u_y and u_z; $\boldsymbol{\Omega}^*$ is the relative angular velocity of separation with projections Ω_x^*, Ω_y^* and Ω_z^*.

For the velocity of the remainder system

$$\mathbf{v}_{S1} = \mathbf{v}_S + \boldsymbol{\Omega} \times \mathbf{SS}_1 + \mathbf{v}^*, \tag{63}$$

and the angular velocity

$$\boldsymbol{\Omega}_1 = \boldsymbol{\Omega} + \boldsymbol{\Omega}_1^*, \tag{64}$$

the kinetic energy of the final (remainder) body in free motion is

$$\begin{aligned} T_{S1} &= \frac{1}{2}(M-m)(\dot{x}_{S1}^2 + \dot{y}_{S1}^2 + \dot{z}_{S1}^2) + \frac{1}{2}(I_{xx1}\Omega_{x1}^2 + I_{yy1}\Omega_{y1}^2 + \\ &I_{zz1}\Omega_{z1}^2 + 2I_{xy1}\Omega_{x1}\Omega_{y1} + 2I_{xz1}\Omega_{x1}\Omega_{z1} + 2I_{zy1}\Omega_{z1}\Omega_{y1}), \end{aligned} \tag{65}$$

where $I_{xx1}, I_{yy1}, I_{zz1}$ are the axial moments of inertia; $I_{xy1}, I_{xz1}, I_{zy1}$ are the centrifugal moments of inertia; $\mathbf{v}^*$ and $\boldsymbol{\Omega}_1$ are the unknown velocity and angular velocity of the remainder body.

The total kinetic energy of the system after the separation is

$$T_2 = T_{S1} + T_{S2}. \tag{66}$$

Taking into account (58) and using the relations (59) and (62), the following system of equations is obtained

$$\begin{aligned}(M-m)\dot{x}_S &= (M-m)\dot{x}_{S1} + m(u_x + \Omega_y z_{S2} - \Omega_z y_{S2}),\\ (M-m)\dot{y}_S &= (M-m)\dot{y}_{S1} + m(u_y + \Omega_z x_{S2} - \Omega_x z_{S2}),\\ (M-m)\dot{z}_S &= (M-m)\dot{z}_{S1} + m(u_z + \Omega_x y_{S2} - \Omega_y x_{S2}),\end{aligned}$$

$$\begin{aligned}0 &= -(I_{xx}\Omega_x + I_{xy}\Omega_y + I_{xz}\Omega_z) + I_{xx1}\Omega_{1x} + I_{xy1}\Omega_{1y} + I_{xz1}\Omega_{1z}\\ &\quad + I_{xx2}(\Omega_x + \Omega_x^*) + I_{xy2}(\Omega_y + \Omega_y^*) + I_{xz2}(\Omega_z + \Omega_z^*)\\ &\quad + m[-(\Omega_z x_{S2} - \Omega_x z_{S2})z_{S2} + (\Omega_x y_{S2} - \Omega_y x_{S2})y_{S2}\\ &\quad - z_{S2}(u_y + \dot{y}_S) + y_{S2}(u_z + \dot{z}_S)]\\ &\quad + (M-m)[-(\Omega_z x_{S1} - \Omega_x z_{S1})z_{S1} + (\Omega_x y_{S1} - \Omega_y x_{S1})y_{S1}\\ &\quad - z_{S1}(v_y + \dot{y}_S) + y_{S1}(v_z + \dot{z}_S)],\\ 0 &= -(I_{yy}\Omega_y + I_{xy}\Omega_x + I_{zy}\Omega_z) + I_{yy1}\Omega_{1y} + I_{xy1}\Omega_{1x} + I_{zy1}\Omega_{1z}\\ &\quad + I_{yy2}(\Omega_y + \Omega_y^*) + I_{xy2}(\Omega_x + \Omega_x^*) + I_{zy2}(\Omega_z + \Omega_z^*)\\ &\quad + m[(\Omega_y z_{S2} - \Omega_z y_{S2})z_{S2} - (\Omega_x y_{S2} - \Omega_y x_{S2})x_{S2}\\ &\quad + z_{S2}(u_x + \dot{x}_S) - x_{S2}(u_z + \dot{z}_S)]\\ &\quad + (M-m)[(\Omega_y z_{S1} - \Omega_z y_{S1})z_{S1} - (\Omega_x y_{S1} - \Omega_y x_{S1})x_{S1}\\ &\quad + z_{S1}(v_x + \dot{x}_S) - x_{S1}(v_z + \dot{z}_S),\\ 0 &= -(I_{zz}\Omega_z + I_{xz}\Omega_x + I_{zy}\Omega_y) + I_{zz1}\Omega_{1z} + I_{xz1}\Omega_{1x} + I_{zy1}\Omega_{1y}\\ &\quad + I_{zz2}(\Omega_z + \Omega_z^*) + I_{xz2}(\Omega_x + \Omega_x^*) + I_{zy2}(\Omega_y + \Omega_y^*)\\ &\quad + m[-y_{S2}(\Omega_y z_{S2} - \Omega_z y_{S2})\\ &\quad + m[x_{S2}(\Omega_z x_{S2} - \Omega_x z_{S2}) - y_{S2}(u_x + \dot{x}_S) + x_{S2}(u_y + \dot{y}_S)]\\ &\quad + (M-m)[-y_{S1}(\Omega_y z_{S1} - \Omega_z y_{S1}) + x_{S1}(\Omega_z x_{S1} - \Omega_x z_{S1})\\ &\quad - y_{S1}(v_x + \dot{x}_S) + x_{S1}(v_y + \dot{y}_S)].\end{aligned} \tag{67}$$

Using the relations

$$\begin{aligned}&\Omega_{2x} = \Omega_x + \Omega_x^*, \quad \Omega_{2y} = \Omega_y + \Omega_{2y}^*, \quad \Omega_{2z} = \Omega_z + \Omega_z^*,\\ &\dot{x}_{S1} = \dot{x}_S + v_x + (\Omega_y z_{S1} - \Omega_z y_{S1}), \quad \dot{y}_{S1} = \dot{y}_S + v_y + (\Omega_z x_{S1} - \Omega_x z_{S1}),\\ &\dot{z}_{S1} = \dot{z}_S + v_z + (\Omega_x y_{S1} - \Omega_y x_{S1}), \quad \dot{x}_{S2} = \dot{x}_S + u_x + (\Omega_y z_{S2} - \Omega_z y_{S2}),\\ &\dot{y}_{S2} = \dot{y}_S + u_y + (\Omega_z x_{S2} - \Omega_x z_{S2}), \quad \dot{z}_{S2} = \dot{z}_S + u_z + (\Omega_x y_{S2} - \Omega_y x_{S2}),\end{aligned} \tag{68}$$

the Eqs. (67) are transformed into

$$
\begin{aligned}
M\dot{x}_S &= m\dot{x}_{S2} + (M-m)\dot{x}_{S1},\\
M\dot{y}_S &= m\dot{y}_{S2} + (M-m)\dot{y}_{S1},\\
M\dot{z}_S &= m\dot{z}_{S2} + (M-m)\dot{z}_{S1},\\
I_{xx}\Omega_x + I_{xy}\Omega_y + I_{xz}\Omega_z &= I_{xx1}\Omega_{1x} + I_{xy1}\Omega_{1y} + I_{xz1}\Omega_{1z}\\
&\quad + I_{xx2}\Omega_{2x} + I_{xy2}\Omega_{2y}\\
&\quad + I_{xz2}\Omega_{2z} + (M-m)(y_{S1}\dot{z}_{S1} - z_{S1}\dot{y}_{S1})\\
&\quad + m(y_{S2}\dot{z}_{S2} - z_{S2}\dot{y}_{S2}),\\
I_{xy}\Omega_x + I_{yy}\Omega_y + I_{yz}\Omega_z &= I_{xy1}\Omega_{1x} + I_{yy1}\Omega_{1y}\\
&\quad + I_{yz1}\Omega_{1z} + I_{xy2}\Omega_{2x} + I_{yy2}\Omega_{2y}\\
&\quad + I_{yz2}\Omega_{2z} + m(z_{S2}\dot{x}_{S2} - x_{S2}\dot{z}_{S2})\\
&\quad + (M-m)(z_{S1}\dot{x}_{S1} - x_{S1}\dot{z}_{S1}),\\
I_{xz}\Omega_x + I_{yz}\Omega_y + I_{zz}\Omega_z &= I_{xz1}\Omega_{1x} + I_{yz1}\Omega_{1y} + I_{zz1}\Omega_{1z}\\
&\quad + I_{xz2}\Omega_{2x} + I_{yz2}\Omega_{2y}\\
&\quad + I_{zz2}\Omega_{2z} + (M-m)(x_{S1}\dot{y}_{S1} - y_{S1}\dot{x}_{S1})\\
&\quad + m(x_{S2}\dot{y}_{S2} - y_{S2}\dot{x}_{S2}).
\end{aligned}
\tag{69}
$$

Introducing the projections of the velocities, angular velocities and position vectors

$$
\begin{aligned}
\mathbf{v}_S &= \dot{x}_S\mathbf{i}+\dot{y}_S\mathbf{j}+\dot{z}_S\mathbf{k}, & \mathbf{v}_{S1} &= \dot{x}_{S1}\mathbf{i}+\dot{y}_{S1}\mathbf{j}+\dot{z}_{S1}\mathbf{k},\\
\mathbf{v}_{S2} &= \dot{x}_{S2}\mathbf{i}+\dot{y}_{S2}\mathbf{j}+\dot{z}_{S2}\mathbf{k}, & &\\
\boldsymbol{\Omega} &= \Omega_x\mathbf{i}+\Omega_y\mathbf{j}+\Omega_z\mathbf{k}, & \boldsymbol{\Omega}_1 &= \Omega_{1x}\mathbf{i}+\Omega_{1y}\mathbf{j}+\Omega_{1z}\mathbf{k},\\
\boldsymbol{\Omega}_2 &= \Omega_{2x}\mathbf{i}+\Omega_{2y}\mathbf{j}+\Omega_{2z}\mathbf{k}, & &\\
\mathbf{SS}_1 &= x_{S1}\mathbf{i}+y_{S1}\mathbf{j}+z_{S1}\mathbf{k}, & \mathbf{SS}_2 &= x_{S2}\mathbf{i}+y_{S2}\mathbf{j}+z_{S2}\mathbf{k},
\end{aligned}
\tag{70}
$$

as well as the inertia tensors

$$
\mathbf{I}_S = \begin{bmatrix} I_{xx} & I_{xy} & I_{xz}\\ I_{yx} & I_{yy} & I_{yz}\\ I_{zx} & I_{zy} & I_{zz}\end{bmatrix}, \qquad \mathbf{I}_{S1} = \begin{bmatrix} I_{xx1} & I_{xy1} & I_{xz1}\\ I_{yx1} & I_{yy1} & I_{yz1}\\ I_{zx1} & I_{zy1} & I_{zz1}\end{bmatrix}, \tag{71}
$$

$$
\mathbf{I}_{S2} = \begin{bmatrix} I_{xx2} & I_{xy2} & I_{xz2}\\ I_{yx2} & I_{yy2} & I_{yz2}\\ I_{zx2} & I_{zy2} & I_{zz2}\end{bmatrix},
$$

into Eq. (21) and Eq. (23), we obtain the above mentioned Eq. (69). The solutions of (58) are equal to those obtained from (21) and Eq. (23)

without external forces and torques. The main advantage of the suggested analytical procedure is its simplicity for practical use in comparison to the classical method based on the general principles of dynamics which have the vectorial form (Cveticanin, 2009_1).

4.1 Increase of the kinetic energy

Analyzing the relation (58) it can be concluded that the kinetic energy of the body before separation T_1 and the sum of the kinetic energies of the remainder and separated bodies T_2 differs. The difference between the kinetic energies T_1 and T_2 is the result of transformation of the deformation energy of the body into kinetic energies of the separated and remainder bodies during separation.

Theorem 4.1. *In the perfectly plastic separation of a body the increase of the kinetic energy is equal to the sum of the kinetic energies corresponding to the relative velocities and angular velocities of the remainder and separated bodies*

$$\Delta T = [\frac{1}{2}(M-m)(\mathbf{v}^*)^2 + \frac{1}{2}\mathbf{I}_{S1}(\mathbf{\Omega}_1^*)^2] + [\frac{1}{2}m\mathbf{u}^2 + \frac{1}{2}\mathbf{I}_{S2}(\mathbf{\Omega}^*)^2]. \tag{72}$$

Proof Using the relations (55) and (56) the difference between the kinetic energy before separation T_1 and after separation T_2 is

$$\begin{aligned} T_2 - T_1 &= [\frac{1}{2}(M-m)\mathbf{v}_{S1}\mathbf{v}_{S1} + \frac{1}{2}\mathbf{I}_{S1}\mathbf{\Omega}_1\mathbf{\Omega}_1 \\ &\quad +(\frac{1}{2}m\mathbf{v}_{S2}\mathbf{v}_{S2} + \frac{1}{2}\mathbf{I}_{S2}\mathbf{\Omega}_2\mathbf{\Omega}_2)] - \frac{1}{2}M\mathbf{v}_S\mathbf{v}_S + \frac{1}{2}\mathbf{I}_S\mathbf{\Omega}\mathbf{\Omega}, \end{aligned} \tag{73}$$

i.e.

$$\begin{aligned} T_2 - T_1 &= [\frac{1}{2}(M-m)\mathbf{v}_{S1}\mathbf{v}_{S1} - M\mathbf{v}_S\mathbf{v}_S + \frac{1}{2}m\mathbf{v}_{S2}\mathbf{v}_{S2}] + \frac{1}{2}M\mathbf{v}_S\mathbf{v}_S \\ &\quad +[\frac{1}{2}\mathbf{I}_{S1}\mathbf{\Omega}_1\mathbf{\Omega}_1 - \mathbf{I}_S\mathbf{\Omega}\mathbf{\Omega} + \frac{1}{2}\mathbf{I}_{S2}\mathbf{\Omega}_2\mathbf{\Omega}_2] + \frac{1}{2}\mathbf{I}_S\mathbf{\Omega}\mathbf{\Omega}. \end{aligned} \tag{74}$$

After some transformation the relation (74) reads

$$\begin{aligned} T_2 - T_1 &= \frac{1}{2}(M-m)(\mathbf{v}_{S1} - \mathbf{v}_S)(\mathbf{v}_{S1} - \mathbf{v}_S) + \frac{1}{2}m(\mathbf{v}_{S2} - \mathbf{v}_S)(\mathbf{v}_{S2} - \mathbf{v}_S) \\ &\quad -(M-m)(\mathbf{SS}_1 \times \mathbf{v}_{S1})\mathbf{\Omega} - m(\mathbf{SS}_2 \times \mathbf{v}_{S2})\mathbf{\Omega} \\ &\quad +\frac{1}{2}\mathbf{I}_{S1}(\mathbf{\Omega}_1 - \mathbf{\Omega})(\mathbf{\Omega}_1 - \mathbf{\Omega}) + \frac{1}{2}\mathbf{I}_{S2}(\mathbf{\Omega}_2 - \mathbf{\Omega})(\mathbf{\Omega}_2 - \mathbf{\Omega}). \end{aligned} \tag{75}$$

Introducing the relations (63) and (60) and also (64) and (61) for the velocities $\mathbf{v}_{S1}$ and $\mathbf{v}_{S2}$ and angular velocities $\boldsymbol{\Omega}_1$ and $\boldsymbol{\Omega}_2$ into (75) we obtain

$$
\begin{aligned}
T_2 - T_1 \quad = \quad & \frac{1}{2}(M-m)(\mathbf{SS}_1 \times \boldsymbol{\Omega} + \mathbf{v}^*)(\mathbf{SS}_1 \times \boldsymbol{\Omega} + \mathbf{v}^*) \\
& + \frac{1}{2}m(\mathbf{SS}_2 \times \boldsymbol{\Omega} + \mathbf{u})(\mathbf{SS}_2 \times \boldsymbol{\Omega} + \mathbf{u}) \\
& - (M-m)(\mathbf{SS}_1 \times (\mathbf{v}_S + \mathbf{SS}_1 \times \boldsymbol{\Omega} + \mathbf{v}^*))\boldsymbol{\Omega} \\
& - m(\mathbf{SS}_2 \times (\mathbf{v}_S + \mathbf{SS}_2 \times \boldsymbol{\Omega} + \mathbf{u}))\boldsymbol{\Omega} \\
& + \frac{1}{2}\mathbf{I}_{S1}(\boldsymbol{\Omega}_1^*)(\boldsymbol{\Omega}_1^*) + \frac{1}{2}\mathbf{I}_{S2}(\boldsymbol{\Omega}^*)(\boldsymbol{\Omega}^*). \qquad (76)
\end{aligned}
$$

Since the time of separation is negligibly short and the displacements of mass centers and the angle positions of the bodies during the separation are also negligibly small we assume that the positions of mass centers and angle positions of the bodies remain constant during the separation. Using this assumption that S is the mass center of the body and S_1 and S_2 are the mass centers of the remainder and separated body and after some calculation the relation (76) is simplified to

$$
\begin{aligned}
\Delta T \quad = \quad & T_2 - T_1 = \frac{1}{2}(M-m)\mathbf{v}^*\mathbf{v}^* + \frac{1}{2}m\mathbf{u}\mathbf{u} \\
& + \frac{1}{2}\mathbf{I}_{S1}\boldsymbol{\Omega}_1^*\boldsymbol{\Omega}_1^* + \frac{1}{2}\mathbf{I}_{S2}\boldsymbol{\Omega}^*\boldsymbol{\Omega}^*. \qquad (77)
\end{aligned}
$$

The theorem is proved.

Remark 4.2. For the perfectly plastic direct central impact of two perfectly inelastic bodies moving translatory Carnot proved the following theorem (see Starzhinskii, 1982): There is the loss of kinetic energy which is equal to the kinetic energy corresponding to the loss of velocities of the two bodies in impact. Carnot's theorem talks about the loss of the kinetic energy during impact, and the suggested theorem (72) about the increase of the kinetic energy during separation. The perfectly plastic impact of two perfectly inelastic bodies is opposite to the perfectly plastic separation of the body into separated and remainder bodies.

4.2 The separation of a pendulum

Let us consider a pendulum with mass M and length L rotating around a fixed point O (Fig.3).

The aim is to obtain the angular velocity of the remainder pendulum after the separation of a part whose mass is m and length l.

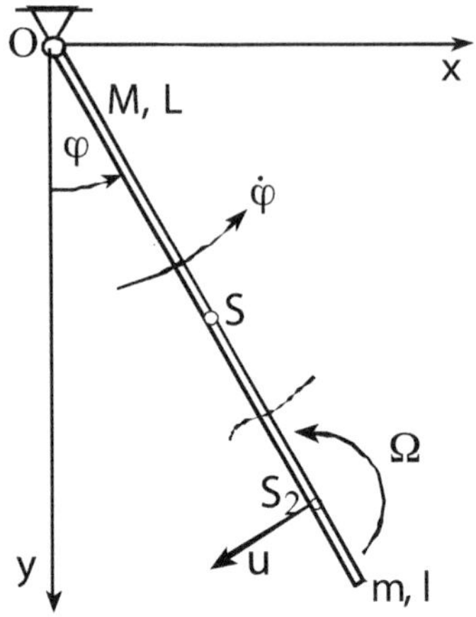

Figure 3. Model of the separated pendulum.

For the angular velocity of rotation $\dot{\varphi}$, the kinetic energy of the motion is

$$T_1 = \frac{1}{2}\frac{ML^2}{3}\dot{\varphi}^2. \tag{78}$$

The part of the pendulum is separating with the velocity $\mathbf{u}$ and the angular velocity Ω and it is moving in the xy plane. Its kinetic energy is

$$T_{S2} = \frac{1}{2}mv_{S2}^2 + \frac{1}{2}\frac{ml^2}{12}\dot{\varphi}_2^2, \tag{79}$$

where

$$\dot{\varphi}_2 = \dot{\varphi} + \Omega, \qquad \mathbf{v}_{S2} = \mathbf{v}_{S2t} + \mathbf{u}, \qquad v_{S2t} = \dot{\varphi}(OS_2) = \dot{\varphi}(L - \frac{l}{2}). \tag{80}$$

Substituting (80) into (79), we obtain

$$T_{S2} = \frac{1}{2}m[u^2 + (\dot{\varphi}+\Omega)^2(L-\frac{l}{2})^2 - 2(\dot{\varphi}+\Omega)(u_x y_{S2} - u_y x_{S2})] + \frac{1}{2}\frac{ml^2}{12}(\dot{\varphi}+\Omega)^2, \tag{81}$$

where

$$x_{S2} = (L - \frac{l}{2})\sin\varphi, \qquad y_{S2} = (L - \frac{l}{2})\cos\varphi. \tag{82}$$

The kinetic energy of the remainder pendulum is

$$T_{S1} = \frac{1}{2}\frac{(M-m)(L-l)^2}{3}\dot{\varphi}_1^2, \tag{83}$$

where $\dot{\varphi}_1$ is an unknown angular velocity. The derivation of the kinetic energies (78), (81) and (83) for the generalized velocity $\dot{\varphi}$ is done according to (58). Then, the angular velocity of the remainder pendulum is determined

$$\dot{\varphi}_1 = \frac{ML^2\dot{\varphi} - \frac{ml^2}{4}(\dot{\varphi}+\Omega) - 3m(\dot{\varphi}+\Omega)(L-\frac{l}{2})^2 + 3m(u_x y_{S2} - u_y x_{S2})}{(M-m)(L-l)^2}. \tag{84}$$

In order to prove the obtained result (84), the method based on the angular momentum of the system is considered. The angular momentum of the pendulum for the fixed point O is

$$L_0 = L_S + M r_S v_S = \frac{1}{3} M L^2 \dot{\varphi}, \tag{85}$$

where $L_S = I_S \dot{\varphi}$ is the angular momentum of the pendulum with the moment of inertia $I_S = \frac{1}{12} M L^2$, the position r_S and the velocity of the mass center v_S

$$r_S = \frac{L}{2}, v_S = \frac{L}{2}\dot{\varphi}. \tag{86}$$

The angular momentum of the separated body is

$$\begin{aligned} L_2 \mathbf{k} &= L_{S2}\mathbf{k} + m r_{S2} v_{S2} \mathbf{k} + \mathbf{r}_{S2} \times m\mathbf{u} \qquad (87)\\ &= [\frac{1}{12} m l^2 \dot{\varphi}_2 + m(L - \frac{l}{2})(\dot{\varphi}+\Omega) + m(x_{S2}u_y - y_{S2}u_x)]\mathbf{k}, \end{aligned}$$

for

$$\begin{aligned} L_{S2} &= I_{S2}\dot{\varphi}_2 = \frac{1}{12} m l^2 (\dot{\varphi}+\Omega), \qquad (88)\\ m r_{S2} v_{S2} &= m(L-\frac{l}{2})(\dot{\varphi}+\Omega), \qquad \mathbf{r}_{S2} \times m\mathbf{u} = m\mathbf{k}(x_{S2}u_y - y_{S2}u_x), \end{aligned}$$

where $\mathbf{k}$ is the unit vector perpendicular to the xy plane. The angular momentum of the remainder pendulum is

$$L_0^r = L_{S1} + (M-m) r_{S1} v_{S1} = \frac{1}{3}(M-m)(L-l)^2 \dot{\varphi}_1, \tag{89}$$

where

$$L_{S1} = I_{S1}\dot{\varphi}_1 = \frac{1}{12}(M-m)(L-l)^2\dot{\varphi}_1, \qquad r_{S1} v_{S1} = \frac{(L-l)^2}{4}\dot{\varphi}_1. \tag{90}$$

Using the assumption that the angular momenta of the system before and after mass separation are equal to each other, the following equation is obtained

$$\frac{ML^2\dot{\varphi}}{3} = \frac{(M-m)(L-l)^2}{3}\dot{\varphi}_1 + \frac{ml^2}{12}(\dot{\varphi}+\Omega) + m[(\dot{\varphi}+\Omega)(L-\frac{l}{2})^2 - (u_x y_{S2} - u_y x_{S2})]. \quad (91)$$

Comparing the eq.(84) with (91), it is obvious that they are in agreement.

4.3 Conclusion

The proposed analytical procedure for obtaining the velocity and angular velocity of the remainder body after the mass separation is based on the principles of momentum and angular momentum of the body before the separation and the system of bodies after separation. By using the derivatives in the velocity of the kinetic energy of the whole body and of the system of bodies after separation, the required quantities are found. These quantities depend on the velocity and angular velocity of the separated mass, mass of the body before separation and also on the mass of the separated body. In addition, these values depend on the moment of inertia of the whole body, moment of inertia of the separated body and on the position of the separated part. The procedure given in this paper is of a general type and applicable to solving all the cases where discontinual mass variation occurs. The developed analytical method is much more suitable for engineering applications than the direct use of the general vectorial principles of dynamics. The results obtained with the developed analytical method are equal to those obtained by using the classical procedure.

The kinetic energy of the system increases during separation in general. The increase of the kinetic energy of the separated and remainder bodies during the perfectly plastic separation is equal to the kinetic energies of relative motion of the two bodies.

As it is stated in the previous Chapter, the dynamics of body separation represents an inverse process of the perfectly plastic impact of the two perfectly inelastic bodies. During the separation, which lasts for a very short time interval τ, the body undergoes a relaxation which causes the energy of deformation (potential energy) to be transformed into kinetic energies of the separated and the remainder bodies. This additional kinetic energy causes the relative motion of the separated and remainder bodies. The separation forces and torques which act between the separated and remainder bodies give the separation impulses and moments of separation impulses. We regard the separated and remainder body as one complex system. Then the separation impulses and moments between these bodies are internal within that system. The external forces and torques are assumed to be negligible in comparison to those caused by body separation. It is this fact that the law of conservation of the momentum and angular momentum of the system can be applied. The linear momentum of the body before the separation and the sum of the linear momenta of two parts after the separation remain invariable. Also, the angular momentum of the body before the separation and the sum of angular momenta of two parts after the separation remain invariable.

5 Dynamics of the body with continual mass variation

Based on the mathematical expressions of the principle of the momentum and of the angular momentum, the differential equations of motion of the continaul mass variation of a body are obtained.

Namely, according to (6) and (20) and the principle of the momentum, it is

$$\Delta \mathbf{K} \equiv M\Delta \mathbf{v}_{S1} \pm m(\mathbf{v}_{S2} - \mathbf{v}_{S1}) = \mathbf{F}_r \Delta t, \tag{92}$$

where the velocity variation is

$$\Delta \mathbf{v}_{S1} = \mathbf{v}_{S1} - \mathbf{v}_S. \tag{93}$$

Substituting the relation (17) into (22), the principle of the angular momentum has the form

$$\Delta \mathbf{L}_O \equiv \mathbf{r}_S \times \Delta K \mp \mathbf{SS}_2 \times m(\mathbf{v}_{S1} - \mathbf{v}_{S2}) + \Delta \mathbf{L}_S \pm \mathbf{L}_{S2} = (\mathbf{M}_0^{Fr} + \mathfrak{M})\Delta t, \tag{94}$$

where

$$\Delta \mathbf{L}_S = \mathbf{L}_{S1} - \mathbf{L}_S, \tag{95}$$

and

$$\mathbf{L}_{S2} = \mathbf{I}_{S2} \mathbf{\Omega}_2. \tag{96}$$

Introducing the notation for the adding or separating mass and moment of inertia and its absolute velocity as

$$m = \Delta M, \qquad \mathbf{I}_{S2} = \Delta \mathbf{I}, \qquad \mathbf{v}_{S2} = \mathbf{u}, \tag{97}$$

and also

$$\mathbf{v}_{S1} = \mathbf{v}, \tag{98}$$

the relations (92) and (94) are transformed

$$M\Delta \mathbf{v} = \mathbf{F}_r \Delta t \mp \Delta M(\mathbf{u} - \mathbf{v}), \tag{99}$$

$$\Delta \mathbf{L}_S = (\mathbf{M}_0^{Fr} + \mathfrak{M})\Delta t - \mathbf{r}_S \times \Delta K \pm \mathbf{SS}_2 \times \Delta M(\mathbf{v} - \mathbf{u}) \mp \Delta \mathbf{I} \mathbf{\Omega}_2. \tag{100}$$

By introducing the moment of external forces $\mathbf{M}_S^{Fr}$ for the mass center of the body S the connection between the two resultant moments $\mathbf{M}_O^{Fr}$ and $\mathbf{M}_S^{Fr}$ for the two points O and S is

$$\mathbf{M}_O^{Fr} = \mathbf{M}_S^{Fr} + \mathbf{r}_S \times \mathbf{F}_r. \tag{101}$$

Multiplying the Eq. (20) with the position vector $\mathbf{r}_S$ it is

$$\mathbf{r}_S \times \Delta K = \mathbf{r}_S \times \mathbf{F}_r. \tag{102}$$

Substituting (101) and (102) into (100) yields

$$\Delta \mathbf{L}_S = (\mathbf{M}_S^{Fr} + \mathfrak{M})\Delta t \pm \mathbf{SS}_2 \times \Delta M(\mathbf{v} - \mathbf{u}) \mp \Delta \mathbf{I}\boldsymbol{\Omega}_2. \tag{103}$$

Dividing the Eqs. (99) and (103) with the infinitesimal time Δt, it is

$$M\frac{\Delta \mathbf{v}}{\Delta t} = \mathbf{F}_r \mp \frac{\Delta M}{\Delta t}(\mathbf{u} - \mathbf{v}), \tag{104}$$

$$\frac{\Delta \mathbf{L}_S}{\Delta t} = (\mathbf{M}_S^{Fr} + \mathfrak{M}) \pm \mathbf{SS}_2 \times \frac{\Delta M}{\Delta t}(\mathbf{v} - \mathbf{u}) \mp \frac{\Delta \mathbf{I}}{\Delta t}\boldsymbol{\Omega}_2. \tag{105}$$

For the limit condition, when the infinitesimal time tends to zero, the relations (104) and (105) transform into

$$M\frac{d\mathbf{v}}{dt} = \mathbf{F}_r + (\mp\left|\frac{dM}{dt}\right|)(\mathbf{u} - \mathbf{v}), \tag{106}$$

$$\frac{d\mathbf{L}_S}{dt} = (\mathbf{M}_S^{Fr} + \mathfrak{M}) + \mathbf{SS}_2 \times (\mp\left|\frac{dM}{dt}\right|)(\mathbf{u} - \mathbf{v}) + (\mp\left|\frac{d\mathbf{I}}{dt}\right|)\boldsymbol{\Omega}_2, \tag{107}$$

i.e.,

$$M\frac{d\mathbf{v}}{dt} = \mathbf{F}_r + \frac{dM}{dt}(\mathbf{u} - \mathbf{v}), \tag{108}$$

$$\frac{d\mathbf{L}_S}{dt} = (\mathbf{M}_S^{Fr} + \mathfrak{M}) + \mathbf{SS}_2 \times \frac{dM}{dt}(\mathbf{u} - \mathbf{v}) + \frac{d\mathbf{I}}{dt}\boldsymbol{\Omega}_2. \tag{109}$$

Remark 5.1. It must be emphasized that the sign 'minus' or 'plus' ($\mp$) in front of the elementary mass dM and moment of inertia $d\mathbf{I}$ have to be eliminated in the equations which describe the motion of the continually mass variable bodies. Namely, the sign of the first time derivative of the mass (dM/dt) and of the moment of inertia ($d\mathbf{I}/dt$) are negative, if the mass and the moment of inertia are decreasing in time (mass separation), and positive, if the mass and the moment of inertia are increasing in time (mass addition). The sign of these first derivatives is automatically obtained during the calculation and it availables the elimination of the mentioned signs in the formulas in front of the time derivatives of the mass and moment of inertia.

For $\mathbf{L}_S = \mathbf{I}\boldsymbol{\Omega}$ where $\mathbf{I} \equiv \mathbf{I}_S$ is the tensor of moment of inertia, and after some simple modification the differential equations of motion of the body with continual mass variation follow as

$$\frac{d}{dt}(M\mathbf{v}) = \mathbf{F}_r + \frac{dM}{dt}\mathbf{u}, \tag{110}$$

$$\frac{d}{dt}(\mathbf{I}\boldsymbol{\Omega}) = (\mathbf{M}_S^{Fr} + \mathfrak{M}) + \mathbf{SS}_2 \times \frac{dM}{dt}(\mathbf{u}-\mathbf{v}) + \frac{d\mathbf{I}}{dt}\boldsymbol{\Omega}_2. \tag{111}$$

For (see for example Goldstein, 1980)

$$\frac{d}{dt}(\mathbf{I}\boldsymbol{\Omega}) = \boldsymbol{\Omega}\frac{d\mathbf{I}}{dt} + \mathbf{I}\frac{d\boldsymbol{\Omega}}{dt} + \boldsymbol{\Omega} \times \mathbf{I}\boldsymbol{\Omega}, \tag{112}$$

the differential equations of motion transform into

$$M\frac{d\mathbf{v}}{dt} = \mathbf{F}_r + \frac{dM}{dt}(\mathbf{u}-\mathbf{v}), \tag{113}$$

$$\mathbf{I}\frac{d\boldsymbol{\Omega}}{dt} + \boldsymbol{\Omega} \times \mathbf{I}\boldsymbol{\Omega} = (\mathbf{M}_S^{Fr} + \mathfrak{M}) + \mathbf{SS}_2 \times \frac{dM}{dt}(\mathbf{u}-\mathbf{v}) + \frac{d\mathbf{I}}{dt}(\boldsymbol{\Omega}_2 - \boldsymbol{\Omega}). \tag{114}$$

The last terms in the Eqs. (113) and (114) represent the reactive force

$$\boldsymbol{\Phi} = \frac{dM}{dt}(\mathbf{u}-\mathbf{v}), \tag{115}$$

and the reactive torque

$$\mathbb{R} = \frac{d\mathbf{I}_S}{dt}(\boldsymbol{\Omega}_2 - \boldsymbol{\Omega}), \tag{116}$$

which exist due to variation of mass and moment of inertia of the body. The reactive force $\boldsymbol{\Phi}$ gives the moment due to point S and it is

$$\mathbf{M}_S^{\boldsymbol{\Phi}} = \mathbf{SS}_2 \times \boldsymbol{\Phi}. \tag{117}$$

Substituting (115) - (117) into (113) and (114), we have

$$M\frac{d\mathbf{v}}{dt} = \mathbf{F}_r + \boldsymbol{\Phi}, \tag{118}$$

$$\mathbf{I}\frac{d\boldsymbol{\Omega}}{dt} + \boldsymbol{\Omega} \times \mathbf{I}\boldsymbol{\Omega} = \mathbf{M}_S^{Fr} + \mathfrak{M} + \mathbf{M}_S^{\boldsymbol{\Phi}} + \mathbb{R}. \tag{119}$$

The first equation defines the translational motion and the second the rotation around the mass center S.

For practical reasons it is convenient to rewrite the vector differential equations (118) and (119) into the scalar ones. Introducing the fixed coordinate system $Oxyz$, the components u, v, w of the velocity $\mathbf{v}$, the components u_2, v_2, w_2 of the velocity $\mathbf{u}$ and F_x, F_y, F_z the components of the resultant $\mathbf{F}_r$ the vector differential equation of translational motion is given with three scalar equations

$$M\frac{du}{dt} = F_x + \Phi_x, \qquad M\frac{dv}{dt} = F_y + \Phi_y, \qquad M\frac{dw}{dt} = F_z + \Phi_z, \qquad (120)$$

The terms on the right side of (120)

$$\Phi_x = \frac{dM}{dt}(u_2 - u), \qquad \Phi_y = \frac{dM}{dt}(v_2 - v), \qquad \Phi_z = \frac{dM}{dt}(w_2 - w), \quad (121)$$

are called the projections of the reactive force. The reactive force is the consequence of mass variation. This result is published in the paper Cveticanin & Kovacic, 2007.

For the reference system $S\xi\eta\zeta$ fixed to the body with the origin in the center of mass of the body the inertial tensor $\mathbf{I}$ has nine components, but only six of them are independent: $I_{\xi\xi}, I_{\eta\eta}, I_{\varsigma\varsigma}$, are the moments of inertia and $I_{\xi\eta}, I_{\xi\varsigma}, I_{\eta\varsigma}$and also $I_{\eta\xi}, I_{\varsigma\xi}, I_{\varsigma\eta}$ are the products of inertia. If the axes are principal and products of inertia are zero simultaneously the inertial tensor $\mathbf{I}$ has only three principal moments of inertia $I_{\xi\xi}, I_{\eta\eta}, I_{\varsigma\varsigma}$. The angular velocity $\mathbf{\Omega}$ has three components p, q, r in this frame. If p_2, q_2, r_2 are the components of the angular velocity $\mathbf{\Omega}_2$, $M_\xi^{\mathbf{\Phi}}$, $M_\eta^{\mathbf{\Phi}}$ and $M_\varsigma^{\mathbf{\Phi}}$ are the body-axis components of $\mathbf{M}_S^{\mathbf{\Phi}}$, M_ξ, M_η and M_ζ are the body-axis components of $\mathbf{M}_S^{Fr}$ and $\mathfrak{M}_\xi, \mathfrak{M}_\eta$ and $\mathfrak{M}_\zeta$ are the projections of the torque the vector $\mathfrak{M}$, the equation for rotational motion (119) is given with three scalar equations

$$\begin{aligned}
I_{\xi\xi}\frac{dp}{dt} + (I_{\varsigma\varsigma} - I_{\eta\eta})qr &= M_\xi + \mathfrak{M}_\xi + M_\xi^{\mathbf{\Phi}} + \mathfrak{R}_\xi, \\
I_{\eta\eta}\frac{dq}{dt} + (I_{\xi\xi} - I_{\varsigma\varsigma})pr &= M_\eta + \mathfrak{M}_\eta + M_\eta^{\mathbf{\Phi}} + \mathfrak{R}_\eta, \\
I_{\varsigma\varsigma}\frac{dr}{dt} + (I_{\eta\eta} - I_{\xi\xi})pq &= M_\varsigma + \mathfrak{M}_\zeta + M_\varsigma^{\mathbf{\Phi}} + \mathfrak{R}_\zeta.
\end{aligned} \qquad (122)$$

The terms on the right side of (122)

$$\mathfrak{R}_\xi = \frac{dI_{\xi\xi}}{dt}(p_2 - p), \qquad \mathfrak{R}_\eta = \frac{dI_{\eta\eta}}{dt}(q_2 - q), \qquad \mathfrak{R}_\zeta = \frac{dI_{\varsigma\varsigma}}{dt}(r_2 - r), \quad (123)$$

are called the projections of the reactive torque. The reactive torque is the consequence of variation of moment of inertia of the body.

The Eqs. (120) and (122) are the six scalar differential equations of motion of the body with variable mass. For the both cases, when mass is separating or adding, the differential equations of motion have the same form (120) and (122), but the signs of separating mass dM/dt and separating moment of inertia dI/dt are negative, while the signs of adding mass dM/dt and adding moment of inertia dI/dt are positive.

5.1 Discussion of the differential equations of motion

1) Comparing the relations (118) and (119) with those for the free motion of a body with constant mass (see for example Starzhinskii, 1982)

$$M\frac{d\mathbf{v}}{dt} = \mathbf{F}_r, \qquad \mathbf{I}\frac{d\boldsymbol{\Omega}}{dt} + \boldsymbol{\Omega} \times \mathbf{I}\boldsymbol{\Omega} = \mathbf{M}_S^{Fr} + \mathfrak{M}, \tag{124}$$

it is evident that due to variation of the mass and moment of inertia some additional terms exist which represent the reactive force $\boldsymbol{\Phi}$, its moment $\mathbf{M}_S^{\boldsymbol{\Phi}}$ and the reactive torque $\mathbb{R}$.

2) For the case when the relative velocity and angular velocity of mass and moment of inertia variation is zero, the differential equations of motion have the form (124), but M and $\mathbf{I}$ are time variable. Namely, due to the fact that $\mathbf{u} = \mathbf{v}$ and $\boldsymbol{\Omega}_2 = \boldsymbol{\Omega}$, the reactive force and torque and also the corresponding moment are zero, i.e.,

$$\boldsymbol{\Phi} = 0, \qquad \mathbb{R} = 0, \qquad \mathbf{M}_S^{\boldsymbol{\Phi}} = 0. \tag{125}$$

3) For the case when the absolute velocity $\mathbf{u} = 0$ and the angular velocity $\boldsymbol{\Omega}_2 = 0$ of added or separated mass are zero, the differential equations of motion transform into

$$\frac{d}{dt}(M\mathbf{v}) = \mathbf{F}_r, \tag{126}$$

$$\frac{d}{dt}(\mathbf{I}\boldsymbol{\Omega}) = \mathbf{M}_S^{Fr} + \mathfrak{M} + \mathbf{M}_S^{\boldsymbol{\Phi}^*}, \tag{127}$$

where the modified reactive force and its moment are

$$\boldsymbol{\Phi}^* = \mathbf{v}\frac{dM}{dt}, \qquad \mathbf{M}_S^{\boldsymbol{\Phi}^*} = \mathbf{SS}_2 \times \boldsymbol{\Phi}^*. \tag{128}$$

The sign of the moment of the reactive force is negative.

Cornelisse *et al.*, 1979, introduce the additional assumption: the moment of the reactive force is sufficiently small in comparison with other values in the system and can be omitted. The differential equations of motion (126) and (127) simplify into

$$\frac{d}{dt}(M\mathbf{v}) = \mathbf{F}_r, \qquad \frac{d}{dt}(\mathbf{I\Omega}) = \mathbf{M}_S^{Fr} + \mathfrak{M}. \tag{129}$$

In the paper of Meshchersky (1896) this special case is considered for the rotating body around a fixed axle

$$\frac{d}{dt}(I_\zeta r) = M_\zeta^{Fr} + \mathfrak{M}_\zeta. \tag{130}$$

where I_ζ is the time variable moment of inertia, r is the angular velocity, M_ζ^{Fr} is the moment of the resultant force due to the axle ς and $\mathfrak{M}_\zeta$ is the torque rotating the body around the axle ς.

4) Mathematical model of the mass variable body with translatory motion is equal to that of the particle with variable mass and it is

$$M\frac{d\mathbf{v}}{dt} = \mathbf{F}_r + \mathbf{\Phi}. \tag{131}$$

This relation was for the first time introduced by Meshchersky. Based on this equation the modern rocket dynamics is developed.

5) If the mass variable body rotates around a fixed axle, the system of differential equations of motion (120) and (122) simplify into only one

$$I_\varsigma\frac{dr}{dt} = M_\varsigma + \mathfrak{M}_\zeta + \mathfrak{R}_\zeta, \tag{132}$$

The reactive torque $\mathfrak{R}_\zeta$ depends on the variation of the moment of inertia for the rotation axle and the relative angular velocity which is the difference between the angular velocity of body rotation and the angular velocity of the separated or added mass. If the relative angular velocity of the mass variation is zero, the differential equation of body rotation is

$$I_\varsigma\ddot{\varphi} = M_\varsigma + \mathfrak{M}_\zeta. \tag{133}$$

where $\dot{\varphi} = r$.and the moment of inertia is time dependent.

6) For the straightforward motion of a mass variable body the differential equation is according to (131)

$$M\frac{dv}{dt} = F_r + \Phi, \tag{134}$$

where

$$\Phi = \frac{dM}{dt}(u - v). \tag{135}$$

Comparing (132) and (134) it can be concluded that the differential equations have the same form, where the reactive force Φ for the translatory motion corresponds to the reactive torque $\Re_\varsigma$.

Remark 5.2. Using the principle of solidification Bessonov, 1967, obtained the differential equations of free motion of the body. Bessonov does not take the reactive torque into consideration.

5.2 Band is winding up on a drum

Let us consider the in-plane motion of a drum on which the band is winding up (Fig.4). The differential equations of the motion are according to (120) and (122)

$$\begin{aligned} \frac{d}{dt}(M\dot{x}_{S1}) &= F_x + \frac{dM}{dt}v_{xb}, \\ \frac{d}{dt}(M\dot{y}_{S1}) &= F_y + \frac{dM}{dt}v_{yb}, \\ \frac{d}{dt}(I_{S1}\dot{\varphi}) &= M_{S1} + M_{S1}^{\mathbf{\Phi}} + \frac{dI_{S1}}{dt}\Omega_b, \end{aligned} \tag{136}$$

where

$$M_{S1}^{\mathbf{\Phi}} = \frac{dM}{dt}[(\mathbf{S}_1\mathbf{S}_2)_x(v_{yb} - \dot{y}_{S1}) - (\mathbf{S}_1\mathbf{S}_2)_y(v_{xb} - \dot{x}_{S1}), \tag{137}$$

$(\mathbf{S}_1\mathbf{S}_2)_x$ and $(\mathbf{S}_1\mathbf{S}_2)_y$ are projections of the position vector of the point of mass addition due to the mass centre S_1, Ω_b is the angular velocity of the winding band, v_{xb} and v_{yb} are the projections of the linear velocity of the winding band, M is the mass of the drum with band, I_{S1} is moment of inertia of the drum with band, $\dot{x}_{S1}$ and $\dot{y}_{S1}$ are the projections of the velocity of the mass center of the drum with band and $\dot{\varphi}$ is the angular velocity of the drum with band (Cveticanin & Kovacic, 2007).

The technical requirement for winding up of the band is the absolute velocity of the band v_b to be constant. Only for that condition the rolling up of the band on the drum is accurate without crumpling of the band or its plucking. The band is moving translatory with velocity v horizontally, parallel to y-axle in Fig.4. The projections of the band velocity are

$$v_{xb} = 0, \qquad v_{yb} = v. \tag{138}$$

The rolling up of one band layer is discussed. The angle of rolling up of the band is in the interval from $\varphi = 0$ to $\varphi = 2\pi$.

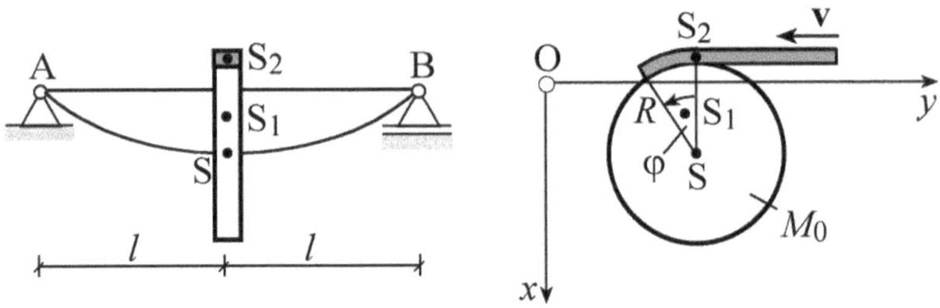

Figure 4. The model of the rotor on which the band is winding up.

The geometric and physical properties of the drum with band If the mass of the drum with unrolled band is M_0 and the rolling mass is M_r

$$M_r = \mu\varphi, \tag{139}$$

where

$$\mu = Rhb\rho, \tag{140}$$

h is the thickness, b is the width and ρ is the density of band, the mass M variation is a linear function of the angle φ

$$M = M_0 + M_r = M_0 + \mu\varphi. \tag{141}$$

The position of mass centre of the drum with unrolled mass is

$$SS_1 = \frac{M_r}{M}(SS'), \tag{142}$$

where SS' is the distance of the mass centre of the unrolled mass on the drum

$$SS' = R\frac{\sin(\varphi/2)}{\varphi/2}. \tag{143}$$

According to the relations (141), (142) and (143) the distance between the mass centre of the whole system and the rotation centre is obtained

$$SS_1 = \frac{2\mu}{M_0 + \mu\varphi}\sin(\frac{\varphi}{2}). \tag{144}$$

The moment of inertia of the drum with the unrolled mass is J_0 and the moment of inertia of the band which is rolling up is

$$J_r = \int_0^{\varphi} R^2 dM_r = \int_0^{\varphi} R^3 hb\rho d\varphi = j\varphi, \tag{145}$$

where $j = R^3 hb\rho = \mu R^2$ is the unit moment of inertia. The total moment of inertia is obtained by superposition the both moments of inertia

$$J_S = J_0 + j\varphi. \tag{146}$$

Applying the Steiner theorem the moment of inertia for the parallel axis settled in the mass centre is obtained

$$J_{S1} = J_S - M(SS_1)^2. \tag{147}$$

Forces acting on the system During winding up of the band the following forces act: the elastic force of the shaft, the damping torque, the reactive force and the reactive torque.

The elastic force of the shaft is projected in the fixed coordinate system

$$F_x = -cx_S = -c(x_{S1} - SS_1 \cos\frac{\varphi}{2}), \qquad F_y = -cy_S = -c(y_{S1} - SS_1 \sin\frac{\varphi}{2}), \tag{148}$$

where c is the rigidity of the shaft.

According to (136) and (138) the projections of the reactive force Φ and the reactive torque $\Im$ are obtained

$$\Phi_x = \frac{dM}{dt}(-\dot{x}_{S1}), \qquad \Phi_y = \frac{dM}{dt}(v - \dot{y}_{S1}), \qquad \Re = \frac{dI_{S1}}{dt}(-\dot{\varphi}).$$

If the rotational damping torque acts

$$M_D = -D\dot{\varphi}, \tag{149}$$

where D is the damping coefficient, and the moment of the reactive force according to S_1 is considered, the differential equations of the plane motion is obtained

$$\begin{aligned}
M\ddot{x}_{S1} + cx_{S1} &= c(SS_1)\cos\frac{\varphi}{2} + \frac{dM}{dt}(-\dot{x}_{S1}),\\
M\ddot{y}_{S1} + cy_{S1} &= c(SS_1)\sin\frac{\varphi}{2} + \frac{dM}{dt}(v - \dot{y}_{S1}),\\
J_{S1}\ddot{\varphi} + D\dot{\varphi} &= \frac{dJ_{S1}}{dt}(-\dot{\varphi}) + x_{S1}c(SS_1)\sin\frac{\varphi}{2} - y_{S1}c(SS_1)\cos\frac{\varphi}{2}\\
&\quad + \frac{dM}{dt}(v - \dot{y}_{S1})(\mathbf{S}_1\mathbf{S}_2)_x - \frac{dM}{dt}(-\dot{x}_{S1})(\mathbf{S}_1\mathbf{S}_2)_y.
\end{aligned} \tag{150}$$

It is worth to be mentioned that due to winding up of the band the symmetry of the disc is disturbed and the products of inertia J_{xz} and J_{xz}

are not zero. As the thickenss of the band which is winding up is quite small, the mentioned products of inertia are also small and can be neglected.

For the case when the mass centre and the geometric centre of the disc are quite close and the eccentricity is $SS_1 \approx 0$, we have the coordinates $x_{S1} \approx x_S$ and $y_{S1} \approx y_S$. Omitting the terms with SS_1 in the equation (150) and using the relations (141), (146) and (147) the differential equations of the plane motion of the disc, with variable mass and without the unbalance force, are obtained

$$M\ddot{x}_S + cx_S = -\mu\dot{\varphi}\dot{x}_S, \qquad M\ddot{y}_S + cy_S = \mu\dot{\varphi}(v - \dot{y}_S), \tag{151}$$

$$J_S\ddot{\varphi} + D\dot{\varphi} = -j\dot{\varphi}^2 + \mu\dot{\varphi}(v - \dot{y}_S)(\mathbf{SS}_2)_x + \mu\dot{\varphi}\dot{x}_S(\mathbf{SS}_2)_y. \tag{152}$$

The system of differential equations (151) and (152) is nonlinear.

Shaft is rigid If the shaft of the drum with winding up band is rigid the motion of the system transforms to a rotation around the rigid axle ($x_S = y_S = 0$)

$$(J_0 + j\varphi)\ddot{\varphi} + D\dot{\varphi} = -j\dot{\varphi}^2 + \mu\dot{\varphi}vR. \tag{153}$$

Introducing the new variable $u(\varphi) = \dot{\varphi}$ the differential equation (153) is transformed to the Bernoulli equation

$$(J_0 + j\varphi)\frac{du}{d\varphi} + ju = (\mu vR - D), \tag{154}$$

whose solution for the initial condition $\dot{\varphi}(0) = \Omega_b$ has the form

$$\dot{\varphi} = \frac{J_0\Omega_b + (\mu vR - D)\varphi}{J_0 + j\varphi}. \tag{155}$$

The relation (155) describes the variation of the angular velocity of the drum when the absolute velocity v of the winding band is constant. The angular velocity of the drum decreases during winding up of a layer.

Integrating the differential equation (155) for the initial angle $\varphi(0) = 0$ the time history of angle variation is obtained

$$\varphi + (\frac{J_0}{j} - \frac{J_0\Omega_b}{\mu vR - D})\ln\left|1 + \varphi\frac{\mu vR - D}{J_0\Omega_b}\right| = \frac{\mu vR - D}{j}t. \tag{156}$$

This form of solution is not convenient for discussion. Introducing the new variable

$$r = 1 + \frac{\mu vR - D}{J_0\Omega_b}\varphi, \tag{157}$$

the equation (156) is

$$r + \frac{D}{j\Omega_b}\ln r = 1 + \frac{(\mu v R - D)^2}{J_0 j\Omega_b}t. \tag{158}$$

Let us introduce the new function

$$w = -\ln r - \frac{f_1}{k_1}, \tag{159}$$

where

$$f_1 = 1 + \frac{(\mu v R - D)^2}{J_0 j\Omega_b}t, \quad k_1 = -\frac{\mu v R - D - j\Omega_b}{j\Omega_b}. \tag{160}$$

After substituting (159) into (156) and some transformation the obtained result is

$$w\exp(w) = x, \tag{161}$$

where

$$x = \frac{1}{k_1}\exp(-\frac{f_1}{k_1}). \tag{162}$$

The solution $w(x)$ of (161) is the Lambert's w function (see Weisstein, 2007)

$$w(x) \equiv lambertw(\frac{1}{k_1}\exp(-\frac{f_1}{k_1})). \tag{163}$$

Substituting into (159) the solution for r is obtained

$$r = -k_1(lambertw(-\frac{1}{k_1}\exp(-\frac{f_1}{k_1}))) \equiv -k_1 w(x), \tag{164}$$

which gives the implicit solution for (156)

$$\varphi = J_0(\frac{1}{j} - \frac{\Omega_b}{\mu v R - D})w - \frac{J_0\Omega_b}{\mu v R - D}. \tag{165}$$

For the case when damping is neglected and assuming that $v = \Omega_b R$ the relation (156) is simplified and the angle time function is linear

$$\varphi = \Omega_b t. \tag{166}$$

Shaft is elastic Let us transform the differential equations (151) introducing the variables

$$x_S = x(\varphi), \qquad \dot{x}_S = \frac{dx}{d\varphi}\dot{\varphi}, \qquad \ddot{x}_S = \frac{d^2x}{d\varphi^2}\dot{\varphi}^2 + \frac{dx}{d\varphi}\ddot{\varphi}, \tag{167}$$

$$y_S = y(\varphi), \qquad \dot{y}_S = \frac{dy}{d\varphi}\dot{\varphi}, \qquad \ddot{y}_S = \frac{d^2y}{d\varphi^2}\dot{\varphi}^2 + \frac{dy}{d\varphi}\ddot{\varphi}, \tag{168}$$

The obtained system of differential equations of plane motion is

$$(M_0 + \mu\varphi)\dot{\varphi}^2 \frac{d^2x}{d\varphi^2} + \frac{dx}{d\varphi}[\mu\dot{\varphi}^2 + (M_0 + \mu\varphi)\ddot{\varphi}] + cx = 0, \tag{169}$$

$$(M_0 + \mu\varphi)\dot{\varphi}^2 \frac{d^2y}{d\varphi^2} + \frac{dy}{d\varphi}[\mu\dot{\varphi}^2 + (M_0 + \mu\varphi)\ddot{\varphi}] + cy = \mu\dot{\varphi}v, \tag{170}$$

$$J_S\ddot{\varphi} + (D - \mu vR)\dot{\varphi} + j\dot{\varphi}^2 = -\mu\dot{\varphi}^2 R\frac{dy}{d\varphi}. \tag{171}$$

Substituting (155) into (169) and (170), assuming that μ/M_0, j/J_0, μ/J_0 and $D/J_0\Omega_b$ are small parameters, the simplified differential equations are formed

$$\frac{d^2x}{d\varphi^2} + 2\delta\frac{dx}{d\varphi} + \omega^2(\varphi)x = 0, \tag{172}$$

$$\frac{d^2y}{d\varphi^2} + 2\delta\frac{dy}{d\varphi} + \omega^2(\varphi)y = \frac{\mu}{M_0}R, \tag{173}$$

$$\ddot{\varphi} + \frac{D - \mu Rv}{J_S}\dot{\varphi} + \frac{j}{J_S}\dot{\varphi}^2 = -\frac{\mu}{J_0}R\dot{\varphi}^2\frac{dy}{d\varphi}, \tag{174}$$

where

$$2\delta = 2\frac{\mu}{M_0} - \frac{j}{J_0} - \frac{D}{J_0\Omega_b}, \qquad \omega^2(\varphi) \equiv \omega^2 = k^2(1 - A\varphi),$$

$$k^2 = \frac{\omega_1^2}{\Omega_b^2}, \qquad \omega_1^2 = \frac{c}{M_0}, \qquad A = 3\frac{\mu}{M_0} - 2\frac{j}{J_0} - \frac{D}{J_0\Omega_b}.$$

To obtain the approximate analytic solutions of (172) - (174) the Bogolubov-Mitropolski method is modified for the non-homogenous rheo-linear differential equations.

Omitting the terms on the right side of the equation (174) as small values the approximate solution of (174) corresponds to the case of rigid shaft (155). Substituting (155) into (173) the solution of the differential equation (173) is assumed as

$$y = a(\varphi)\exp(-\delta\varphi)\cos\Psi(\varphi) + \frac{1}{\omega^2(\varphi)}\frac{\mu}{M_0}R$$
$$\equiv a\exp(-\delta\varphi)\cos\Psi + \frac{1}{\omega^2(\varphi)}\frac{\mu}{M_0}R, \tag{175}$$

where

$$\Psi(\varphi) = \int \omega(\varphi)d\varphi + \alpha(\varphi), \tag{176}$$

and the first derivative of the function y

$$\frac{dy}{d\varphi} = (-\delta a \cos\Psi - a\omega(\varphi)\sin\Psi)\exp(-\delta\varphi) - (\frac{\mu}{M_0})^2 \frac{R}{k^2(1-A\varphi)^2}, \tag{177}$$

with

$$\frac{da}{d\varphi}\cos\Psi - a\frac{d\alpha}{d\varphi}\sin\Psi = 0. \tag{178}$$

Eliminating the second order small term in (177) the relation transforms to

$$\frac{dy}{d\varphi} \approx (-\delta a \cos\Psi - a\omega(\varphi)\sin\Psi)\exp(-\delta\varphi). \tag{179}$$

Using the relations (175) - (179) the differential equation (173) is transformed into a system of two first order differential equations

$$\frac{da}{d\varphi} = -\frac{a}{\omega}\frac{d\omega}{d\varphi}\sin^2\Psi, \qquad \frac{d\alpha}{d\varphi} = \frac{1}{2\omega}\frac{d\omega}{d\varphi}\sin 2\Psi. \tag{180}$$

It is at this point the averaging procedure $\frac{1}{2\pi}\int\limits_0^{2\pi}(\bullet)d\Psi$ is introduced and the relations (180) are simplified into

$$\frac{da}{d\varphi} = -\frac{a}{2\omega}\frac{d\omega}{d\varphi}, \qquad \frac{d\alpha}{d\varphi} = 0. \tag{181}$$

For the initial conditions

$$\varphi = 0, \qquad a = a_0, \qquad \alpha = \alpha_0, \tag{182}$$

the solution of the equation (173) in the first approximation is

$$y_S = y(\varphi) = \frac{a_0}{\sqrt[4]{1-A\varphi}}\exp(-\delta\varphi)\cos(k\sqrt{(1-A\varphi)} + \alpha_0) + \frac{R}{k^2}\frac{\mu}{M_0}. \tag{183}$$

According to the suggested procedure the solution of the equation (172) is

$$x_S = x(\varphi) = \frac{b_0}{\sqrt[4]{1-A\varphi}}\exp(-\delta\varphi)\cos(k\sqrt{(1-A\varphi)} + \beta_0), \tag{184}$$

where b_0 and β_0 are initial amplitude and phase. The parameter values have to satisfy the relation

$$3\frac{\mu}{M_0} - 2\frac{j}{J_0} - \frac{D}{J_0\Omega_b} < \frac{1}{2\pi}. \tag{185}$$

The motion of the rotor center depends on the ratio between the small parameters μ/M_0, j/J_0 and $D/J_0\Omega_b$. For small value of the rotational damping and higher velocity of the rolling band the vibrations decrease.

Using the obtained solution (183) the correction for the angle velocity (155) can be denoted. Due to the fact that $\dot{y}_S$ tends to zero for technical reasons the relation (155) is guessed to be accurate enough.

5.3 Conclusion

During the process of continual mass variation, the mass and moment of inertia of the rigid body vary due to adding or separating of mass in the short infinitesimal time interval: mass but also the form and the volume of the body are continually varying in time. It causes the body mass center position variation and also the change of the moment of inertia and the products of moment of inertia. Due to mass and moment of inertia variation the reactive force and reactive torque act. Namely, the absolute velocity and angular velocity of addition or separation differs in general from the velocity of mass center and angular velocity of the initial body and it causes the impact to occur. As the mass variation is continual the impact is substituted with a "reactive force" and "reactive torque" which continually act on the body. The force and torque depend on the absolute velocity of mass centre and angular velocity of the separated or added body.

For the rolling up of the band on the drum mass and moment of inertia of the drum with band is varying. Mass and moment of inertia depend on the angle position of the wounded band. Due to geometry variation of the drum with band the mass center position inside the system is varying, too. This variation seems to be small and is neglected in our consideration. During winding up of the band on the drum the impact occurs due to difference of velocity and angular velocity of the band and drum. It causes the vibrations of the mass centre of drum. The vibrations of drum mass centre depend on the amount on the band winding up on the drum: the higher the amount of band on the drum the smaller the vibrations. The damping property of drum also has an influence on the vibrations: the higher the damping the smaller the vibrations.

The band is winding up with constant velocity. This requires the angle velocity of drum to vary. The angle velocity variation is the function of the moment of inertia of the band which is winding up and also of the damping properties of the system: for higher damping the angle velocity decreases faster than for the smaller damping; the larger the moment of inertia of the winding up band the slower the decrease of the angular velocity. This result is of technical importance for regulating of the rotation of the drum.

6 Lagrange's equations of the body with continual mass variation

Let us rewrite the Eqs. (110) and (111) into the form

$$\frac{d}{dt}(M\mathbf{v}) = \mathbf{F}_r + \mathbf{\Phi}_a, \tag{186}$$

$$\frac{d}{dt}(\mathbf{I\Omega}) = \mathbf{M}_S^{Fr} + \mathfrak{M} + \mathbf{M}_S^{\Phi} + \mathbb{R}_a. \tag{187}$$

where

$$\mathbf{\Phi}_a = \frac{dM}{dt}\mathbf{u}, \qquad \mathbb{R}_a = \frac{d\mathbf{I}}{dt}\mathbf{\Omega}_2, \tag{188}$$

and the reactive force $\mathbf{\Phi}$ and the moment of the reactive force $\mathbf{M}_S^{\Phi}$ are given with Eqs. (115) and (117), respectively. Multiplying the Eq. (186) with the virtual displacement $\delta\mathbf{r}$, and the Eq. (187) with the virtual angle $\delta\mathbf{\Psi}$ and by adding them, it follows

$$(\mathbf{F}_r+\mathbf{\Phi}_a)\delta\mathbf{r} + (\mathbf{M}_S^{Fr}+\mathfrak{M}+\mathbf{M}_S^{\Phi}+\mathbb{R}_a)\delta\mathbf{\Psi} - M\ddot{\mathbf{r}}\delta\mathbf{r} - \dot{M}\dot{\mathbf{r}}\delta\mathbf{r} - \mathbf{I}\ddot{\mathbf{\Psi}}\delta\mathbf{\Psi} - \dot{\mathbf{I}}\dot{\mathbf{\Psi}}\delta\mathbf{\Psi} = 0, \tag{189}$$

where $\dot{\mathbf{\Psi}} = \mathbf{\Omega}$ and $(\cdot)^{\cdot} \equiv d(\cdot)/dt$, $(\cdot)^{\cdot\cdot} \equiv d^2(\cdot)/dt^2$.The relation (189) describes the D'Alambert-Lagrange principle for the body with continual mass variation: The total virtual work of all active forces and torques (including the non-ideal constraint reactions), of the reactive force and torque, of the moment of the reactive force and of the inertial force and torque is equal to zero for any virtual displacement and virtual angle of the body. Mathematically, it is

$$\delta A^I + \delta A^{\phi a} + \delta A^{Ra} + \delta A + \delta A^{M\phi} = 0, \tag{190}$$

where

$$\begin{aligned}
\delta A^I &= -(M\ddot{\mathbf{r}}+\dot{M}\dot{\mathbf{r}}\delta\mathbf{r})\delta\mathbf{r} - (\mathbf{I}\ddot{\mathbf{\Psi}} + \dot{\mathbf{I}}\dot{\mathbf{\Psi}})\delta\mathbf{\Psi},\\
\delta A &= \mathbf{F}_r\delta\mathbf{r} + (\mathbf{M}_S^{Fr} + \mathfrak{M})\delta\mathbf{\Psi},\\
\delta A^{Ra} &= \mathbb{R}_a\delta\mathbf{\Psi}\\
\delta A^{M\phi} &= \mathbf{M}_S^{\Phi}\delta\mathbf{\Psi}\\
\delta A^{\phi a} &= \mathbf{\Phi}_a\delta\mathbf{r}.
\end{aligned} \tag{191}$$

us introduce $i = 1, 2, ..., 6$ independent generalized coordinates q_i. The virtual displacement and the virtual angle are determined by formulas

$$\delta\mathbf{r} = \sum_{i=1}^{6} \frac{\partial\mathbf{r}}{\partial q_i}\delta q_i, \qquad \delta\mathbf{\Psi} = \sum_{i=1}^{6} \frac{\partial\mathbf{\Psi}}{\partial q_i}\delta q_i, \tag{192}$$

where δq_i is the variation of the generalized coordinate q_i. Substituting (192) into (190) and after some modification we have

$$\sum_{i=1}^{6}(Z_i + Q_i^{\phi a} + Q_i^{Ra} + Q_i + Q_i^{M\phi})\delta q_i = 0, \tag{193}$$

where the generalized inertial force Z_i, generalized force of the part of the reactive force $Q_i^{\phi a}$ and reactive torque Q_i^{Ra}, generalized force of the active forces and torques and reactions of non-ideal constraints Q_i and the generalized force of the moment of the reactive force $Q_i^{M\phi}$ are calculated according to following formulas

$$\begin{aligned} Z_i &= -[\frac{d}{dt}(M\mathbf{v})\frac{\partial \mathbf{r}}{\partial q_i} + \frac{d}{dt}(\mathbf{I\Omega})\frac{\partial \mathbf{\Psi}}{\partial q_i}], \\ Q_i^{\phi a} &= \mathbf{\Phi}_a\frac{\partial \mathbf{r}}{\partial q_i} \qquad\qquad Q_i^{Ra} = \mathbb{R}_a\frac{\partial \mathbf{\Psi}}{\partial q_i}, \\ Q_i &= \mathbf{F}_r\frac{\partial \mathbf{r}}{\partial q_i}+(\mathbf{M}_S^{Fr} + \mathfrak{M})\frac{\partial \mathbf{\Psi}}{\partial q_i}, \qquad Q_i^{M\phi} = \mathbf{M}_S^{\Phi}\frac{\partial \mathbf{\Psi}}{\partial q_i}. \end{aligned} \tag{194}$$

The generalized inertial force Z_i is rewritten as

$$Z_i = -\frac{d}{dt}[(M\mathbf{v})\frac{\partial \mathbf{r}}{\partial q_i} + (\mathbf{I\Omega})\frac{\partial \mathbf{\Psi}}{\partial q_i}] + [(M\mathbf{v})\frac{d}{dt}\frac{\partial \mathbf{r}}{\partial q_i} + (\mathbf{I\Omega})\frac{d}{dt}\frac{\partial \mathbf{\Psi}}{\partial q_i}]. \tag{195}$$

The position vector $\mathbf{r}$ and the angle vector $\mathbf{\Psi}$ depend on the generalized coordinates q_i and time t

$$\mathbf{r} = \mathbf{r}(q_i, t), \qquad \mathbf{\Psi} = \mathbf{\Psi}(q_i, t). \tag{196}$$

As the generalized coordinates also depend on time, the velocity and angular velocity have the form

$$\mathbf{v} \equiv \frac{d\mathbf{r}}{dt} = \frac{\partial \mathbf{r}}{\partial t} + \sum_{i=1}^{6}\frac{\partial \mathbf{r}}{\partial q_i}\dot{q}_i, \qquad \mathbf{\Omega} \equiv \frac{d\mathbf{\Psi}}{dt} = \frac{\partial \mathbf{\Psi}}{\partial t} + \sum_{i=1}^{6}\frac{\partial \mathbf{\Psi}}{\partial q_i}\dot{q}_i. \tag{197}$$

Now we take the partial derivatives with respect to $\dot{q}_i$

$$\frac{\partial \mathbf{r}}{\partial q_i} = \frac{\partial \dot{\mathbf{r}}}{\partial \dot{q}_i} = \frac{\partial \mathbf{v}}{\partial \dot{q}_i}, \qquad \frac{\partial \mathbf{\Psi}}{\partial q_i} = \frac{\partial \dot{\mathbf{\Psi}}}{\partial \dot{q}_i} = \frac{\partial \mathbf{\Omega}}{\partial \dot{q}_i}. \tag{198}$$

On the other hand, taking the partial derivatives of both the sides of equalities (197) with respect to q_i we obtain

$$\frac{\partial \mathbf{v}}{\partial q_i} = \frac{\partial^2 \mathbf{r}}{\partial t \partial q_i} + \sum_{j=1}^{6}\frac{\partial^2 \mathbf{r}}{\partial q_i \partial q_j}\dot{q}_j, \qquad \frac{\partial \mathbf{\Omega}}{\partial q_i} = \frac{\partial^2 \mathbf{\Psi}}{\partial t \partial q_i} + \sum_{j=1}^{6}\frac{\partial^2 \mathbf{\Psi}}{\partial q_i \partial q_j}\dot{q}_j. \tag{199}$$

Directly, the time derivative of $(\partial \mathbf{r}/\partial q_i)$ and $(\partial \mathbf{\Psi}/\partial q_i)$ is

$$\frac{d}{dt}\left(\frac{\partial \mathbf{r}}{\partial q_i}\right) = \frac{\partial^2 \mathbf{r}}{\partial t \partial q_i} + \sum_{j=1}^{6} \frac{\partial^2 \mathbf{r}}{\partial q_i \partial q_j}\dot{q}_j, \qquad \frac{d}{dt}\left(\frac{\partial \mathbf{\Psi}}{\partial q_i}\right) = \frac{\partial^2 \mathbf{\Psi}}{\partial t \partial q_i} + \sum_{j=1}^{6} \frac{\partial^2 \mathbf{\Psi}}{\partial q_i \partial q_j}\dot{q}_j. \tag{200}$$

The left sides of the Eqs. (199) and (200) are equal, and consequently,

$$\frac{\partial \mathbf{v}}{\partial q_i} = \frac{d}{dt}\left(\frac{\partial \mathbf{r}}{\partial q_i}\right), \qquad \frac{\partial \mathbf{\Omega}}{\partial q_i} = \frac{d}{dt}\left(\frac{\partial \mathbf{\Psi}}{\partial q_i}\right). \tag{201}$$

Applying (198) and (201), the generalized inertial force (195) is

$$Z_i = -\frac{d}{dt}[(M\mathbf{v})\frac{\partial \mathbf{v}}{\partial \dot{q}_i} + (\mathbf{I\Omega})\frac{\partial \mathbf{\Omega}}{\partial \dot{q}_i}] + [(M\mathbf{v})\frac{\partial \mathbf{v}}{\partial q_i} + (\mathbf{I\Omega})\frac{\partial \mathbf{\Omega}}{\partial q_i}]. \tag{202}$$

After some modification, Eq. (202) transforms into

$$Z_i = -\frac{d}{dt}\frac{\partial T}{\partial \dot{q}_i} + \frac{\partial T}{\partial q_i}, \tag{203}$$

where T is the kinetic energy

$$T = \frac{1}{2}M\mathbf{vv} + \frac{1}{2}\mathbf{I\Omega\Omega}. \tag{204}$$

Substituting (203) into (193) the general equation of dynamics for mass variation is

$$\sum_{i=1}^{6}\left[\frac{d}{dt}\left(\frac{\partial T}{\partial \dot{q}_i}\right) - \frac{\partial T}{\partial q_i} - \left(Q_i + Q_i^{\phi a} + Q_i^{Ra} + Q_i^{M\phi}\right)\right]\delta q_i = 0 \tag{205}$$

Since the coordinates q_i are independent so are the variations δq_i and therefore condition (205) implies

$$\frac{d}{dt}\left(\frac{\partial T}{\partial \dot{q}_i}\right) - \frac{\partial T}{\partial q_i} = Q_i + Q_i^{\phi a} + Q_i^{Ra} + Q_i^{M\phi}, \qquad i = 1, 2, ..., 6. \tag{206}$$

The Lagrange's equations of motion of the variable mass body is given by Bessonov, , by using the method of solidification. Bessonov's equations are not general, as he assumed that the absolute angular velocity of the added body is zero and the generalized force Q_i^{Ra} is omitted.

7 Vibration of the body with continual mass variation

As the special type of motion the oscillation of the body with time variable mass is considered. The motion is bounded, periodical and with monotone change of the direction of motion around the equilibrium position or it is added to the steady state motion of the body. In this Section the vibration of the one-degree-of-freedom oscillator with variable mass will be considered

Based on the Eq. (120) with (121) and introducing the generalized coordinate x, the mathematical model of the one-degree-of-freedom oscillator with time variable mass is

$$M\ddot{x} = F_x + \frac{dM}{dt}(u_x - \dot{x}), \tag{207}$$

where F_x is the resultant force and u_x is the absolute velocity of the adding or separating particle in x direction. In general, the resultant force is a function of the displacement x, velocity $\dot{x}$ and time t

$$M\ddot{x} = F_x(x, \dot{x}, t) + \frac{dM}{dt}(u_x - \dot{x}). \tag{208}$$

If an elastic force of odd parity acts, i.e.,

$$F_e(-x) = F_e(-x), \tag{209}$$

the differential equation (208) is as follows

$$M\ddot{x} + F_e(x) = F_x(x, \dot{x}, t) + \frac{dM}{dt}(u_x - \dot{x}). \tag{210}$$

Eq. (210) describes the vibration of the time variable one-degree-of-freedom system.

7.1 Oscillator with strong nonlinear deflection

Let us consider the oscillator where:

1. Mass variation is slow and depends on the 'slow time' $\tau = \varepsilon t$ where $\varepsilon << 1$ is a small parameter

$$M = m(\tau). \tag{211}$$

2. The elastic force depends on the nonlinear deflection $x\left|x\right|^{\alpha-1}$, where the order of nonlinearity $\alpha \in \mathbb{R}_+$ is the positive rational number written as a termination decimal or as an exact fraction, $\alpha \in \mathbb{Q}_+ = \left\{\frac{m}{n} > 0 : m \in \mathbb{Z}, n \in \mathbb{Z}, n \neq 0\right\}$ and $\mathbb{Z}$ is integer.

3. The absolute velocity of the adding or separating mass is zero, i.e., $u_x = 0$.

4. The additional force which acts on the oscillator is small and is the function of the deflection x and velocity $\dot{x}$: $F_x = \varepsilon f_1(x, \dot{x})$.

The mathematical model for such an oscillator is

$$m(\tau)\ddot{x} + k_\alpha x\,|x|^{\alpha-1} = -\varepsilon\frac{dm(\tau)}{d\tau}\dot{x} + \varepsilon f_1(x,\dot{x}), \tag{212}$$

i.e.

$$\ddot{x} + \omega^2(\tau)x\,|x|^{\alpha-1} = \varepsilon f(\tau,x,\dot{x}), \tag{213}$$

where

$$\varepsilon f(\tau,x,\dot{x}) = -\frac{\varepsilon}{m(\tau)}\frac{dm(\tau)}{d\tau}\dot{x} + \frac{\varepsilon}{m(\tau)}f_1(x,\dot{x}). \tag{214}$$

and

$$\omega(\tau) = \sqrt{\frac{k_\alpha}{m(\tau)}}. \tag{215}$$

Generating solution For $\varepsilon = 0$ the generating equation of (213) is

$$\ddot{x} + \omega_0^2 x\,|x|^{\alpha-1} = 0, \tag{216}$$

with initial conditions

$$x(0) = x_0, \qquad \dot{x}(0) = 0, \tag{217}$$

where $\omega_0^2 \equiv \omega^2(0)$ =const.Integrating (216) and using the initial conditions, the first integral of the energy type is obtained (Cveticanin, 2008)

$$\frac{\dot{x}^2}{2} + \frac{c_1^2}{\alpha+1}\,|x|^{\alpha+1} = \frac{c_1^2}{\alpha+1}\,|x_0|^{\alpha+1}. \tag{218}$$

The both terms on the left side are positive and the motion is periodic. (see for example Cveticanin, 2009_2; Mickens, 2010; Cveticanin & Pogany, 2012). For the new variable $|x| = |x_0|\,|u|^{1/(\alpha+1)}$ the period of vibration is

$$T_{ex} = \frac{4\,|x_0|^{(1-\alpha)/2}}{c_1\sqrt{2(\alpha+1)}}\int_0^1 (1-|u|)^{-1/2}u^{-\alpha/(\alpha+1)}du. \tag{219}$$

Introducing the Euler beta function $B(m,n)$ (see Rosenberg, 1963)

$$B(m,n) = \int_0^1 (1-|u|)^{n-1}u^{m-1}du, \tag{220}$$

the relation (219) can be rewritten as follows

$$T_{ex} = \frac{4\left|x_0\right|^{(1-\alpha)/2}}{c_1\sqrt{2(\alpha+1)}}B(\frac{1}{\alpha+1},\frac{1}{2}). \tag{221}$$

Due to

$$B(m,n) = \frac{\Gamma(m)\Gamma(n)}{\Gamma(m+n)}, \tag{222}$$

the exact period is

$$T_{ex} = \frac{4\left|x_0\right|^{(1-\alpha)/2}}{c_1\sqrt{2(\alpha+1)}}\frac{\Gamma(\frac{1}{\alpha+1})\Gamma(\frac{1}{2})}{\Gamma(\frac{3+\alpha}{2(\alpha+1)})}, \tag{223}$$

where Γ is the Euler gamma function (Abramowitz & Stegun, 1979). Using $\Gamma(\frac{1}{2}) = \sqrt{\pi}$ (see Abramowitz & Stegun, 1979) the period expression is finally

$$T_{ex} = \frac{1}{c_1\left|x_0\right|^{(\alpha-1)/2}}\left(\frac{2\sqrt{2}\pi}{\sqrt{(\alpha+1)}}\right)\left(\frac{\Gamma(\frac{1}{\alpha+1})}{\sqrt{\pi}\Gamma(\frac{3+\alpha}{2(\alpha+1)})}\right). \tag{224}$$

Using the property of the conservative oscillator (216), for the condition $\dot{x} = 0$, according to the relation (218), the extremal amplitudes during a period of vibration are obtained as $x_{\max} = x_0$ and $\mathrm{x}_{\min} = -x_0$.

Using the exact period (224) and the statement of the constant amplitude of vibration, it is obvious that the approximate solution of (216) would satisfy the relation

$$x(0) = x(T_{ex}) = ... = x(nT_{ex}) = x_0. \tag{225}$$

Finally, the approximate solution has to satisfy not only the initial conditions (217) but also the requirements of the exact amplitude and frequency, i.e. the relation (225) for (224).

Using the procedure published in Cveticanin, 2012, the approximate solution of (212) is obtained.

The solution of the differential equation (216) with constant parameters is assumed as a cosine function

$$x = A\cos\psi, \tag{226}$$

where

$$\psi = \theta + \Omega t, \tag{227}$$

A and θ are arbitrary constants and Ω is the exact frequency of vibration

$$\Omega = \frac{2\pi}{T_{ex}} = \Omega_\alpha \omega_0 A^{\frac{\alpha-1}{2}}. \tag{228}$$

Substituting (226) and its first and second time derivatives

$$\dot{x} = -A\Omega \sin\psi, \tag{229}$$
$$\ddot{x} = -A\Omega^2 \cos\psi, \tag{230}$$

the equation (216) is approximately satisfied

$$\frac{d^2}{dt^2}(A\cos\psi) + \omega_0^2 (A\cos\psi)^{\frac{\alpha-1}{2}} \approx 0. \tag{231}$$

According to the generating solution (226) with (227) of the Eq. (216) and its first time derivative (229), the trial solution of the Eq.(213) and its time derivative are assumed as

$$x = A(t)\cos\psi(t), \tag{232}$$
$$\dot{x} = -A(t)\Omega(\tau, A(t))\sin\psi(t), \tag{233}$$

where

$$\dot{\psi} = \dot{\theta} + \Omega(\tau, A(t)), \tag{234}$$

$A(t)$, $\theta(t)$ and $\psi(t)$ are time variable functions and according to (228) it is

$$\Omega(\tau, A(t)) = \Omega_\alpha \omega(\tau)(A(t))^{\frac{\alpha-1}{2}}. \tag{235}$$

Calculating the first time derivative of (232) and equating it with expression (233), it follows

$$\dot{A}(t)\cos\psi(t) - A(t)\dot{\theta}(t)\sin\psi(t) = 0. \tag{236}$$

Substituting (232), (233) and the time derivative of (233) into (213) and using the relation (231), we obtain

$$-\dot{A}\Omega\sin\psi - A\dot{A}\frac{\partial\Omega}{\partial A}\sin\psi - \varepsilon A\frac{\partial\Omega}{\partial\omega}\frac{\partial\omega}{\partial\tau}\sin\psi - A\Omega\dot{\theta}\cos\psi$$
$$= \varepsilon f(\tau, A\cos\psi, -A\Omega\sin\psi), \tag{237}$$

where $\omega \equiv \omega(\tau), A \equiv A(t), \theta \equiv \theta(t), \psi \equiv \psi(t)$. Hence, two first order differential equations (236) and (237) replace the second order differential equation (213). Solving (236) and (237) with respect to $\dot{A}$ and $\dot{\theta}$, we have

$$\dot{A}(\Omega - A\frac{\partial\Omega}{\partial A}\sin^2\psi) = -\varepsilon A\frac{\partial\Omega}{\partial\omega}\frac{\partial\omega}{\partial\tau}\sin^2\psi - \varepsilon f(\tau, A\cos\psi, -A\Omega\sin\psi)\sin\psi, \tag{238}$$

$$A\dot{\theta}(\Omega - A\frac{\partial\Omega}{\partial A}\sin^2\psi) = -\varepsilon A\frac{\partial\Omega}{\partial\omega}\frac{\partial\omega}{\partial\tau}\sin\psi\cos\psi$$
$$-\varepsilon f(\tau, A\cos\psi, -A\Omega\sin\psi)\cos\psi. \quad (239)$$

Averaging the differential equations in the period 2π we obtain the following equations

$$\dot{A} = -\frac{2\varepsilon A}{(5-\alpha)\omega}\frac{d\omega}{d\tau} - \frac{4\varepsilon A^{\frac{1-\alpha}{2}}}{(5-\alpha)\omega\Omega_\alpha}\frac{1}{2\pi}\int_0^{2\pi} f(\tau, A\cos\psi, -A\Omega\sin\psi)\sin\psi d\psi, \quad (240)$$

$$A\dot{\theta} = -\frac{2\varepsilon A^{\frac{1-\alpha}{2}}}{(5-\alpha)\omega\Omega_\alpha}\frac{1}{2\pi}\int_0^{2\pi} f(\tau, A\cos\psi, -A\Omega\sin\psi)\cos\psi d\psi, \quad (241)$$

and from (234) and (235)

$$\dot{\psi} = \Omega_\alpha\omega A^{\frac{\alpha-1}{2}} - \frac{2\varepsilon A^{\frac{1-\alpha}{2}}}{(5-\alpha)\omega\Omega_\alpha}\frac{1}{2\pi}\int_0^{2\pi} f(\tau, A\cos\psi, -A\Omega\sin\psi)\cos\psi d\psi. \quad (242)$$

Solving the averaged differential equation (240) and substituting the obtained solution for A into (242) the approximate function ψ is obtained which gives the solution (232).

Small linear damping force acts For the special case when beside the reactive force also the linear damping force acts

$$F_x = -\varepsilon b\dot{x}, \quad (243)$$

where εb is the small damping coefficient, the function f is

$$f = -\left(\frac{b}{m} + \frac{1}{m}\frac{dm}{d\tau}\right)\dot{x}. \quad (244)$$

As the mass variation is slow and the damping coefficient is small, the reactive and damping force are also small in comparison to the elastic force. Substituting (215) and (244) into (240) and (242) the differential equation (212) transforms into a system of two averaged first order differential equations

$$\frac{\dot{A}}{A} = -\frac{\varepsilon}{(5-\alpha)m}\left(\frac{dm}{d\tau}\right) - \frac{2\varepsilon b}{(5-\alpha)\,m}, \quad (245)$$

$$\dot{\psi}(t) = \Omega_\alpha A^{\frac{\alpha-1}{2}} \sqrt{\frac{k_\alpha}{m}}, \tag{246}$$

In general, the averaged amplitude variation is the solution of (245)

$$A = A_0 \left(\frac{m_0}{m}\right)^{\frac{1}{5-\alpha}} \exp\left(-\frac{2\varepsilon b}{5-\alpha}\int \frac{dt}{m}\right), \tag{247}$$

which gives the phase angle function

$$\psi = \psi_0 + \Omega_\alpha \sqrt{k_\alpha} \left(A_0 m_0^{\frac{1}{5-\alpha}}\right)^{\frac{\alpha-1}{2}} \int m^{-\frac{\alpha-1}{2(5-\alpha)}-\frac{1}{2}} \left(\exp\left(-\frac{2\varepsilon b}{5-\alpha}\int \frac{dt}{m}\right)\right)^{\frac{\alpha-1}{2}} dt. \tag{248}$$

The amplitude and the phase of vibration vary in time due to damping, but also due to mass variation. The order of nonlinearity has a significant influence on the velocity of amplitude and phase increase or decrease.

Linear mass variation Let us consider the case when the mass variation is linear, as it is suggested by Yuste (1991)

$$m = m_0 + m_1\tau = m_0 + \varepsilon m_1 t, \tag{249}$$

where m_1 is a constant and ε is a small parameter. According to (245), we obtain the differential equation for the amplitude variation

$$\frac{\dot{A}}{A} = -\frac{\varepsilon(2b + m_1)}{(5-\alpha)m} \tag{250}$$

a) For the special parameter values, when $m_1/b = -2$ the amplitude of vibration is constant i.e.,

$$A = A_0 = const. \tag{251}$$

and the relation (248) transforms into

$$\psi = \psi_0 + \frac{2}{\varepsilon m_1}\Omega_\alpha A_0^{\frac{\alpha-1}{2}} \sqrt{k_\alpha}\left(m^{1/2} - m_0^{1/2}\right) \tag{252}$$

For this special case in spite of the action of the linear damping the amplitude of vibration is constant due to the fact that the linear mass separation makes the compensation to the effect of damping. Using the series expansion of the function ψ we have

$$\psi = \psi_0 + 2\Omega_\alpha A_0^{\frac{\alpha-1}{2}} \sqrt{\frac{k_\alpha}{m_0}} t. \tag{253}$$

The approximate value of the period of vibration is independent on the mass variation and damping coefficient, and is given as follows

$$T = \frac{2\pi}{2\Omega_\alpha A_0^{\frac{\alpha-1}{2}} \sqrt{\frac{k_\alpha}{m_0}}}. \tag{254}$$

The approximate period value depends only on the order of nonlinearity.

b) For $m_1/b \neq -2$ the amplitude-time and phase-time functions are

$$A = A_0 \left(\frac{m}{m_0}\right)^{-\frac{1}{5-\alpha}\left(1+\frac{2b}{m_1}\right)}, \tag{255}$$

and

$$\psi = \frac{\sqrt{m_0 k_\alpha}}{\varepsilon m_1} \frac{A_0^{\frac{\alpha-1}{2}} \Omega_\alpha}{\frac{1}{2} - \frac{\alpha-1}{2(5-\alpha)}\left(1 + \frac{2b}{m_1}\right)} \left(\left(\frac{m}{m_0}\right)^{\frac{1}{2}-\frac{\alpha-1}{2(5-\alpha)}\left(1+\frac{2b}{m_1}\right)} - 1\right) + \psi_0, \tag{256}$$

which give the approximate solution (232)

$$x = A_0 \left(\frac{m}{m_0}\right)^{-\frac{1}{5-\alpha}(1+\frac{2b}{m_1})}$$
$$\cos\left(\frac{\sqrt{m_0 k_\alpha}}{\varepsilon m_1} \frac{A_0^{\frac{\alpha-1}{2}} \Omega_\alpha}{\frac{1}{2} - \frac{\alpha-1}{2(5-\alpha)}(1 + \frac{2b}{m_1})} \left(\left(\frac{m}{m_0}\right)^{\frac{1}{2}-\frac{\alpha-1}{2(5-\alpha)}(1+\frac{2b}{m_1})} - 1\right) + \psi_0\right). \tag{257}$$

The amplitude and phase variation depend on the relation m_1/b, parameter m_1 and order of nonlinearity α.

Let us consider a numerical example were the order of nonlinearity is $\alpha = 4/3$, the rigidity $k_{4/3} = 1$ and the mass decrease is $m = 1 - 0.01t$, where $m_0 = 1$, $m_1 = 1$ and $\varepsilon = 0.01$. The differential equation of motion is

$$\ddot{x} + \frac{x\,|x|^{1/3}}{1 - 0.01t} = 0.01\,(1 - b)\,\dot{x}, \tag{258}$$

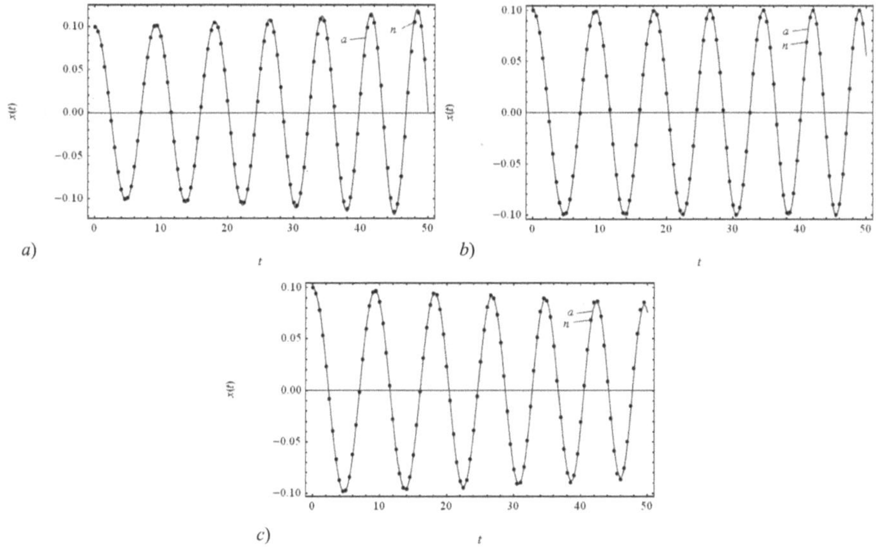

Figure 5. The $x - t$ diagrams obtained analytically (a – full line) and numerically (n – dot line) for: a) $b = 0$, b) $b = 1/2$ and c) $b = 1$.

where b is the damping coefficient. For the initial conditions $x(0) = A_0 = 0.1$ and $\dot{x}(0) = 0$ the analytical solution (257) has the form

$$x = \frac{0.1}{(1 - 0.01t)^{0.27273(1-2b)}} \cos\left(\frac{66.028}{0.5\,(1 - 0.0909(1-2b))}\left(1 - (1 - 0.01t)^{0.5(1-0.0909(1-2b))}\right)\right). \tag{259}$$

In Fig.5 the approximate solution (259) and the numerical solution of (258), obtained by using of the Runge-Kutta procedure, are plotted. The $x - t$ diagrams for various values of the damping parameter b are shown.

It can be concluded that for $b = 1/2$ the amplitude of vibration is constant as it is previously stated (see Eq. (248)). For the case when the damping is neglected ($b = 0$), due to mass decrease and existence of the reactive force, the amplitude of vibration increases. For certain damping ($b = 1$) which is higher than the limit value ($b = 1/2$) the amplitude of vibration decreases. The analytical solution is in a very good relation to the numeric one in spite of the long time interval of consideration.

Linear oscillator For the linear oscillator when $\alpha = 1$

$$A = A_0 \left(\frac{m_0}{m} \right)^{\frac{1}{4}(1+\frac{2b}{m_1})} \tag{260}$$

and

$$\psi = \frac{2\sqrt{k_1}}{\varepsilon m_1} \left(\sqrt{m} - \sqrt{m_0} \right) + \psi_0 \tag{261}$$

If the damping parameter is zero, i.e., $b = 0$, the amplitude variation is $A = A_0 \left(\frac{m_0}{m} \right)^{1/4}$. Using the series expansion of the functions in (261) the approximate frequency of vibration is $\sqrt{k_1/m_0}$ which corresponds to the systems with constant mass and without damping.

Influence of the reactive force Let us analyze the influence of the reactive force described with the function

$$f = -\frac{1}{m} \frac{dm}{d\tau} \dot{x}. \tag{262}$$

Substituting (262) into (247) and (248) we obtain the variation of the amplitude

$$A = A_0 \left(\frac{m_0}{m} \right)^{\frac{1}{5-\alpha}}, \tag{263}$$

and of the phase angle function

$$\psi = \psi_0 + \Omega_\alpha \sqrt{k_\alpha} \left(A_0 m_0^{\frac{1}{5-\alpha}} \right)^{\frac{\alpha-1}{2}} \int m^{-\frac{\alpha-1}{2(5-\alpha)} - \frac{1}{2}} dt. \tag{264}$$

For the certain order of nonlinearity α the amplitude of vibration increases with decreasing of the mass in time. If the mass increases, the amplitude of vibration decreases for the oscillator of the certain degree of nonlinearity. For the linear oscillator, when $\alpha = 1$, the amplitude variation is, as previously published by Bessonov 1967,

$$A = A_0 \left(\frac{m_0}{m} \right)^{1/4}, \tag{265}$$

and for the pure cubic oscillator with cubic nonlinearity (see Cveticanin, 1992)

$$A = A_0 \left(\frac{m_0}{m} \right)^{1/2}. \tag{266}$$

If the mass increases, the amplitude decreases faster for higher order of nonlinearity. If the mass decreases the amplitude increases faster for smaller order of nonlinearity.

For the case when the relative velocity of the adding or separated mass is zero and the reactive force is zero, or for the case when the reactive force is sufficiently small and can be omitted, the amplitude and phase angle functions are according to (240), (242) and (215)

$$\dot{A} = \frac{A\dot{m}}{(5-\alpha)m}, \tag{267}$$

and

$$\dot{\psi} = \Omega_\alpha \sqrt{\frac{k_\alpha}{m}} A^{\frac{\alpha-1}{2}}. \tag{268}$$

Integrating the relations (267) and (268) for the initial amplitude A_0, phase angle ψ_0 and mass m_0, it follows

$$A = A_0 \left(\frac{m}{m_0}\right)^{\frac{1}{5-\alpha}}, \tag{269}$$

$$\psi = \psi_0 + \Omega_\alpha \sqrt{k_\alpha} \left(A_0 m_0^{-\frac{1}{5-\alpha}}\right)^{\frac{\alpha-1}{2}} \int (m)^{\frac{\alpha-1}{2(5-\alpha)}-\frac{1}{2}} dt. \tag{270}$$

Due to (232) and the relations (269) and (270), the approximate solution is

$$x = A_0 \left(\frac{m}{m_0}\right)^{\frac{1}{5-\alpha}} \cos\left(\psi_0 + \Omega_\alpha \sqrt{k_\alpha} \left(A_0 m_0^{-\frac{1}{5-\alpha}}\right)^{\frac{\alpha-1}{2}} \int (m)^{\frac{\alpha-1}{2(5-\alpha)}-\frac{1}{2}} dt\right). \tag{271}$$

Analyzing the relation (269) it is obvious that for the same order of nonlinearity α, the amplitude of vibration increases by increasing of the mass. Besides, for the same mass variation, the amplitude increases faster for higher orders of nonlinearity.

7.2 Conclusion

Due to previous consideration it can be concluded:

1. The vibration of the oscillator with monotone time variable parameter has time variable amplitude and phase. The free vibrations for all of the oscillators with a strong nonlinearity of any order and with the certain monotone slow time variable parameters are qualitatively the same independently on the order of the nonlinearity. The order of nonlinearity quantitatively changes the amplitude and the phase of vibrations but has no influence on the character of vibrations. Namely, for certain parameter variation the higher the order of nonlinearity, the faster or slower is the amplitude and phase increase or decrease. The tendency of increase or decrease of amplitude and phase i.e., frequency of vibration variation is not

directed by the order of nonlinearity but with the type of time parameter variation.

2. It is evident that in the oscillator with variable mass for the special relation between the coefficient of damping and parameter of mass variation (which affects the reactive force) the amplitude of vibration is constant, but the phase angle varies independently on the order of nonlinearity.

3. The approximate solution of the nonlinear differential equation with strong nonlinearity of any order (integer or non-integer) and time variable parameter can be obtained analytically.

4. The approximate analytic method for solving the differential equation based on the exact solution of the corresponding differential equation with constant parameters and strong nonlinearity of any order (integer or non-integer) gives very accurate results in comparison to the numerical one.

5. The solving method based on the approximate solution with exact period of vibration of the corresponding oscillator with constant parameter gives very convenient results for the oscillator with time variable parameters. For technical purpose the solution is accurate enough and appropriate for practical use. This solution has the form of trigonometric function and satisfies the requirements for simplicity and usefulness for application in techniques.

The vibrations of the mass variable systems are widely investigated by Leach, 1983; Abdalla, 1986_1; Abdalla, 1986_2; Crespo *et al.*, 1990, Xie *et al.*, 1995; Sanchez-Otiz & Salas-Brits, 1995; Flores *et al.*, 2003., too. The results published in this Chapter and in the mentioned papers are applied in dynamics of mechanisms and rotors with time variable mass (see Cveticanin, 1998_2).

8 Bibliography

M.S. Abdalla. Canonical treatment of harmonic oscillator with variable mass. *Physical Review A,* 33 (5):2870-2876, 1986_1.

M.S. Abdalla. Time-dependent harmonic oscillator with variable mass under the action of a driving force. *Physical Review A,* 34 (6):4598-4605, 1986_2.

M. Abramowitz, I.A. Stegun. *Handbook of Mathematical Functions with Formulas, Graphs and Mathematical Tables.* Nauka, Moscow, 1979. (in Russian)

N.G. Apykhtin, V.F. Jakovlev. On the motion of dynamically controlled systems with variable masses. *Prikladnaja Matematika i Mehanika,* 44 (3):427-433, 1980.

A.G. Azizov. On the motion of a controlled system of variable mass.

Prikladnaja Matematika i Mehanika, 50 (4):567-572, 1986.

A.P. Bessonov. *Osnovji dinamiki mehanizmov s peremennoj massoj zvenjev.* Nauka, Moscow, 1967.

A. Cayley. On a class of dynamical problems. *Proceeding of the Royal Society of London*, III: 506-511, 1857.

A. Cayley. On a class of dynamical problems. *Phylosophical Magazine and Journal of Science*, XV:306-310, 1858.

A. Cayley. *The collected mathematical papers*, IV(225):7-11, 1859.

J.W. Cornelisse, H.F.R. Schoyer, K.F. Wakker. *Rocket Propulsion and Spaceflight Dynamics.* Pitman, London, 1979.

G. Crespo, A.N. Proto, A. Plastino, D. Otero. Information-theory approach to the variable-mass harmonic oscillator. *Physical Review A,* 42 (6):3608-3617, 1990.

L. Cveticanin. Vibrations of a textile machine rotor. *Journal of Sound and Vibration*, 97 (2):181-187, 1984.

L. Cveticanin. The stability of a textile machine rotor with increasing mass. *Mechanism and Machine Theory*, 23 (4):275-278, 1988.

L. Cveticanin. Stability of a clamped-free rotor with variable mass for the case of radial rubbing. *Journal of Sound and Vibrations*, 129 (3):489-499, 1989.

L. Cveticanin. The oscillations of a textile machine rotor on which the textile is wound up. *Mechanism and Machine Theory*, 26 (3):253-260, 1991.

L. Cveticanin. The influence of the reactive force on a nonlinear oscillator with variable parameter. *Journal of Vibrations and Acoustics, Trans. ASME*, 114 (4):578-580, 1992.

L. Cveticanin. An approximative solution of a coupled differential equation with variable parameter. *Trans. ASME, J. of Applied Mechanics*, 60(1):214-217, 1993_1.

L. Cveticanin. Conservation laws in systems with variable mass. *Journal of Applied Mechanics, Trans. ASME,* 60 (4):954-958, 1993_2.

L. Cveticanin. The influence of the reactive force on the motion of the rotor on which the band is winding up. *Journal of Sound and Vibration*, 167 (2):382-384, 1993_3.

L. Cveticanin. Some conservation laws for orbits involving variable mass and lineat damping. *Journal of Guidane, Control and Dynamics*, 17 (1):209-211, 1994.

L. Cveticanin. Approximate solution of a time-dependent differential equation. *Meccanica*, 30:665-671, 1995.

L. Cveticanin. On the stability of rheo-linear rotor systems based on some new first integrals. *Mechanics, Research Communications, Basic and Applied*, 23 (5):519-530, 1996_1.

L. Cveticanin. The influence of the reactive force on the stability of motion for one-degree-freedom mechanisms with variable mass. *Machine Vibration*, 5:224-228, 1996_2.

L. Cveticanin. Self-excited vibrations of the variable mass rotor/fluid system. *Journal of Sound and Vibration*, 212 (4):685-702, 1998_1.

L. Cveticanin. *Dynamics of machines with variable mass.* Gordan and Breach Science Publishers, London, 1998_2.

L. Cveticanin. Dynamic buckling of a single-degree-of-freedom system with variable mass. *European Journal of Mechanics, A/Solids*, 20 (4):661-672, 2001.

L. Cveticanin. A qualitative analysis of the quasi-linear one-degree-of-freedom system. *European Journal of Mechanics, A/Solids,* 23:667-675, 2004.

L. Cveticanin. Particle separation form a four particle system. *European Journal of Mechanics A/Solids*, 26:270-285, 2007.

L. Cveticanin, I. Kovacic. On the dynamics of bodies with continual mass variation. *Trans ASME, Journal of Applied Mechanics,* 74:810-815, 2007.

L. Cveticanin, Dj. Djukic. Motion of body with discontinual mass variation. *Nonlinear Dynamics*, 52 (3):249-261, 2008.

L. Cveticanin. Dynamics of body separation – Analytical procedure. *Nonlinear Dynamics*, 55 (3):269-278, 2009_1.

L. Cveticanin. Oscillator with fraction order restoring force. *Journal of Sound and Vibration*, 320, (4-5):1064-1077, 2009_2.

L. Cveticanin, T. Pogany. Oscillator with a sum of non-integer order non-linearities. *Journal of Applied Mathematics*, vol. 2012, art. no. 649050, 20 pages, 2012.

L. Cveticanin. Oscillator with non-integer order nonlinearity and time variable parameters. *Acta Mechanica*, 223 (7):1417-1429, 2012.

S. Djerassi. An algorithm for simulation of motions of 'variable mass' systems. *Advances in the Astronautical Sciences,* 99 (1):461-474, 1998.

Ch. Dufour. Sur l'acceleration seculaire du mouvement de la lune. *Comptes rendus des Seances de l'Ac. des Sc.*, LXII:840-842, 1866.

F.O. Eke, T.C. Mao. On the dynamics of variable mass systems. *The International Journal of Mechanical Engineering Education*, 30 (2), 2002.

J. Flores, G. Solovey, S. Gill. Variable mass oscillator. *American Journal of Physics,* 71 (7):721-725, 2003.

Z.-M. Ge, Y.H. Cheng. Extended Kane's equations for nonholonomic variable mass system. *Journal of Applied Mechanics, Trans. ASME,* 49 (2):429-431, 1982.

Z.-M. Ge. Equations of motion of nonlinear nonholonomic variable mass system with applications. *Journal of Applied Mechanics, Trans. ASME,* 51 (2):435-437, 1984.

H. Goldstein. *Classical Mechanics.* Addison-Wesley, Reading, 1980.

L. N., Grudtsyn. Plane perturbed motion of a material point of variable mass. *PMM*, 36 (1):172-174, 1972.

H. Gylden. Die Bahnbewegungen in einem Systeme von zwei Koerpern in dem Falle das Massen Veraenderungen unterworfen sind. *Astronomische Nachrichten*, 109 (2593):1-6, 1884.

J.E. Howard. Particle dynamics with variable mass and charge. *Physics Letters, Section A: general, Atomic and Solid State Physics*, 366 (1-2):91-96, 2007.

A.O. Ignat'yev. On the instability of an equilibrium point of a linear oscillator with variable parameters. *Prikladnaja Matematika i Mehanika*, 55 (4):701-703, 1991.

Ya.F. Kayuk, A. Tilavov. Motion of an elastically suspended solid of variable mass. *Prikladnaja Mekhanika*, 23 (15):102-109, 1987.

Ya.F. Kayuk, A. Akhmedov. Spatial motion of an elastically suspended cylindrical body of variable mass. *Prikladnaja Mekhanika*, 28 (7):62-69, 1992.

Ya.F. Kayuk, V.I. Denisenko. Motion of a mechanical system with variable mass - inertial characteristics. *International Applied Mechanics*, 40 (7):814-820, 2004.

P.G.L. Leach. Harmonic oscillator with variable mass. *Journal of Physics A: General Physics.* 16 (14), art.no. 019:3261-3269, 1983.

C. Leubnert, P. Krumm. Lagrangians for simple systems with variable mass. *European Journal of Physics*, 11 (1) art.no.005:31-34, 1990.

T. Levi Civita. Sul moto di un corpo di massa variable. *Rendiconti del Lincel*: 329-333, 621-622, 1928.

L.G. Luk'yanov. Conservative two-body problem with variable masses. *Astronomy Letters*, 31 (8):563-568, 2005.

J.J. McPhee, R.N. Dubey. Dynamic analysis and computer simulation of variable-mass multi-rigid-body sysems. *International Journal for Numerical Methods in Engineering*, 32 (8):1711-1725, 1991.

L. Meirovitch. general motion of a variable-mass flexible rocket with internal flow. *Journal of Spacecraft and Rockets*, 7 (2):186-195, 1970.

I.V. Meshcherski. Odin chasnij sluchaj zadachi Gouldena. *Astronomische Nachrichten*, 132(3153):9, 1893.

I.V. Meshchersky. Dinamika tochki peremennoj massji. *Magistarskaja disertacija*, Peterburgski Universitet, Petersburg, 1897.

I.V. Meshcherskij. *Rabotji po mehanike tel peremennoj massji.* Gos.Izd. tehniko-teoret.lit, Moscow, 1952.

R.E. Mickens. *Truly Nonlinear Oscillations.* World Scientific, Singapore, 2010.

A. Oppalzer. Ueber die Ursache, welche den Unterschied zwischen der theoretischen berechneten Secularacceleration in der Laenge des Mondes und der thatsaechlichen bedingen kann. *Astronomische Nachrichten*, 108 (2573):67-72, 1884.

C.P. Pesce. The application of Lagrange equations to mechanical systems with mass explicitly dependent on position. *Journal of Applied Mechanics, Trans. ASME,* 70 (5):751-756, 2003.

R. Rosenberg. *The ateb(h)-functions and their properties. Quarterly of Applied Mathematics*, 21 (1):37-47, 1963.

G.I. Sanchez-Ortiz, A.L. Salas-Brito. Chaos in a variable mass relaxation oscillator model for the leaky tap. *Physica D: Nonlinear Phenomena,* 89 (1-2):151-168, 1995.

V.M. Starzhinskii. *An Advanced Course of Theoretical Mechanics.* Mir Publishers, Moscow, 1982.

T. Tran, F.O. Eke. Effects of internal mass flow on the attitude dynamics of variable mass systems. *Advances in the Astronautical Sciences*, 119, (Issue SUPPL.):1297-1316, 2005.

S.M. Wang, F.O. Eke. Rotational dynamics of axisymmetric variable mass systems. *Journal of Applied Mechanics*, 62 (4):970-974, 1995.

E.W. Weisstein. *Lambert W-function.* From MathWorld-AWolfram Web Resource, http://mathworld.wolfram.com/LambertW-Function.html

G.-Q. Xie, S.-W. Qian, Z.-Y. Gu. Separation of variables treatment of the time-dependent damped harmonic oscillator with an arbitrary varying mass and with a force quadratic in the velocity under the action of an arbitrary time-varying force. *Physics Letters A,* 207 (1-2):11-16, 1995.

S.B. Yuste. On Duffing oscillators with slowly varying parameters. *International Journal of Non-Linear Mechanics*, 26 (5):671-677, 1991.

Mechanics of multi-component media with exchange of mass and non-classical supplies

Dmitry Indeitsev and Yulia Mochalova
Institute of Problems in Mechanical Engineering of the Russian Academy of Sciences, Saint-Petersburg, Russia

Abstract An approach for description of internal evolution processes in materials basing on the Euler equations and the mass balance equations containing source terms is proposed. Dynamics of the complex materials such as structured liquids in nanochannels, metals with dissolved hydrogen and various impurities are discussed within a two-component continuum model. The effect of stress state on the internal structure of the materials is investigated.

1 Motivation and some examples

The classical equations of continuum include the equations of momentum and mass balance and the equation of state. As a rule, the mass balance equation after linearization is not used for further solution of problems. However, in some cases the mass balance equation plays a very important role. Primarily these are problems examining materials with a complex internal structure – materials with various impurities, structured liquids in nanochannels, metals with dissolved hydrogen. In these cases, we must begin the research with the mass balance equations and the analysis of source terms.

One of possible approaches, which allows us to consider the influence of internal degrees of freedom on structural reconstructions at a material, is the description of unknown particle kinematics of continuous media by phenomenological transport equations of substance (for example Fick' laws). The diffusion equations, that describe the relative motion of particles within a representative volume, should complement the basic equations of motion of a continuous media. This usually leads to the following difficulties.

First, the diffusion transfer of mass (which can change the internal structure of the material) can depend on the stress state of the material. This leads to necessity to choose diffusion coefficients using experimental data and thus makes it is impossible to build a general mathematical model describing materials with complex structures.

H. Irschik, A. K. Belyaev (Eds.), *Dynamics of Mechanical Systems with Variable Mass*,
CISM International Centre for Mechanical Sciences DOI 10.1007/ 978-3-7091-1809-2_4

Secondly, it is known that the mass diffusion transfer inside material can cause changes in the internal structure of the material. It is not derived directly from the phenomenological equations. Usually to describe this phenomenon, we have to introduce artificial parameters associated with the concentration change in the equation of state of the material.

The aim of those lectures is to propose a different approach which allows us to describe internal evolution processes in the material. The approach is based on usage the Euler equations and the mass balance equations containing source terms. Choosing by some means the source terms that determine the mass transfer between a moving substance and the medium, we can derive the equation of state of the substance. Besides the effect of stress state on evolution processes in the material (the diffusion of impurities) taken into account by introduction to the basic equations of dissipative term with a coefficient depending on the spherical part of the strain tensor. Then the classical evolution equations arise as a particular case within our approach.

In this section we show how this approach may be used to some model problems. In 1.2 we give a brief exposition of the law of particles conservation. An example of a chemical adsorption of substances is given in 1.3. Selecting the source terms in the mass balance equation allows us to control the process of adsorption. In 1.4 we discuss the connection of rheological models of materials and continuum models and shows how the choice of the source terms can affect the equation of state of the material. In 1.5 using continuum mechanics we obtain diffusion equation and introduce the resistance force to the diffusion flux being proportional to its velocity. The diffusion coefficient depends on the normal deformation basic media.

1.1 Particle balance and mass balance equation

Let $\eta(\boldsymbol{r}, t)$ be the number density at a given point $\boldsymbol{r}$ of an inertial reference system. Specifying by $\mathrm{d}N$ the number of particles per unit volume $\mathrm{d}V$ we can write

$$\mathrm{d}N = \eta(\boldsymbol{r}, t)\mathrm{d}V, \quad \eta \geq 0.$$

The mass density $\rho(\boldsymbol{r}, t)$ and the number density are connected by the relation

$$\rho = m\eta, \tag{1}$$

where m is the mass of one particle. Let V be a volume in the reference frame and the boundary of V be a closed surface $S = \partial V$. It is assuming that the total number of particles in medium remains unchanged and we can formulate the following particle balance equation

$$\frac{\mathrm{d}}{\mathrm{d}t}\int_V \eta(\boldsymbol{r},t)\mathrm{d}V = -\int_S \eta\boldsymbol{n}\cdot\boldsymbol{v}\mathrm{d}S = -\int_V \nabla\cdot(\eta\boldsymbol{v})\mathrm{d}V, \tag{2}$$

where $\boldsymbol{v}$ is the velocity of particles. In the local form equation(2) can be written

$$\frac{\partial\eta}{\partial t} + \nabla\cdot(\eta\boldsymbol{v}) = 0. \tag{3}$$

Using (1) we can get the mass balance equation

$$\frac{\partial\rho}{\partial t} + \nabla\cdot(\rho\boldsymbol{v}) = 0. \tag{4}$$

If the density of particles can be changed, the particle balance equation should be modified as follows

$$\begin{aligned}\frac{\mathrm{d}}{\mathrm{d}t}\int_V \eta(\boldsymbol{r},t)\mathrm{d}V &= \int_V \chi(\boldsymbol{r},t)\mathrm{d}V - \int_S \eta\boldsymbol{n}\cdot\boldsymbol{v}\mathrm{d}S \\ &= \int_V \big[\chi(\boldsymbol{r},t) - \nabla\cdot(\eta\boldsymbol{v})\big]\mathrm{d}V.\end{aligned} \tag{5}$$

Here the function χ is the rate of production (destruction) of particles at a point of the reference frame. Then equations (3) and (4) take the form

$$\frac{\partial\eta}{\partial t} + \nabla\cdot(\eta\boldsymbol{v}) = \chi \tag{6}$$

or

$$\frac{\partial\rho}{\partial t} + \nabla\cdot(\rho\boldsymbol{v}) = J. \tag{7}$$

Here the functions χ and $J = m\chi$ are co-called source teams and characterize the rate of mass production (destruction) of particles.

Now we present some examples where equation (7) is used for description of various phenomena.

1.2 Adsorption of impurities

The first example is a chemical adsorption of substances (see Whitham (1974)). The situation is that a fluid carrying dissolved substances or particles (impurities) flows through a fixed bed and impurities being carried is partially adsorbed on the fixed solid material in the bed. The fluid flow is idealized to have a constant velocity v. Then if ρ_f is the density of the

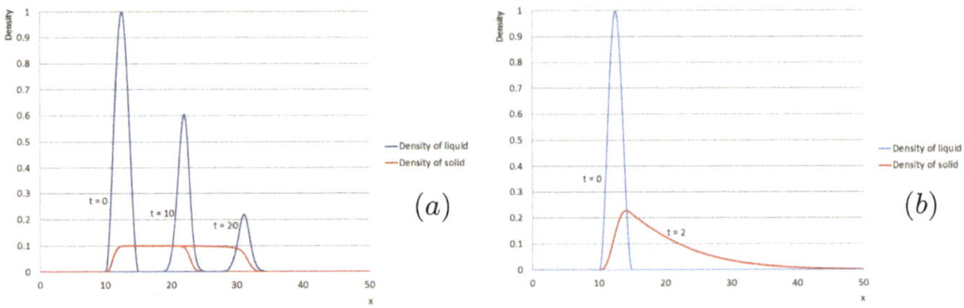

Figure 1. The density of the adsorbed substance: (a) the uniform distribution $v = 1$, $k_1 = 4$, $k_2 = 0$, $A = 0.1$, $B = 1$; (b) the uneven distribution $v = 20$, $k_1 = 4$, $k_2 = 0$, $A = 0.7$, $B = 1$.

material carried in the fluid, and ρ_s is the density deposited on the solid, the conservation equations have the forms

$$\frac{\partial \rho_s}{\partial t} = J, \tag{8}$$

$$\frac{\partial \rho_f}{\partial t} + \frac{\partial}{\partial x}(\rho_f v) = -J, \tag{9}$$

where the source term is as follows:

$$J = k_1(A - \rho_s)\rho_f - k_2\rho_s(B - \rho_f). \tag{10}$$

The first term of (10) represents deposition from the fluid to the solid at a rate proportional to the amount in the fluid, but limited by the amount being already on the solid up to the capacity A. The second term is the reverse transfer from the solid to the fluid. The numerical solution of the system (8)–(10) is carried out. Depending on the system parameters, we have the uniform distribution of the adsorbed substance (Figure 1a) and the non-uniform distribution of the adsorbed substance (Figure 1b).

Let us find the analytical solution under certain simplifications. The system (8)–(10) can be rewritten as

$$\frac{\partial}{\partial t}(\rho_f + \rho_s) + v\,\frac{\partial \rho_f}{\partial x} = 0,$$
$$\frac{\partial \rho_s}{\partial t} = k_1\rho_f(A - \rho_s) - k_2\rho_s(B - \rho_f).$$

For relatively slow changes in the densities and relatively high reaction rates k_1, k_2, the second equation is taken in the approximate quasi-equilibrium form in which the $\partial \rho_s / \partial t$ is neglected and

$$\rho_s(\rho_f) = \frac{k_1 A}{k_2 B + (k_1 - k_2)\rho_f}.$$

Substituting this expression into the first equation of the system (8)–(10) yields

$$\frac{\partial \rho_f}{\partial t} + c(\rho_f)\frac{\partial \rho_f}{\partial x} = 0, \quad c(\rho_f) = \frac{v}{1 + \partial \rho_s / \partial \rho_f}.$$

Thus, the density changes propagate at the speed of $c(\rho_f)$. If the densities concerned are small, the value of $c(\rho_f)$ is approximately equal to

$$c = \frac{k_2 B}{k_1 A + k_2 B} v.$$

The propagation speed depends on the reaction rates involved, being slower for substances with larger attraction toward the solid.

1.3 Rheological models of materials. The equations of state and source terms

The concept of a rheological model of a material is given by Reiner (1958); Palmov (1998). Rheological models are often used to describe the materials with complex internal structure, in particular dispersed systems of two or three phases. Rheology considers such materials as homogeneous, the mechanical properties of which coincide with the properties of real materials. A mathematical model of the mechanical properties of the material is given by the constitutive equation (the equation of state). To compose the constitutive equations for materials with complex rheological properties each basic property material is modeled by suitable rheological element. For example the elasticity is simulated by an elastic spring (the Hooke element), the viscosity is simulated by a viscous damper (the Newton element) and the plasticity is simulated by a dry friction damper (the St.Venant element). By combining the fundamental rheological elements either in series or parallel we form the rheological model with the complex material properties.

Our aim is to show the connection of rheological models of materials and continuum models. Consider as an example the Maxwell material. This viscoelastic material whose rheological model consists of a Hooke element and a Newton element in series. The constitutive equations for this material in the simplest one-dimensional case is as follows:

$$\dot{\varepsilon} = \frac{\dot{\sigma}}{E} + \frac{\sigma}{\mu}, \tag{11}$$

where ε is the strain, σ is the mean normal stress, E is the elastic modulus, μ is the viscosity coefficient. It is shown that the rheological equation of state (11) can be obtained from the mass conservation law (7). The rheological material has a complex internal structure and the mass exchange between the material and the medium is possible, for example, due to change in a phase state of the material. Suppose that the material is initially at rest with the density ρ_0 and the perturbation quantity $\widetilde{\rho} = \rho - \rho_0$ and the velocity v are small. Then, linearizing equation (7) and expressing the velocity in terms of the displacement u, we obtain

$$\frac{\partial \widetilde{\rho}}{\partial t} + \rho_0 \frac{\partial^2 u}{\partial x \partial t} = J.$$

Assuming $\varepsilon = \partial u / \partial x$, it yields

$$\frac{\partial \widetilde{\rho}}{\partial t} + \rho_0 \frac{\partial \varepsilon}{\partial t} = J. \tag{12}$$

The mean normal stress is assumed to be a function of relative mass density ρ_0/ρ

$$\sigma = k \left(1 - \frac{\rho}{\rho_0} \right). \tag{13}$$

The coefficient k links the stress to a change of mass density. Substituting equation (13) into (12), we find

$$\dot{\varepsilon} = \frac{\dot{\sigma}}{k} + \frac{J}{\rho_0}. \tag{14}$$

In order to obtain the equation of state of the material we need to define the source term J. It is done as follows:

$$J = -\alpha \widetilde{\rho}, \quad \alpha \geq 0,$$

where the coefficient α determines the rate of exchange processes between the material and the medium (in this case dissipation of the material) and, taking into account equation (13) we arrive at the following expression

$$J = \frac{\alpha \rho_0}{k} \sigma,$$

Then, substituting the last expression in equation (14), we get

$$\dot{\varepsilon} = \frac{\dot{\sigma}}{k} + \frac{\alpha}{k} \sigma. \tag{15}$$

The constitutive equation (15) describes the Maxwell material by using a mass balance equation. The equation (15) coincides with the rheological constitutive equation (11), when $E = k$ and $\mu = k/\alpha$. It is noted that if there are no exchange processes between the material and the medium $\alpha = 0$ ($J = 0$), then (15) is the equation of state describing the Hooke material.

The rheological constitutive equation for the Kelvin–Voigt material (viscoelastic material with the rheological model that represents a parallel connection of a Hooke element and a Newton element) can be obtained from a continuous two-component model of the media.

1.4 Diffusion equation

Consider a flux of particles moving in a media (the penetration of a substance, such as a liquid, gas, or vapor, through a solid or another liquid; motion special liquids). Assume that we can neglect the exchange of particles between the medium and diffusion flux. The diffusion equation is generally obtained from the law of mass conservation (4), assuming that the mass diffusion flux is proportional to the density gradient (Fick's first law)

$$\rho \boldsymbol{v} = -\kappa_D \nabla \rho,$$

and the corresponding diffusion equation (Fick's second law) is

$$\frac{\partial \rho}{\partial t} = \kappa_D \nabla^2 \rho.$$

The constant κ_D is the diffusion coefficient. It depends on the properties of the media and the type of the diffusion liquid. We obtain the diffusion equation using the system of equations describing the motion of a homogeneous substance (liquid). For this purpose we consider fluxes in which the entropy of a liquid element is constant. Then, the system of equations of a flux motion is determined by the law of mass conservation, described by equation (4), and the equation of dynamics is as follows:

$$\rho \frac{\mathrm{d}\boldsymbol{v}}{\mathrm{d}t} = \rho \boldsymbol{F} - \nabla p, \tag{16}$$

where p is the pressure at the point of the diffusion flux, $\boldsymbol{F}$ is the mass external force. Equations (4) and (16) should be supplemented by an equation of state, defining the density as a function of pressure

$$\rho = \rho(p). \tag{17}$$

The specific form of the equation of state is related with the nature of matter flux. It introduces into equation (16) a dissipative term, i.e. we assume that the force of interaction between matter flux and the medium is proportional to the velocity of the flux and equation (16) can be written as

$$\rho \frac{\mathrm{d}\boldsymbol{v}}{\mathrm{d}t} = \rho \boldsymbol{F} - \nabla p - \beta \boldsymbol{v}. \tag{18}$$

The force $\beta \boldsymbol{v}$ is obtained by linearization of the known formula used in hydraulics, see Loitsyansky (1987), where the square-law of resistance is given. The coefficient β, which can depend on the stress state of the medium (strain field), is defined below. Supposing that the forces of inertia can be considered as negligibly small (examples of such flows, see Batchelor (1967)), $\boldsymbol{v}$ is defined by equation (18) and substituting it into equation (3), we can be obtained

$$\frac{\partial \rho}{\partial t} = \nabla \cdot \Big(\frac{1}{\beta} \nabla p \Big) - \rho \nabla \cdot \boldsymbol{F} - \boldsymbol{F} \cdot \nabla \rho. \tag{19}$$

Usually the mass external force occurs under the force of gravity $\nabla \cdot \mathbf{F} = 0$ and the last term in equation (19) is negligibly small, then

$$\frac{\partial \rho}{\partial t} = \nabla \cdot \Big(\frac{1}{\beta} \nabla p \Big). \tag{20}$$

Suppose that the substance is initially at rest with pressure p_0 and density ρ_0. Assuming the density values of the perturbation $\widetilde{\rho} = \rho - \rho_0$ and pressure $\widetilde{p} = p - p_0$ are small, we can assume that the equation of state has the form

$$\widetilde{p} = c_0^2 \, \widetilde{\rho}, \tag{21}$$

where c_0 is speed of propagation of sound waves in the matter flux. Substituting equation (21) into (20), we obtain

$$\frac{\partial \rho}{\partial t} = \nabla \cdot \Big(\frac{c_0^2}{\beta} \nabla \rho \Big), \quad \beta = k \frac{\rho_0}{D(\varepsilon)}, \tag{22}$$

where the coefficient k is a dimensionless number which depends on physical state of the medium, $D = D(\varepsilon)$ is the size of through passage section, which can depend on a stress state of the medium. Indeed, for structures under uniaxial compression the size of flow section can change and to determine the dependence $D = D(\varepsilon)$ it is necessary to consider the state of stress the basic medium. Thus, the obtained equation (22) is the diffusion equation of the flux of matter, taking into account the influence of the stress state of the medium. In the absence of deformation of the medium $D = D_0$, equation (22) coincides with the classical diffusion equation (19), where

$$\varkappa_D = k\frac{c_0^2 \rho_0}{D_0}. \tag{23}$$

Introduction of the dissipative terms and the resistance coefficient β are discussed further in detail to describe the motion of fluid flow in nanochannels (Section 3) and the motion of mobile hydrogen particles (Section 4).

1.5 Conclusion to section 1

The approach which allows us to describe internal evolution processes in the material with the help of the Euler equations and the mass balance equations containing source terms is proposed. Now it is of interest to extend the results to models of two-component continuum. According to the remarks from above the lectures are organized as follows. A mechanical two-component model of the solid of complex structure is presented in Section 2. This model is used in Sections 3 and 4. In Section 3 we propose a mathematical model of a fluid flow in a two-dimensional nanochannel, which is caused by the motion of one of the confining walls parallel to the other immovable wall. The two-component model of the material, in which atomic hydrogen dissolved, is constructed in Section 4.

2 Two-component model of medium

The rational mechanics of continuous medium ignores such an important physical property of any real material as its discrete structure. It is clear that the model of a solid in the framework of the rational mechanics should have a complex structure in order to reflect the properties of discrete structure of the matter. Such a complex structure is determined by the presence of internal degrees of freedom and the influence of dynamics of the material. The presence of these degrees of freedom can result in change of the basic macroparameters which are usually used for description of the material by means of the classical equations of continuum mechanics.

As shown by Sobolev (1991); Sobolev (1997), one of the approaches to description of the continuous media behaviour is to introduce two-component models. These models allow one to explain some physical phenomena which have not been properly understood. In particular, these are the question a fluid flow behaviour in nanochanels, problem of hydrogen diffusion in metals.

The classical approaches usually introduce additional parameters in the constitutive equations. In this case the required phenomenological relations

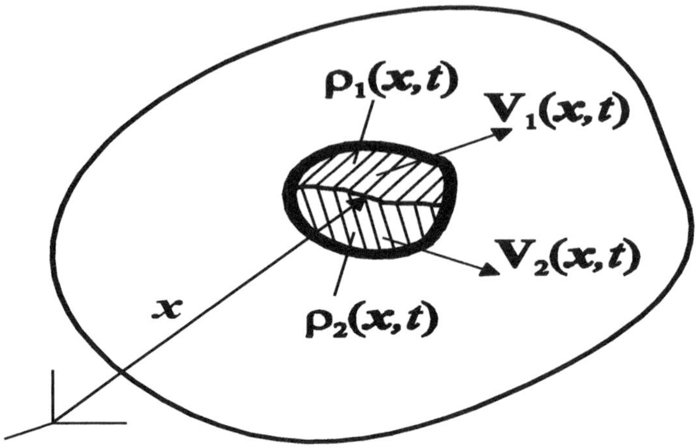

Figure 2. A schematics of the two-component model.

allow determining these new variables. The basic equations for the two-component model introduced in this section emphasize the essential role of the internal structure of the material and enable description of the above-mentioned physical phenomena.

2.1 The basic assumptions and equations

We postulate a model of the material with a carrying medium whose components are particles described by the displacement vector $\boldsymbol{u}_1(x,t)$. An additional set of particles interacting with each other and with the carrying medium is attached to the carrying medium. The absolute displacement of particles of this additional medium is given by vector $\boldsymbol{u}_2(x,t)$. Both sets are supposed to be mutually penetrating continuous media. In other words, we introduce the concept of the material point that has a complex structure and consists of two components. In the expressions for displacements the argument $\boldsymbol{x}$ is the position vector of the material point in actual configuration, i.e. Euler's description is taken, see Figure 2.

Physically, the different components of the material occupy different spatial volumes. In this regard there arises a question of the conditions of interaction of parts of the introduced solid on their internal boundaries. An axiomatic construction of the model reduces to assignment of interaction force $\boldsymbol{R}$ and the mass exchange J between the components. Realization of these representation results in a two-component model.

The law of mass conservation in the local form for each component and for the overall material is supposed to hold:

$$\frac{\partial \rho_1}{\partial t} + \nabla \cdot (\rho_1 \boldsymbol{v}_1) = J, \tag{24}$$

$$\frac{\partial \rho_2}{\partial t} + \nabla \cdot (\rho_2 \boldsymbol{v}_2) = -J, \tag{25}$$

$$\frac{\partial \rho}{\partial t} + \nabla \cdot (\rho \boldsymbol{v}) = 0.$$

Here ρ_1, ρ_2, ρ are densities of components and the overall material, respectively. The right hand side of equations in (24), (25) have the source terms J which point out the possibility of exchange of particles between the components.

By virtue of the law of momentum conservation we have

$$\rho \boldsymbol{v}(x,t) = \rho_1 \boldsymbol{v}_1(x,t) + \rho_2 \boldsymbol{v}_2(x,t).$$

Here and in what follows we assume the following expression for the density of the overall material: $\rho = \rho_1 + \rho_2$.

Velocities of the components and the center of mass of the material point are expressed as follows:

$$\boldsymbol{v}_i(x,t) = \frac{\mathrm{d}_i \boldsymbol{u}_i(x,t)}{\mathrm{d}t}, \quad i = 1, 2, \quad \boldsymbol{v}(x,t) = \frac{\mathrm{d}\boldsymbol{u}(x,t)}{\mathrm{d}t}.$$

Here

$$\frac{\mathrm{d}_i}{\mathrm{d}t} = \frac{\partial}{\partial t} + \boldsymbol{v}_i(x,t) \cdot \nabla, \quad \frac{\mathrm{d}}{\mathrm{d}t} = \frac{\partial}{\partial t} + \boldsymbol{v}(x,t) \cdot \nabla$$

denote the material derivatives.

The motion of the overall material point is governed by the law of dynamics in the local form

$$\nabla \cdot \boldsymbol{\tau} + \rho \boldsymbol{F} - J\big(\boldsymbol{v}_1 - \boldsymbol{v}_2\big) = \rho \frac{\mathrm{d}\boldsymbol{v}}{\mathrm{d}t}.$$

The mass external force $\boldsymbol{F}$ can be given by

$$\rho \boldsymbol{F} = \rho_1 \boldsymbol{F}_1 + \rho_2 \boldsymbol{F}_2.$$

The equation of dynamics is convenient to rewrite in the form of two equations

$$\nabla \cdot \boldsymbol{\tau}_1 + \rho_1 \boldsymbol{F}_1 + \boldsymbol{R} - J\boldsymbol{v}_1 = \rho_1 \frac{\mathrm{d}\boldsymbol{v}_1}{\mathrm{d}t}, \quad \nabla \cdot \boldsymbol{\tau}_2 + \rho_2 \boldsymbol{F}_2 - \boldsymbol{R} + J\boldsymbol{v}_2 = \rho_2 \frac{\mathrm{d}\boldsymbol{v}_2}{\mathrm{d}t} \tag{26}$$

where $\boldsymbol{R}$ is the force of interaction of two components of the material of complex structure. The interaction force $\boldsymbol{R}$ has an expression which is explicitly determined by the specific structure of the medium under consideration. In addition to this, the overall stress tensor of the material point is supposed to be the sum of the stress tensors of separate components

$$\boldsymbol{\tau} = \boldsymbol{\tau}_1 + \boldsymbol{\tau}_2.$$

Let us note that the above equations for the two-component body (the two-component medium) are in agreement with the equations of mechanics of continuous heterogeneous media developed for modeling diverse mixtures (Nigmatulin (1990)).

The basic equations of moment and mass balance can be complemented by the equations of energy balance and the second law of thermodynamics for the each component. These equations are given by Indeitsev and Naumov (2009), where the problem of propagation of mechanical and temperature pulses in solids are studied. In the present paper we do not consider a heat exchange between the components.

Now we construct two-component model for the fluid flows in nanochannels.

3 Features of fluid behaviour in nanochannels

New equations that describe the behavior of fluids in nanochannels and take into account the molecular structure of the fluid and results of real and numerical experiments are presented. The Poiseuille flows are considered. The obtained results show that it is possible to describe the structural transformations in thin layers by using the continuum mechanics methods. New degrees of freedom of the material are introduced via the second continuum that makes up for the role of the forming new phase of a state. In the models considered here, the properties of the new phase are determined by the influence of rigid boundaries with a different structure.

3.1 Introduction

Fluid flows in micro- and nanochannels are of great interest from both the viewpoint of fundamental science and practical applications (Drummond and Israelachvili (2001)). By virtue of this, modeling of such a flow became one of most quickly developing trends in hydrodynamics. The topicality of this modeling is also supported by the results of numerous experiments that have been conducted during last two decades (see, for example Gourdon and Israelachvili (2003); Thomson and Robbins (1990)) and revealed great differences between the behavior of fluids in volumes with a size of 50 molecular diameters or smaller and the predictions of classical continuum theories. These experiments showed a substantial increase of the effective viscosity of the fluid in such volumes as compared with its macroscopic value. Classical hydrodynamics, that does not allow for atomic (molecular) structure of fluid, does not present adequate description of fluid flows in nanochannels with a width of 50 molecular diameters or smaller. It is well known that the classical Poiseuille flow is described by the Navier–Stokes equations and the velocity profile in this case has a parabolic shape. Nevertheless, the fact that some fluids flowing in rather narrows gaps begin to feel the boundaries, which leads to restructuring of the profile, have long attracted the attention of researchers. Moreover, in their numerous studies Deryagin and Zheleznyi (1974) found that fluids at the boundary with a solid body form layers with an ordered structure that determines special features of fluids in these layers.

To study this flow, we suggest new equations that describe the behavior of fluids and take into account the molecular structure of the fluid and results of real and numerical experiments. Numerical experiments were conducted by computer modeling and by the molecular dynamics (MD) method formulated by Allen and Tilesly (1989).

3.2 Proposed model and main equations

To describe the fluid flow in the channel, we use the two-component model (see Abramyan (2010)). We assume that the fluid in the channel is affected by the walls, i.e., has a possibility to be structured. The medium outside the channel is a usual molecular viscous fluid. The motion of the latter in the interior of the channel filled with a certain structured medium is similar to the flow through a 'sieve' whose 'feed through' cell dimensions significantly depends on the density of the ordered phase. We assume that, in the process of the fluid flow, the main resistance force is the reaction of fluid particle interaction with the structure cells, which is proportional to the difference of velocities of particles of the interacting components. At rest, without any applied external loads, the channel is filled with a medium which is ordered under the action of the channel walls. It is natural to assume that this phenomenon is inhomogeneous over the layer thickness, namely, the medium particles in the central part of the layer experience lesser influence of the walls than the particles on the boundary with the surfaces. We consider some specific cases in which the influence of the walls is such that, as a rule, the structures near the walls are more concentrated than those in the middle. The stressed state of the ordered medium is modeled as the pure shear stress.

The so-called molecular fluid is fed into the channel, and this fluid interacts with the structure. This interaction force mainly depends on the density of the ordered structure, and hence on the so-called flow section (the distance between particles of the ordered medium). The larger the particle density, the lesser the flow section through which the incoming particles can pass, and hence the force of the two media interaction is the largest. Thus, this interaction of two media, like the viscous friction force, depends on the difference of their particle velocities: it is the larger the higher the velocity of motion of one component relative to the other. In this case, it is assumed that this dependence is linear.

Obviously, as the incoming particles of the molecular fluid move with an input velocity greater than a certain value, the fluid has the tendency to pass freely through the immovable structured medium with possible separation of particles of the latter. This means that if the input pressure is sufficiently high, then the fluid medium can 'destroy' the structure where it exists. Otherwise, if the pressure is insufficient, then the velocities of the applied particles are small, the structure density increases, and the flow rate of the constantly incoming fluid decreases. If the incoming fluid particles are sufficiently slow, then the structuring continues until complete sedimentation of fluid particles, i.e., the channel is 'choked up' and the fluid cannot pass through it anymore.

It is important to note that, in the equations of mass balance, there arise source terms determining the rate of transformation of fluid-like particles into solid-like particles and conversely. We assume that the sedimentation rate must be proportional to the particle concentration in the fluid, and the separation rate must be proportional to the structured medium concentration. Obviously, as the number of the fluid particles decreases, the number of solid-like particles increases, and hence the inverse process may occur.

Since an ordered structure is formed for certain pressure and velocities, the frow rate through each cross-section of the channel decreases in time. The main effect considered in this problem is the phenomenon of molecular fluid sedimentation on the structure, which may result in the so-called 'choking' effect.

Let us denote the number density of fluid particles per unit volume by n_f, the number density of solid (structured, precipitated) particles per unit volume by n_s, and $n_f + n_s = 1$ Then $\rho_s = mn_s$ and $\rho_f = mn_f$ are the density of the solid and fluid particles, m is the mass of particle. Assuming that the rate of solid particles are close to zero, we write the basic equations of moment and mass balance in the form

$$\nabla \cdot \Big[-\boldsymbol{I}p + \mu(\nabla \boldsymbol{v} + (\nabla \boldsymbol{v})^T)\Big] = \rho_f \frac{\mathrm{d}\boldsymbol{v}}{\mathrm{d}t} + J\boldsymbol{v} + \boldsymbol{R}, \tag{27}$$

$$\frac{\partial \rho_f}{\partial t} + \nabla \cdot \big(\rho_f \boldsymbol{v}\big) = J, \qquad \frac{\partial \rho_s}{\partial t} = -J. \tag{28}$$

Here p is the pressure of fluid fraction (liquid phase), $\boldsymbol{I}$ is the unit tensor, $\boldsymbol{v} = \boldsymbol{v}_f$ is the velocity vector of the fluid particles, μ is the equivalent viscosity of fluid fraction, J is the rate of sedimentation (adhesion) and separation of fluid particles at the checkpoint of the reference system, $\boldsymbol{R}$ is the force of interaction between the fractions. Using $n_s = 1 - n_f$, equations (28) can be rewritten as

$$\frac{\partial n_f}{\partial t} = J/m,$$
$$\nabla \cdot \big(n_f \boldsymbol{v}\big) = 0.$$

The source terms J has the following form

$$J = \begin{cases} -k_1 n_f, & |\boldsymbol{v}| < v_*, \\ k_2(n - n_f), & |\boldsymbol{v}| > v_*, \end{cases}$$

where k_1 and k_2 are constants obtained experimentally, v_* is a certain critical velocity. The source term is defined by to the above-described scenario of the events. The forces of interaction between the components have the following form:

$$\boldsymbol{R} = \frac{k n_f}{D(n_s)}\, \boldsymbol{v}, \tag{29}$$

where k is the constant obtained from experimental data, $D(n_s)$ is the cell characteristic open area dimension. Taylor series expansion of $D(n_s)$ about the equilibrium point is given by $D(n_s) = D_0 - D_1 n_s$. Then equation (29) has the form

$$\boldsymbol{R} = k\, g(n_f)\, \boldsymbol{v}, \quad g(n_f) = \frac{n_f}{D_0 - D_1(1 - n_f)} \tag{30}$$

where D_0 is the characteristic open area dimension of the structured cell.

The conditions at the initial time moment are chosen as follows (see Figure 3):

$$n_s(0, y) = n_{s0}(y), \quad n_f(0, y) = n_{f0}(y)$$

For the problem of the Poiseuille flow we take the following initial and boundary conditions

$$p\big|_{x=0} = p_0, \quad p\big|_{x=L} = 0, \quad v\big|_{x=0} = v\big|_{x=L} = 0.$$

They correspond to the assumption that the structure is more ordered near the walls and less ordered near the channel center. As follows from expression (30), the quantity $g(n_f)$ depends on the molecular fluid fraction in the entire volume n_f and on the values of the coefficients D_0 and D_1. As $n_f \to 1$, the quantity $g \to 1/D_0$ and the value of the characteristic open area dimension of the structured cell D_0 increases. In this case, the quantity $\boldsymbol{R}$ contained in the equation (27) tends to zero, and the equation itself tends to the classical form.

The solution of the above-posed problem for different sets of parameters confirms the qualitative applicability of the two-component model for describing the effect under study. The choking effect was investigated for different types of initial ordering of the medium, i.e., for different characters of the wall action and for two types of the source term.

Now we consider the obtained diagrams using the mathematical and computer models. The results given below show that the wall material structure itself significantly affects the liquid flow.

In the first computer experiment, we considered a channel whose walls affected by the incoming liquid so that the medium was structured mainly near the walls and significantly less near the center of the channel. Prescribing a certain initial pressure at which the liquid particles were incoming and the other necessary parameters, we observed a regime in which the velocity profiles and the concentration of each of the media had the form shown in Figure 4a and Figure 4b.

In the second computer experiment, we considered a channel whose walls affected the incoming liquid so that the medium was structured very strongly near the input and the walls and significantly less near the center of the channel. The character of the observed regime is shown in Figure 5a for $H = 100$ nm and $L = 200$ nm. The diagrams for the two above-described experiments clearly illustrate the choking regime and show that the input pressure was insufficient in these experiments, which is testified

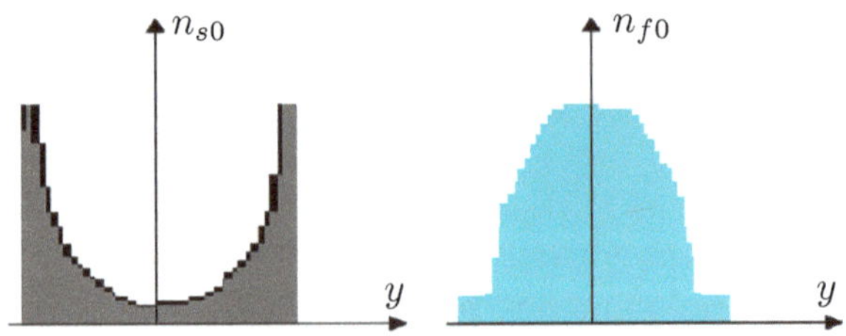

Figure 3. Distribution of particles of the structure and molecular fluid.

by the decrease in the flow rate of the incoming media (see Figure 5b). The computations showed that by increasing the excess pressure in the channel, one can obtain the converse effect, namely, the structured media becomes disordered. We did not calculate how the characteristic open area dimension affects the flow rate.

In the third computer experiment, we considered a channel in which the structured medium distribution was similar to the preceding distribution but the source term in the mass balance equation had the form

$$J = -k_1 \rho_f (A - \rho_s) H(v_* - v) + k_2 \rho_s (B - \rho_f) H(v - v_*) \tag{31}$$

The first term describes the sedimentation (ordering) at a rate proportional to the quantity of the matter in the liquid and bounded by the quantity of the already ordered medium till the saturation A. The second term describes the converse transition. The simulation results are shown in Figure 6. The results obtained confirm that there is a blocking effect, which is illustrated by an increase in solid-like phase concentration and a decrease in liquid flow rate for a certain pressure regime. We note that, under the assumption of strong effect of the walls strongly on the medium in the channel, this model more clearly illustrates the action of these forces on the process of structurization (see Figure 6a).

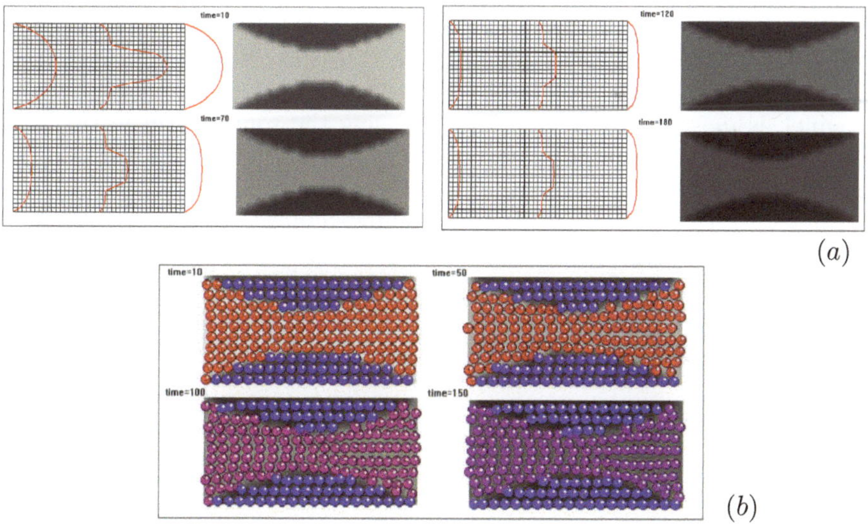

Figure 4. The first experiment. Profiles of velocity and concentration of particles in each component (a), Concentration of particles of each component at different instants of time (b).

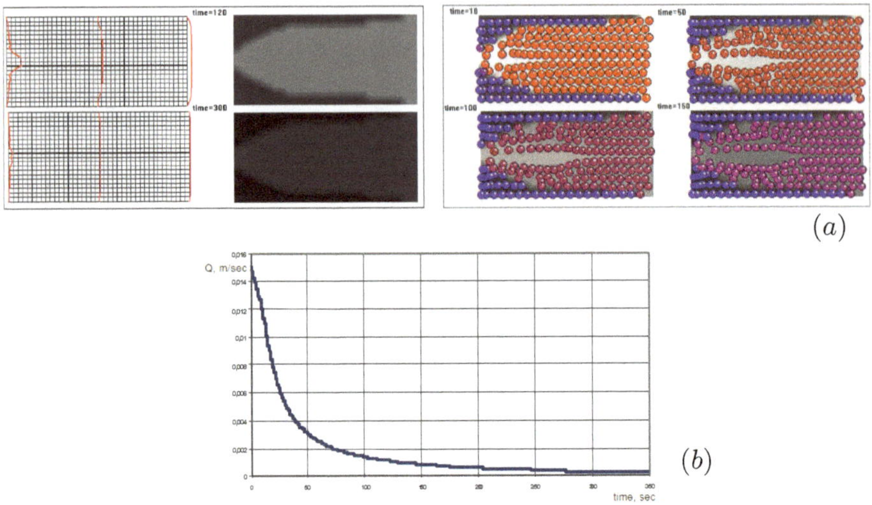

(a)

(b)

Figure 5. The second experiment. Profiles of velocity and concentration of particles in each component(a), flow rate of molecular fluid in the channel (b).

All the above experiments describe the choking effect in a plane channel. This phenomenon, under the assumption that the medium structurization (sedimentation) must decrease from the walls towards the center of the channel, is most precisely described by using the source term in the form (31). The third experiment clearly shows that the particles begin to settle near the already structured medium. Thus, generalizing all the diagrams obtained by using the mathematical and computer models, we see that the claim of this model to describe the choking effect in the channel with significant influence of its walls taken into account is justified completely.

3.3 Conclusion to section 3

We propose a mathematical model of a fluid flow in a plane nanochannel, that is caused by the motion of one of the confining walls parallel to the other, immovable wall. The values of the resistance forces acting on the walls when the distances between them are less than 50 nm, obtained using the above model, are in good agreement with the experimental results and predictions by the MD modeling. The obtained results show that it is possible to describe the structural transformations in thin layers by using the continuum mechanics methods. We introduce new degrees of freedom of the material by using the second continuum, which plays the role of the

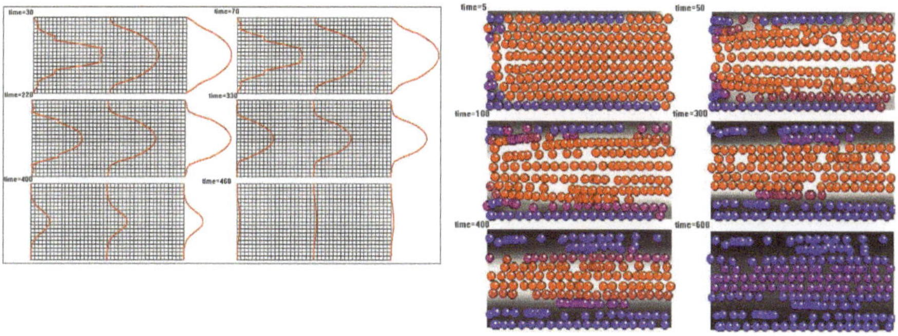

(a)

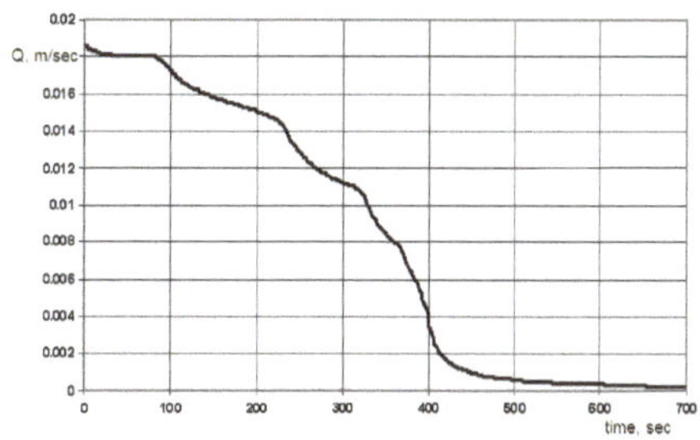

(b)

Figure 6. The third experiment. Profiles of velocity and concentration of particles in each component (a), flow rate of molecular fluid in the channel (b).

arising new phase of state. In the models considered above, the properties of the new phase are determined by the influence of rigid boundaries with a different structure. The solutions thus obtained depend on macroparameters, which can be determined using experimental data. This is a significant distinction of this approach from the earlier approaches, where numerous parameters, which are hard to determine, were introduced in the equations of state.

We note that the two-component model can describe quite well such effects as the flow 'choking' and the 'destroyed' layer reconstruction. This is because the source terms are introduced in the equation of the particle number balance of one or the other component. Depending on the scenario of the events in the material, it is quite possible to control and describe its state by using a suitable source term.

4 Hydrogen diffusion in the crystal structures

Hydrogen embrittlement of materials under load is one of the most important problems of the physics and mechanics of materials. Though the hydrogen concentration in metals can be very low (about 1 atom of hydrogen in 100,000 atoms of the metal matrix), nevertheless its influence on the mechanical properties of the metals can be of crucial importance. The problem of the effect of low hydrogen concentrations in metals on the strength of material nowadays attracts a lot of attention. As a rule, the hydrogen is accumulated in metals during their exploitation. One of the main sources for hydrogen appearance in metals is water (or steam); however, hydrogen diffusion from gas and oil is feasible as well.

In metals, the hydrogen is contained in traps with various bonding energies. It has been established (see Polyanskiy (2005) and references therein) that thermo-mechanical loading results in the hydrogen redistribution over the traps. A number of papers were devoted to the influence of hydrogen on the mechanical properties of metals, see e.g. Ahn (2007). The majority of the papers addressing the effect of hydrogen on the strength of materials utilize primarily phenomenological models and do not discuss the problem of redistribution of hydrogen over the traps. The degradation of mechanical properties in these papers is modeled by means of some empirical dependencies. However, there is an open question: how kinetic processes in the material (such as the redistribution of hydrogen) affect its basic strength properties under static and dynamic loads? The aim of our study is to describe the dynamics of the hydrogenated metal and the influence of internal kinetics on metal macroparameters using the fundamental principles of rational mechanics. The hydrogen diluted in structural materials can be conditionally divided into that with low bonding energy and that with high bonding energy. The hydrogen with low bonding energy is diffuse, and its interaction with material is very weak (mobil hydrogen). The high bonded hydrogen interacts with material very intensively. The mechanical material properties degrade owing to this strong interaction. We suggest a one-dimensional model of two-component continuum, which allows us to describe both the hydrogen diffusion and its interaction with the material and, therefore, to find the equation of state for hydrogen-containing medium. The first component is represented by the crystal lattice of the initial material including stationary hydrogen atoms embedded (attached) in chemical bonds between atoms (which significantly reduce the strength of the bonds), the second component is modeled by free mobile hydrogen atoms dissolved in the material.

4.1 Two-component model

Let us consider the simplest one-dimensional model of a metal rod, in which atomic hydrogen is dissolved. Let N be the total number of particles

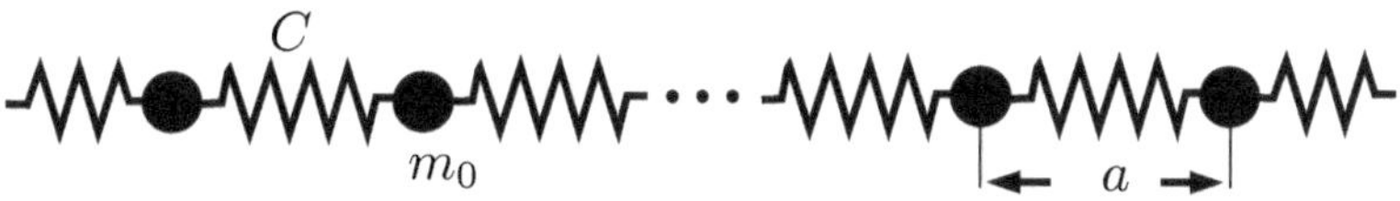

Figure 7. One-dimensional chain of atoms.

per unit volume, N_0 be the number of the particles connected by undamaged bonds per unit volume, N_H^+ be the number of hydrogen particles attached to a lattice per unit volume, N_H^- be the number of mobil hydrogen particles per unit volume. Furthermore, n_0, n_H^+ and n_H^- are the corresponding concentrations of the above-mentioned particles, respectively.

The first component is a lattice structure with bonded hydrogen (in the frame of elastic theory). The relation between the strain ε and the stress σ can be represented as $\sigma = E\varepsilon$, where E is the equivalent module of the lattice defined below. We denote the velocity and the density of the first component by v_1 and $\rho_1 = \rho_0 + \rho_H^+$. Here $\rho_0 = m_0 n_0$, $\rho_H^+ = m_H n_H^+$, m_0, m_H are the mass of lattice atoms and bounded hydrogen atom.

The second component is flow of mobile hydrogen particles of the internal structure of a material (inviscid compressible liquid) and p is the pressure of flow, v_2, $\rho_2 = \rho_H^-$ is the velocity and the density of the second component, $\rho_H^- = m_H n_H^-$.

Equation of state (rheological model). We consider the lattice as a one-dimensional chain consisting of identical particles with a mass of m_0 (mass of atom in the crystal lattice of the material) which are connected with each other by identical nonlinear springs with the lengths a, Figure 7. The equation of movement in the long-wave approximation is as follows (Zhilin (2006))

$$m_0 \ddot{u} = -a[f(a(1 + u'))]'. \tag{32}$$

For small strains $\varepsilon = \partial u/\partial x$ we have the following equation

$$\ddot{u} - \vartheta_0^2 u'' = 0, \qquad \vartheta_0^2 = \sqrt{\frac{C}{m_0}}\, a,$$

because $f[a(1 + \varepsilon)] \approx -Ca\varepsilon$.

It is known that, when hydrogen dissolves in materials, some of the hydrogen atoms are embedded in existing atomic bonds, breaking them and creating new bonds, its stiffness is much smaller than the initial stiffness (see Indeitsev and Semenov (2008)). This effect also takes place for a congestion of lattice defects: dislocations, vacancies, *etc.* By combining the elements

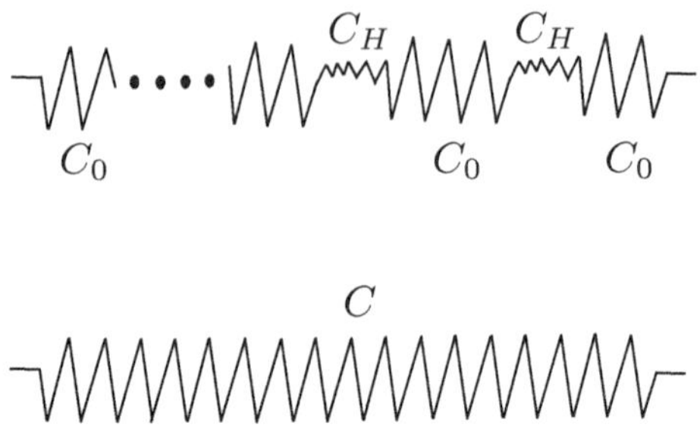

Figure 8. Model of a chain with hydrogen atoms planted on the bonds.

with the old and new bonds in series one obtains the model of the lattice with bonded hydrogen (see Figure 8). It is possible at the assumption $m_0 \gg m_H$. Then the equivalent rigidity of the new bond C can be found from the equation

$$\frac{N}{C} = \frac{N_0}{C_0} + \frac{N_H^+}{C_H}, \qquad N = N_0 + N_H^+,$$

where C_0 is the rigidity of pure material in the absence of hydrogen, C_H of the material with all bonds occupied by hydrogen. The nonlinear force f in equation (32) can be accepted for small strains as

$$f = -Ca\varepsilon = -E\varepsilon.$$

Then using $n_0 = N_0/N$ and $n_H^+ = N_H^+/N$ we obtain the constitutive equation for the lattice structure with bonded hydrogen

$$\sigma = E\varepsilon, \qquad E = \frac{E_0 E_H}{n_0 E_H + n_H^+ E_0}. \tag{33}$$

The equivalent elastic modulus for the lattice E can decrease essentially, since $E_H \ll E_0$ ($C_H \ll C_0$) and depends strongly on the concentration of the attached particles n_H^+ (bonded hydrogen). The number of the lattice-settled hydrogen particles depends on the stress state of the lattice at every point and, generally, on time. The unknown functional dependence of E on $n_H^+(\varepsilon, x, t)$ should be determined from the model of the two-component continuum.

Main equations. Substantive provisions of the theory of two-component continuum can be found in Section 2 (Krivtsov and N.F. Morozov (2001));

therefore, we are presenting only final equations. The equation of dynamics for the lattice structure (the first continuum) is given by

$$\frac{\partial \sigma}{\partial x} = \rho_1 \frac{\partial v_1}{\partial t} + J v_1 + R. \tag{34}$$

Equation (34) has new force terms in the right part. Capturing of hydrogen mobile particles in the lattice practically does not influence the change of inertial characteristics of the lattice structure, i.e. $\rho_H^+ = m_H n_H^+ \ll \rho_0$, but the velocity of the change of those characteristics J results in the occurrence of jet force Jv_1, whose neglecting is impossible. The term R determines the force of interaction between the first and second components and depends on internal processes in the material.

The equation of dynamics for the second component (mobile hydrogen particles) is as follows:

$$-\frac{\partial p}{\partial x} = \rho_2 \frac{\partial v_2}{\partial t} - J v_2 - R, \quad \rho_2 = \rho_H^- = m_H n_H^-. \tag{35}$$

Similarly to a case of compressed liquid, the state equation determining a connection between pressure p and density ρ_H^- takes the form

$$p - p_0 \cong c_H^2 \rho_H^-. \tag{36}$$

The equation of mass balance for the first component is

$$\frac{\partial \rho_1}{\partial t} + \frac{\partial (\rho_1 v_1)}{\partial x} = J,$$

or taking into account an invariance of ρ_0 for ρ_H^+ we have

$$\frac{\partial \rho_H^+}{\partial t} + \frac{\partial (\rho_H^+ v_1)}{\partial x} = J.$$

In terms of the concentration of bonded hydrogen the mass balance equation has the form

$$\frac{\partial n_H^+}{\partial t} + \frac{\partial (n_H^+ v_1)}{\partial x} = J/m_H. \tag{37}$$

Mobil hydrogen obeys the similar equation of mass balance:

$$\frac{\partial \rho_H^-}{\partial t} + \frac{\partial (\rho_H^- v_2)}{\partial x} = -J \quad \text{or} \quad \frac{\partial n_H^-}{\partial t} + \frac{\partial (n_H^- v_2)}{\partial x} = -J/m_H. \tag{38}$$

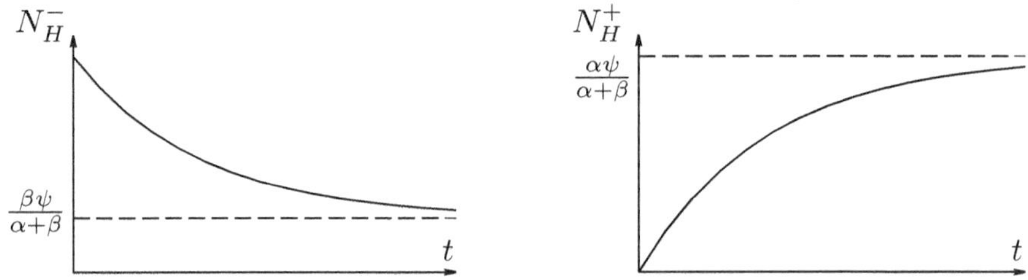

Figure 9. Dependencies of n_H^- and n_H^+ on time.

Interaction between the components. Interaction between the stationary lattice and flow of mobile hydrogen, that has not been attached yet, is described similarly to the flow of compressed liquid, see Loitsyansky (1987). Since the velocity of the hydrogen particles is very low and the suggested approach is linear, we assume that the interaction force is proportional to the difference in the continuum particles velocities:

$$R = k\frac{\rho_H^-}{D(\varepsilon)}\big[v_2 - v_1\big]. \tag{39}$$

Here k is determined in terms of the material properties, $D(\varepsilon)$ is the effective cross section of the flow of the second component Indeitsev and Osipova (2011). The larger is the deformation, the smaller is the quantity D, since it is more difficult for mobile hydrogen to move in strained medium.

Since the concentrations of hydrogen are low, the source terms J has the following form

$$J = \alpha n_H^- - \beta n_H^+ \tag{40}$$

where α and β are positive parameters determined by interaction between bonded and mobile particles of hydrogen. The physical meanings of the coefficients α and β can be defined as follows. For the small velocities v_1 and v_2 the terms $\partial\big(n_H^+ v_1\big)/\partial x$ and $\partial\big(n_H^- v_2\big)/\partial x$ in equations (37) and (38) can be neglected. Than the problem for n_H^- and n_H^+ takes form

$$\frac{\mathrm{d}n_H^+}{\mathrm{d}t} = \alpha n_H^- - \beta n_H^+, \quad \frac{\mathrm{d}n_H^-}{\mathrm{d}t} = -\alpha n_H^- + \beta n_H^+ \tag{41}$$

with the initial conditions

$$n_H^+\big|_{t=0} = 0, \qquad n_H^-\big|_{t=0} = \psi.$$

The solutions of equations (41) are

$$n_H^+ = \frac{\alpha\psi}{\alpha+\beta}\big(1 - e^{-(\alpha+\beta)t}\big), \quad n_H^- = \psi\Big[1 - \frac{\alpha}{\alpha+\beta}\big(1 - e^{-(\alpha+\beta)t}\big)\Big]. \tag{42}$$

The parameters α and β define the rate of hydrogenation of the lattice (and its rate of hydrogen loss) and they should be prescribed, e.g. by using experimental data. The problem (41) describes interchange of hydrogen particles between the components under the condition that the velocity of diffuse hydrogen is small. In this case almost all mobile hydrogen is built in the lattice; hence $\alpha \gg \beta$ (see Figure 9).

The complete set of equations. In this section we summarize the equations derived in previous sections to represent the motion of the metal rod containing dissolved hydrogen. We have

$$\frac{\partial \sigma}{\partial x} = \left(\rho_0 + m_H\, n_H^+\right) \frac{\partial v_1}{\partial t} + J v_1 + R, \quad \sigma = \frac{\varkappa E_0}{n_H^+ + \varkappa n_0}\, \varepsilon, \tag{43}$$

$$-\frac{\partial p}{\partial x} = m_H\, n_H^- \frac{\partial v_2}{\partial t} - J v_2 - R, \quad p - p_0 = m_H c_H^2 n_H^-, \tag{44}$$

$$\frac{\partial \rho_0}{\partial t} + \frac{\partial (\rho_0 v_1)}{\partial x} = 0, \tag{45}$$

$$\frac{\partial n_H^+}{\partial t} + \frac{\partial (n_H^+ v_1)}{\partial x} = J/m_H, \quad \frac{\partial n_H^-}{\partial t} + \frac{\partial (n_H^- v_2)}{\partial x} = J/m_H, \tag{46}$$

$$R = k \frac{m_H\, n_H^-}{D(\varepsilon)} \left[v_2 - v_1\right], \quad J/m_H = \alpha n_H^- - \beta n_H^+. \tag{47}$$

Here $\varkappa = E_H/E_0$. The set of governing equations (43)–(47) is much too complicated for a direct mathematical analysis and we will restrict our attention by the simplest case – static stress state of the hydrogenated metal rod under the uniaxial tension/compression.

4.2 Static stress state

We suppose that the first material component (lattice with bonded hydrogen) is initially at the static stress state, so that the strain ε_0 and the stress σ_0 are related by $\sigma_0 = E\varepsilon_0$. The initial static stage is then disturbed slightly, and we suppose that the perturbation quantities $\widetilde{\sigma}$, $\widetilde{\varepsilon}$, $\widetilde{v}_1$, $\widetilde{n}_H^+$; $\widetilde{n}_H^-$ and $\widetilde{v}_2$ are small in magnitude. Then we find the solution of the problem in the following form

$$\varepsilon = \varepsilon_0 + \tilde{\varepsilon}(x,t), \quad \sigma = \sigma_0 + \widetilde{\sigma}(x,t), \quad v_1 = 0 + \tilde{v}_1, \quad v_2 = v_{20} + \tilde{v}_2 \tag{48}$$

$$n_H^+ = n_{H0}^+ + \widetilde{n}_H^+(x,t), \quad n_H^- = n_{H0}^- + \widetilde{n}_H^-(x,t) \tag{49}$$

Substituting equations (48) and (49) into (43)–(47) we obtain the following linear approximation for the first component

$$\frac{\partial \sigma_0}{\partial x} = 0, \quad \sigma_0 = E_0 \varepsilon_0 \left[1 - \frac{n_{H0}{}^+}{n_{H0}{}^+ + \varkappa n_0} \right], \tag{50}$$

$$\frac{\partial n_{H0}^+}{\partial t} = \alpha n_{H0}^- - \beta n_{H0}^+, \tag{51}$$

for the second component

$$c_H^2 \frac{\partial n_{H0}{}^-}{\partial x} = -k \frac{n_{H0}{}^-}{D(\varepsilon_0)} v_{20}, \tag{52}$$

$$\frac{\partial n_{H0}^-}{\partial t} + \frac{\partial (n_{H0}^- v_{20})}{\partial x} = -\alpha n_{H0}^- + \beta n_{H0}^+. \tag{53}$$

Equations (51) –(53) can be reduced to the equation for concentration of bonded hydrogen n_{H0}^+ that takes form

$$\frac{\partial^2 n_{H0}^+}{\partial t^2} + (\alpha + \beta) \frac{\partial n_{H0}^+}{\partial t} - \frac{c_H^2 D(\varepsilon_0)}{k} \left[\beta \frac{\partial^2 n_{H0}^+}{\partial x^2} + \frac{\partial^3 n_{H0}^+}{\partial x^2 \partial t} \right] = 0. \tag{54}$$

Equation (54) is the equation of the mixed type, it contains terms inherent in the hyperbolic equation and terms of a parabolic kind. It means that at the assignment of the finite initial perturbation one should expect a characteristic front of movement of increase (or decrease) in a bonded hydrogen number density, i.e. an exposed strong dispersion.

Example: approximate analytical solution. Suppose that the lattice structure does not initially contain bonded hydrogen and mobil hydrogen is distributed in the material as follows

$$n_{H0}{}^- \Big|_{t=0} = \frac{\Psi}{2} \Big(1 + \cos \frac{2\pi x}{\lambda} \Big),$$

where λ is the characteristic size of internal structure (such as the distance between the lattice atoms). Then we can analyze equation (54) following the initial conditions

$$n_{H0}{}^+ \Big|_{t=0} = 0, \quad \frac{\partial n_{H0}{}^+}{\partial t} \Big|_{t=0} = \frac{\alpha \Psi}{2} \Big(1 + \cos \frac{2\pi x}{\lambda} \Big). \tag{55}$$

Here Ψ is the limit value of the bonded hydrogen density. We seek a solution in the form

$$n_{H0}{}^+ = \frac{\Psi}{2} \Big(1 + \cos \frac{2\pi x}{\lambda} \Big) q(t), \tag{56}$$

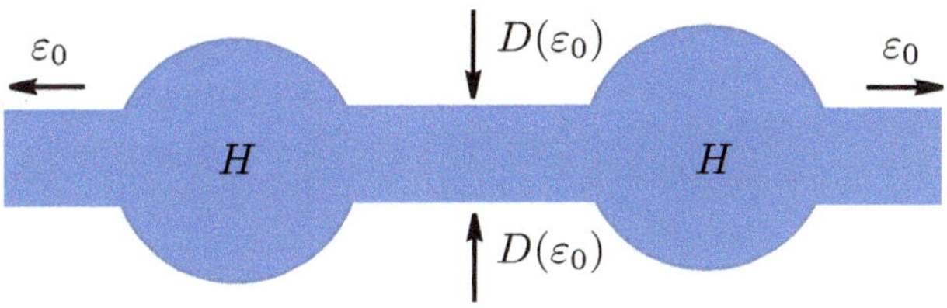

Figure 10. Model of closing of channels and places of a congestion of hydrogen.

and $q(t)$ satisfies the equation:

$$\ddot{q} + \big[\alpha + \beta + \gamma D(\varepsilon_0)\big]\dot{q} + \beta\gamma D(\varepsilon_0)\, q = 0, \quad q\big|_{t=0} = 0, \quad \dot{q}\big|_{t=0} = \alpha, \tag{57}$$

where $\gamma = (2\pi c_H)^2/3k\lambda^2$. Solving equation (57) by assuming that $\beta \ll \alpha$ and $\alpha \ll \gamma D(\varepsilon_0)$, we obtain

$$n_{HO}{}^{+}(x,t) = \frac{\alpha\Psi}{2\big(\alpha + \gamma D(\varepsilon_0)\big)}\Big(1 + \cos\frac{2\pi x}{\lambda}\Big)\Big[1 - \exp\big\{-(\alpha + \gamma D(\varepsilon_0))t\big\}\Big]. \tag{58}$$

Averaging equation (58) with respect to x, we find the simplified expression of N_{HO}^{+}

$$n_{HO}{}^{+} \approx \frac{\alpha\Psi}{\alpha + \gamma D(\varepsilon_0)}. \tag{59}$$

For the small deformations we can assume that the value of the flow cross section D is linearly dependent on ε_0 and

$$D(\varepsilon_0) = D_0 - D_1\varepsilon_0, \quad D_1 > 0. \tag{60}$$

Then the larger is the deformation, the smaller is the value of D. When $\varepsilon_0^* = D_1/D_0$ we have $D = 0$ and any diffusion of mobile hydrogen becomes impossible, so it goes into a bonded state. The behavior of concentration of the bonded hydrogen at a tension of a sample depending on the enclosed strain ε_0 ($\varepsilon_0^1 < \varepsilon_0^2 < ... < \varepsilon_0^*$) is shown in Figure 11a.

Stress-strain diagram. Suppose that we have a limiting concentration of the bounded hydrogen $\Psi \gg \varkappa$, where $\varkappa = E_H/E_0$ (in particular, for steel $\Psi \approx 10^{-6}$, $\varkappa \approx 10^{-7} \div 10^{-8}$). Substituting equations (59) and (60) into (50) yields

$$\sigma_0 \approx \frac{\varkappa E_0 \varepsilon_0}{\varkappa + \alpha\Psi/\big[\alpha + \gamma(D_0 - D_1\varepsilon_0)\big]}. \tag{61}$$

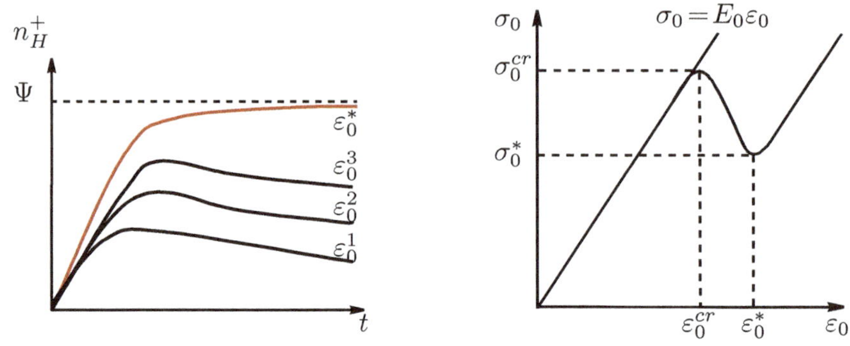

Figure 11. Behavior of concentration of the bonded hydrogen at a tension of a sample depending on the enclosed strain (a), the diagram of strain-stress in view of influence of the bounded hydrogen (b).

Defining the extrema of the function $\sigma_0(\varepsilon_0)$ we get the critical points of the stress–strain diagram

$$\varepsilon_0^{cr} \approx \varepsilon_0^* \left[1 - \sqrt{\frac{\alpha\Psi}{\varkappa\gamma D_0}}\right], \quad \varepsilon_0^* = D_0/D_1.$$

It should be noted that we obtain the equation of state (61) assuming that the induced strains connected to reorganization of internal structure can be neglected compared with the homogeneous static field of the strains ε_0.

Figure 11b shows the qualitative representation of the stress–strain diagram, calculated by equation (61), corresponding to steel and titanium. The part of $\sigma_0(\varepsilon_0)$ dependence with $d\sigma_0/d\varepsilon_0 < 0$ can not be realized. The growth of $\varepsilon_0 > \varepsilon_0^{cr}$ results in hydrogen embrittlement and destruction. However the hydrogen saturation leads to decreasing of the breaking point σ_0^{cr}. Similar $\sigma(\varepsilon)$ curves were observed in the experiments with titanium alloys having large hydrogen concentration and in high-strength steels under various immersion times in NH_4SCN solution (see Takai and Watanuki (2003)).

It should be mention that the dependence (61) is looplike and thereby predicts the first-order phase transition into hydride phase under the load. The stress–strain equation of state (61) agrees well with the corresponding results of the model developed on the basis of statistical mechanics (Indeitsev and Osipova (2011)).

4.3 Conclusion to section 4

The two-component model of the material, in which atomic hydrogen dissolved, has been constructed. It has been shown that the stress-strain equation of state of the hydrogenated metal is shaped like the Van der Waals loop; therefor, brittle hydride regions are nucleated in metal by the mechanism of the first-order phase transition. This allows us to describe the kinetics of hydrogen in metals, to estimate hydrogen transition from the mobile into the bonded state depending on the stress state.

5 Conclusion

Dynamics of the material with complex internal structure has been investigated within a two-component continuum model. The approach which allows us to describe internal evolution processes in materials basing on the Euler equations and the mass balance equations containing source terms has been proposed. The influence of exchange mass between the components on the internal structure of the materials has been investigated. The source terms determining the mass transfer between material components have been defined. Examples – structured liquids in nanochannels, metals with dissolved hydrogen – have been considered.

Acknowledgements. This work is supported by the Russian Foundation of Basic Research (grant number 13-01-00349).

Bibliography

G. B. Whitham, *Linear and Nonlinear Waves*, Wiley, New York, 1974.

M. Reiner, *Rheology*, Springer-Verlag, Gottingen Heidelberg, 1958.

V.A.Palmov, *Vibrations of elastoplastic bodies*, Heidelberg: Springer, 1998.

L.G. Loitsyansky, *Mechanika zidkosti i gaza (Fluid and gas mechanics)*, Moskva, Nauka, 1987.

G. K. Batchelor *Introduce to fluid dynamics*, Cambridge University Press. 1967

S.L. Sobolev, *Transport processes and traveling waves in systems with local nonequilibrium*, In Sov. Phys. Usp., pages 217–229, 1991.

S.L. Sobolev, *Local non-equilibrium transport models*, Phys. Usp., pages 1043–1053, 1997.

R.I. Nigmatulin, *Dynamics of Multiphase Media*, Hemisphere, N.Y., 1990.

D.A. Indeitsev, V.N. Naumov, B.N. Semenov and A.K. Belyaev, Indeitsev D.A. *Thermoelastic waves in a continuum with complex structure*, InZAMM Zeitschrift fur Angewandte Mathematik und Mechanik, page 279–287, 2009.

C. Drummond and J. Israelachvili, *Dynamic Phase Transitions in Confined Lubricant Fluids under Shear*, Phys. Rev., 041506, 2001.

D. Gourdon and J. Israelachvili, *Transitions between Smooth and Complex Stick Slip Sliding of Surfaces*, Phys. Rev., 021602 2003.

P. Thomson and M. O. Robbins, *Shear Flow near Solids: Epitaxial Order and Flow Boundary Conditions*, Phys. Rev., 6830–6837 1990.

B.V. Deryagin, B. V. Zheleznyi, Z.M. Zorin, et al., *Properties of Fluids in ThinQuartz Capillaries*, Surface Forces in Thin Films and Colloid Stability (Nauka, Moscow), pages 90–94, 1974 (in Russian).

M. P. Allen and D. J. Tilesly, *Computer Simulations of Liquids*, Clarendon Press, Oxford, 1989.

A.K. Abramyan, N.M. Bessonov, D.A. Indeitsev, L.V. Mirantsev *Influence of the confining wall structure on the fluid flow in nanochannels*, Mechanics of Solids, pages 379–389, 2010.

D.C. Ahn, P. Sofronis, R.Jr. Dodds *Modeling of hydrogen-assisted ductile crack propagation in metals and alloys*, Int. J. Fract., pages 135–157, 2007.

P.A. Zhilin, *Advanced problems in machanics*, Petersburg: Edition of the Institute For Problems in Mechanical Engineering of Russian Academy of Siences, 2006.

D.A. Indeitsev, B.N. Semenov *About a model of structural-phase transformations under hydrogen influence*, Acta Mech., pages 295–304, 2008.

A.M. Krivtsov and N.F. Morozov, *Anomalies in mechanical characteristics of nanometer-size objects*, Doklady Physics, pages 825–827, 2001.

A.M. Polyanskiy, V.A. Polyanskiy and D.B. Popov-Diumin *Diagnostics of mechanical condition of structural material by method of high-temperuture hydrogen vacuum-extraction*, In Proceedings of the Sixth International Congress on Thermal Stresses, Vienna, Austria, pages 589–592, 2005.

D.A. Indeitsev and E.V. Osipova, *A statistical model of hydride phase formation in hydrogenated metals under loading*, Doclady Physics, pages 523–526, 2011.

K. Takai, R. Watanuki *Hydrogen in trapping ttates innocuous to environmental degradation of high-strength steels*, ISIJ International, pages 520–526, 2003.

Modelling of Fluid-Structure Interaction – Effects of Added Mass, Damping and Stiffness

Andreas Zilian
Université du Luxembourg,
Faculté des Sciences, de la Technologie et de la Communication,
Luxembourg

Abstract Fluid-flow around mechanical structures can sometimes lead to catastrophic failures. Improved modelling of fluid/structure interaction is required for safety and mechanical considerations. In this contribution, concepts for modelling the interaction of structures and fluids are presented. Starting from excitation mechanisms and associated classifications, various model depth approaches are compared. Among them, the use of added coefficients for quasi-steady problems is discussed. On the basis of potential flow theory, different approaches for determining fluid-induced additional mass are established and illustrated using an analytical example. Given the limitations of simplifying the engineering models, the second part of the paper provides a brief overview on computational methods for fluid-structure interaction and presents a monolithic modelling approach using space-time finite elements for discretisation of both fluid and structure. Applications from aero- and hydro-elasticity show the applicability of computational methods for problems involving flow-induced added mass, damping, and stiffness.

1 Flow-induced vibrations

Flow in and around engineered structures can cause structural vibrations that may sometimes become dangerous and destructive. Interactions between fluid flow and structures have become particularly important in light of the many new materials being used for lighter structures and also important for temporary and less robust project states that exist during construction. The resulting designs are often more sensitive to dynamic excitation, but generic predictions of the structural response to flow-induced excitation is often not available.

In civil engineering, the majority of structural designs lead to bluff objects exposed to the flow of fluids such as air and water. However, common

H. Irschik, A. K. Belyaev (Eds.), *Dynamics of Mechanical Systems with Variable Mass*,
CISM International Centre for Mechanical Sciences DOI 10.1007/ 978-3-7091-1809-2_5

civil engineering structures are required to meet criteria such as static resistance and safe response spectra with respect to traffic, machine or earthquake excitation. Motivated by a number of spectacular construction failures that at the time represented the state-of-the-art design, aero-elastic considerations were introduced into the design of modern flow-exposed lightweight structures (Scruton (1969); Sockel (1984); Ruscheweyh (1988)). Further, aero-elastic safety and aerodynamic optimisation and control are becoming critical for the overall design. As a consequence, eligible design rules and estimates of structural safety with respect to flow-induced vibrations are now being requested by engineers. However, the complexity of most fluid-structure interaction phenomena means that it is still very difficult to determine robust and accurate risk estimates that would be applicable to a broad band of operational states (see Bisplinghoff et al. (1955); Försching (1974); Dowell et al. (1989); Blevins (1990)).

The lack of a universal and straightforward model for fluid-structure interaction (FSI) motivated the establishment of a field of experimental and numerical investigation of flow-induced vibrations. Today, numerical simulation is gaining acceptance in wind, coastal, and hydraulic engineering as it provides important local and global estimates of parameters such as the stress and deformation state. Those tools, therefore, enable engineers to gain further insight into the physical mechanisms of coupled problems.

Excitation Mechanisms

Following the classification of fluid-structure interactions that were proposed by Naudascher and Rockwell (1994), three main excitation mechanisms of civil engineering structures can be distinguished: extraneous-flow-induced excitations (EIE), flow-instability-induced excitations (IIE) and movement-induced excitations (MIE). Herein, the structural component may be modelled as single or multiple elastically supported rigid bodies (e.g. sluice gates, air-foil or hydro-foil sections) or flexible structures (e.g. cables, pipes, thin-walled structures, long-span bridges, membrane roofs, or silos). Coupled with surrounding or enclosed fluid flow even these simple structures can produce very complex and unpredictable dynamic behaviour of the overall system. Realistic situations of fluid-structure interaction are often simultaneously driven by several of the excitation mechanisms.

Instability-induced excitation is caused by the intrinsic properties of the fluid and leads to an unstable flow regime around the structure being considered. This self-exciting flow instability produces oscillating forces even if the structure is stationary. A further amplification of the exciting force is possible for fluid-elastic feedback. Moreover, *lock-in* of the frequency of the exciting fluid force and a structural eigenfrequency may be observed in a specific range of flow velocities. Shedding vortices in wind and water flow or unstable free surfaces can result in IIE for a number of civil engineering structures, e.g. cylinders and marine cables in cross-flow or vibrations of weir flaps with unstable overfall stream.

Movement-induced excitation is caused by fluctuating flow forces that result from movements of the vibrating structural component. Small deviations from structural equilibrium induce a re-distribution of impacting fluid forces that further increase the initial disturbances. This effect transfers flow energy to the structural oscillator and is referred to as dynamic instability. Single or multi mode vibrations may occur due to aero-elastic phenomena such as flutter or galloping involving one or multiple bodies. Galloping is a self-excited, transversal oscillation of bluff sections that is initiated at flow speeds greater than critical flow speeds and has been observed, for example, in ice-coated power cables exposed to cross fluid-flow. The frequency of the resulting galloping motion is small compared to the vortex shedding frequency, making the effect distinct from IIE. Flutter may occur if a slender structure undergoes small heaving or pitching vibrations in parallel fluid-flow. Periodic structural motion results in movement-induced vortex shedding at the trailing edge that leads to a phase shift of fluid loading relative to the structural vibration. In civil engineering, flutter can cause severe vibrations of bridge decks, fluid-conveying pipes, facade panels, or cylindrical shells.

Extraneously-induced excitation is caused by periodic or randomly fluctuating flow forces generated by an external energy source. Fluctuations of flow velocities or pressure may arise from periodic pulsations in pumps, uniform ocean waves, turbulent wind, gusts, or chaotic sea motions.

For specific problems, engineers have used empirical data to develop successful strategies to circumvent unwanted flow-induced excitation of structures: symmetry-breaking separation by bevelled trailing edges and shrouds, vortex trapping by concave trailing edges for IIE; or additional damping devices such as dissipative dampers or tuned mass dampers for MIE.

2 Mathematical modelling

A non-negligible number of flow-structure problems in engineering practise are still subject to thorough scientific investigation and meanwhile treated in a rule-of-thumb fashion or on the basis of experience and accumulated expert knowledge. Nevertheless, this expertise serves the key aspect in phenomenological modelling of FSI: careful identification of the driving interactions for later mathematical formulation of the coupled system.

The general physical behaviour of coupled systems can often be mathematically described using differential and algebraic equations. The governing system of equations of the involved continua constitute the conservation laws for mass, momentum and energy as well as the kinematic relationships and the material laws. The integration constants are determined by the boundary conditions. In addition to Dirichlet and Neumann boundary conditions, a formulation of a boundary-coupled multi-field problem requires the definition of appropriate coupling conditions at the coupling interfaces. These interface conditions are established on the basis of the conservation laws and kinematic constraints.

Fully coupled multi-physics. This generic approach –as alternative to the phenomenological approach– allows the field equations for each continuum to be independently formulated, while the coupled nature of the full problem is taken into account by the coupling conditions. The coupled system can be written in terms of the respective primary field functions of the structure u_S (e.g. structural velocity/displacement) and the flow u_F (e.g. fluid velocity, pressure/density). All spatial and temporal derivatives of the involved physical models for both continua are here supposed to be incorporated in the operators $\mathcal{L}_{SS}$ and $\mathcal{L}_{FF}$ while the bi-directional field-coupling is reflected by operators $\mathcal{L}_{SF}$ and $\mathcal{L}_{FS}$, which describe the quality and strength of the interaction. Volume forces are indicated by the time-dependent loading functions f_S and f_F. If it is also assumed that all essential and natural boundary conditions are already incorporated in the typically non-linear field operators, one arrives at the *full* description of the coupled fluid (F) and structure (S) problem:

$$\mathcal{L}_{SS}(u_S) + \mathcal{L}_{SF}(u_F) = f_S(t) \qquad \text{solid equations} \tag{1}$$

$$\mathcal{L}_{FS}(u_S) + \mathcal{L}_{FF}(u_F) = f_F(t) \qquad \text{fluid equations.} \tag{2}$$

An approximate solution of this system of coupled non-linear equations requires a large number of unknowns in order to achieve the accuracy required for capturing the most important interaction phenomena. Typically, discretisation of the structural component involves a degree of freedom on the order of $n_S = 10^3$ while the fluid flow field easily requires $n_F = 10^7$, especially if the laminar-turbulent transition or the turbulent regime is to be considered.

Field elimination. If in the multi-physics problem the behaviour of the structural component is of primary interest, one may (theoretically) perform a static condensation of the fluid. Given the invertibility of the fluid operator, from equation (2) one obtains for the fluid field

$$u_F = \mathcal{L}_{FF}^{-1}(f_F(t) - \mathcal{L}_{FS}(u_S)),$$

which allows its elimination in equation (1)

$$\{\mathcal{L}_{SS} - \mathcal{L}_{SF}\mathcal{L}_{FF}^{-1}\mathcal{L}_{FS}\}(u_S) = f_S(t) - \mathcal{L}_{SF}\mathcal{L}_{FF}^{-1}(f_F(t)). \tag{3}$$

Equation (3) describes the response of a structure in the presence of motion- or deformation-dependent fluid loading and fluctuating fluid forces resulting from flow-intrinsic instabilities (both on the left hand side) as well as time-dependent extraneous excitations (second term on the right hand side). Clearly, the above formulation is only of limited practical interest since the inverse of the fluid operator is unavailable. Nevertheless, the notation reveals the complexity of identifying a suitable reduced-order description for a given fluid-structure interaction problem.

Engineering approach/reduction. A common approach to describing flow-induced loading on structures is to summarize the effects of motion-dependent fluid loading by using so-called *added coefficients* in the modelling. Specifically, one assumes that for quasi-steady processes the above projection can be sufficiently well represented by the operator $\mathcal{A}$

$$-\mathcal{L}_{SF}\mathcal{L}_{FF}^{-1}\mathcal{L}_{FS} \approx \mathcal{A} \tag{4}$$

such that the structure's field equation reduces to

$$\{\mathcal{L}_{SS} + \mathcal{A}\}(u_S, \dot{u}_S, \ddot{u}_S) = f_S(t). \tag{5}$$

The nonlinear operator $\mathcal{A}$ can be approximated by a multi-variate Taylor series expansion in terms of structural displacement u_S, structural velocity $\dot{u}_S$ and structural acceleration $\ddot{u}_S$ - neglecting all mixed terms -

$$\mathcal{A} = \mathcal{A}(u_S, \dot{u}_S, \ddot{u}_S) := \mathcal{A}_k(u_S) + \mathcal{A}_b(\dot{u}_S) + \mathcal{A}_m(\ddot{u}_S)\,, \tag{6}$$

which for the structural oscillator results in expressions for the so-called added fluid stiffness $\mathcal{A}_k$, added fluid damping $\mathcal{A}_b$ and added fluid mass $\mathcal{A}_m$, respectively. Consideration of only first order expansions of the functional $\mathcal{A}$ then leads to the often used form of constant added coefficients, which are generally non-symmetric.

Aerodynamic added mass $\mathcal{A}_m$ captures the surrounding fluid accelerated by the structural vibration and always increases total mass. Added aerodynamic damping $\mathcal{A}_b$ can result in increased reduction of structural vibration amplitudes in addition to structural damping. More importantly, aerodynamic damping - having negative coefficients - constantly introduces flow energy to the oscillator with every cycle (self-excitation), giving rise to dynamic instabilities. Aerodynamic stiffness $\mathcal{A}_k$ may increase or decrease the total stiffness of the system. The presence of negative added fluid stiffness can therefore lead to static divergence if the original structural stiffness is small.

Unfortunately, the quantification of all types of added coefficients is not straightforward. For selective simple interaction problems one may find a set of relevant added coefficients by using analytical approaches and experimental validation. This approach leads to semi-empirical models of fluid-structure interaction using aero-elastic coefficients and their derivatives (Paidoussis (1987); Parkinson (1989)). Alternatively, the quantification of motion-dependent fluid forces (added effects) can be achieved through numerical simulation of the fluid-structure system.

2.1 Modelling with *added coefficients*

The steps and implications of modelling with added coefficients are demonstrated for a simple structure with distributed mass, damping, and stiffness properties. Separating the temporal and spatial components results in the equation of motion for the multi degree-of-freedom oscillator

$$\boldsymbol{\mathcal{M}}\ddot{\mathbf{u}}_S + \boldsymbol{\mathcal{B}}\dot{\mathbf{u}}_S + \boldsymbol{\mathcal{K}}\mathbf{u}_S = \mathbf{f}(t) - \mathbf{f}_F(\ddot{\mathbf{u}}_S, \dot{\mathbf{u}}_S, \mathbf{u}_S) \tag{7}$$

with motion-independent and motion-dependent components in the forcing terms. The first order Taylor series expansion of the flow-induced force $\mathbf{f}_F$

$$\mathbf{f}_F = \frac{\partial \mathbf{f}_F}{\partial \ddot{\mathbf{u}}_S}\ddot{\mathbf{u}}_S + \frac{\partial \mathbf{f}_F}{\partial \dot{\mathbf{u}}_S}\dot{\mathbf{u}}_S + \frac{\partial \mathbf{f}_F}{\partial \mathbf{u}_S}\mathbf{u}_S = \boldsymbol{\mathcal{A}}_m\ddot{\mathbf{u}}_S + \boldsymbol{\mathcal{A}}_b\dot{\mathbf{u}}_S + \boldsymbol{\mathcal{A}}_k\mathbf{u}_S \tag{8}$$

leads to the matrices representing aerodynamic mass, $\boldsymbol{\mathcal{A}}_m$, damping, $\boldsymbol{\mathcal{A}}_b$, and stiffness, $\boldsymbol{\mathcal{A}}_k$, which in the general case are non-symmetric and complex-valued, because the aerodynamic forces are motion-dependent. The overall solution of the resulting equation of structural motion

$$(\boldsymbol{\mathcal{M}} + \boldsymbol{\mathcal{A}}_m)\ddot{\mathbf{u}}_S + (\boldsymbol{\mathcal{B}} + \boldsymbol{\mathcal{A}}_b)\dot{\mathbf{u}}_S + (\boldsymbol{\mathcal{K}} + \boldsymbol{\mathcal{A}}_k)\mathbf{u}_S = \mathbf{f}(t) \tag{9}$$

requires an analysis of the particular and homogeneous case. The solution of the homogeneous equation of motion requires solving the associated quadratic eigenvalue problem

$$\left\{\lambda^2(\boldsymbol{\mathcal{M}} + \boldsymbol{\mathcal{A}}_m) + \lambda(\boldsymbol{\mathcal{B}} + \boldsymbol{\mathcal{A}}_b) + (\boldsymbol{\mathcal{K}} + \boldsymbol{\mathcal{A}}_k)\right\}\hat{\mathbf{u}}_S = \mathbf{0} \tag{10}$$

with partially non-symmetric, and possibly non-positive, complex effective matrices resulting in $2n_S$ complex-valued eigenvalues with right and left eigenvectors.

Static divergence is an instability caused by the presence of flow-induced stiffness that reduces the effective stiffness of the system to zero (e.g. fluid-conveying elastic tube or torsional vibrations of the section in cross-flow).

If the fluid stiffness is a function of the approaching flow velocity $u_{F,\infty}$, the range of critical flow-velocities for which very large structural deformations occur can be determined with the condition $\det(\boldsymbol{\mathcal{K}} + \boldsymbol{\mathcal{A}}_k) = 0$.

The *Dynamic response* of the structural system originates from the forcing term $\mathbf{f}(t)$ that may express periodic (e.g. vortex shedding) or non-periodic (e.g. turbulence, rough sea waves) fluid forces. Such forced vibrations can be calculated using standard approaches. If the fluid force is periodic in time, the structural system may experience parametric resonance: dynamic instability occurs if the frequency of the flow-induced force is an integer multiple of one of the natural frequencies of the structure.

Dynamic divergence can be caused by internal or external flow at high velocities (e.g. flutter of fluid-conveying pipes or transversal galloping). Instabilities controlled by fluid damping (single mode flutter) are associated with the symmetric part of the total damping, $(\boldsymbol{\mathcal{B}} + \boldsymbol{\mathcal{A}}_b)$, which introduces negative fluid damping and causes permanent transfer of flow energy to the structural oscillator (self-excitation). Instabilities controlled by fluid stiffness (coupled mode flutter) are associated with the antisymmetric component of the total stiffness, $(\boldsymbol{\mathcal{K}} + \boldsymbol{\mathcal{A}}_k)$.

From the above discussion it can be concluded that fluid-induced added mass is not causally involved in the various instability phenomena. Nevertheless, the presence of added mass influences the dynamic characteristics of the system as the eigenstates depend on the distribution of total mass. The actual importance of flow-induced mass very much depends on the density of the fluid but also on the effects of variable fluid boundaries (e.g. free surfaces of waves) that may interact with the structural motion making the added fluid mass position-dependent.

2.2 Analytical determination of fluid-induced *added mass*

Determination of added mass consists of a number of important steps: specification of the fluid model to be used, description of structural motion and evaluation of motion-induced fluid forces. The evaluation step involves the computation of fluid force sensitivities on structural rigid body motion and deformation modes from the underlying fluid model equations. This can be achieved by means of approximate solutions or - if the applied fluid model and the flow domain are not too complicated - by means of analytical solutions, as they are presented in (Korotkin, 2009) for ship applications.

Modelling of the fluid flow. In order to allow for an analytical investigation, the fluid here is assumed to be inviscid, incompressible, and irrotational. With the fluid velocity $\mathbf{u}_F$ and both the continuity equation together with the rotation-free condition

$$\nabla \cdot \mathbf{u}_F = 0 \qquad \text{and} \qquad \nabla \times \mathbf{u}_F = 0 \tag{11}$$

allow the introduction of the potential ϕ which fulfils the relation $\mathbf{u}_F = \nabla\phi$ and lead to the Laplace equation of potential flow

$$\Delta\phi = 0 \quad \text{in}\, \Omega \tag{12}$$

with the Neumann boundary condition

$$(\nabla\phi - \dot{\mathbf{u}}_S) \cdot \mathbf{n}_\Sigma = 0 \quad \text{on}\, \Sigma \tag{13}$$

at the fluid-structure interface Σ with the normal vector $\mathbf{n}_\Sigma$ pointing outwards the fluid domain, representing kinematic compatibility between fluid and structure in terms of velocity. Similarly, impermeability along the outer fluid boundary Γ requires

$$\nabla\phi \cdot \mathbf{n}_\Gamma = 0 \quad \text{on}\, \Gamma. \tag{14}$$

With the Bernoulli equation for unsteady flows in the absence of volume forces

$$\frac{\partial\bar{\phi}}{\partial t} + \frac{1}{2}\|\nabla\bar{\phi}\|_2^2 + \frac{p}{\rho_F} = h(t)\,, \tag{15}$$

the pressure state can be expressed by

$$p = -\rho_F \frac{\partial\phi}{\partial t} - \frac{1}{2}\rho_F \|\nabla\phi\|_2^2\,, \tag{16}$$

where ρ_F is fluid density and the level $h(t)$ is incorporated into the potential by the transformation $\bar{\phi} = \phi + \int_{t_0}^{t} h(\tau)\,\mathrm{d}\tau$ such that $\nabla\bar{\phi} = \nabla\phi$. The potential ϕ of the non-oscillatory fluid only depends on time if the boundary conditions are time-dependent. Here, the velocity at the interface Σ may vary in time due to the oscillatory behaviour of the structure, which can be described as a linear combination of its rigid body and flexural modes. For internal flow situations, the incompressibility of the fluid restricts the spectrum of admissible structural deformations.

Description of the fluid-structure interface motion. The motion of the fluid-structure interface Σ, which is equivalent to the motion of the structural boundary, can be represented by a separation ansatz in space (surface coordinate $\mathbf{s}$) and time t using the basis of linear independent deformation modes

$$\mathbf{u}_S(\mathbf{s}, t) = \sum_k \mathbf{v}_k(\mathbf{s}) \cdot g_k(t) \tag{17}$$

where $\mathbf{v}_k(\mathbf{s})$ is the k-th global mode of the interface and $g_k(t)$ is the associated generalised coordinate.

Modal decomposition of fluid response. Since the model equations of the chosen fluid are linear, the fluid velocity potential associated with each interface mode can be expressed by the modal superposition

$$\phi(\mathbf{x},t) = \sum_k \phi_k(\mathbf{x}) \cdot \dot{g}_k(t) \tag{18}$$

and the solution to the governing equation (12) in terms of ϕ_k with mode-specific boundary conditions

$$(\nabla\phi_k - \mathbf{v}_k \dot{g}_k) \cdot \mathbf{n}_\Sigma = 0 \quad \text{on}\,\Sigma \qquad \text{and} \qquad \nabla\phi_k \cdot \mathbf{n}_\Gamma = 0 \quad \text{on}\,\Gamma. \tag{19}$$

Approaches to the derivation of modal added fluid mass. Three different approaches for determining added mass within the framework of modal decomposition of the structural motion are considered:

- Direct evaluation of the resultant fluid force $f_{F;k}$ exerted on the structural mode k can be achieved by integrating the modal projection of the interface fluid pressure, such that

$$\begin{aligned}
f_{F;k} &= \int_\Sigma (\mathbf{v}_k \dot{g}_k \cdot \mathbf{n}_\Sigma)\, p \,\mathrm{d}\Sigma && (20)\\
&= \int_\Sigma (\mathbf{v}_k \dot{g}_k \cdot \mathbf{n}_\Sigma)\, \rho_F \left(-\frac{\partial\phi}{\partial t} - \frac{1}{2}\|\nabla\phi\|_2^2\right) \mathrm{d}\Sigma && (21)\\
&= \int_\Sigma (\nabla\phi_k \cdot \mathbf{n}_\Sigma)\, \rho_F \left(-\sum_n \frac{\partial\phi_n}{\partial t}\right) \mathrm{d}\Sigma && (22)\\
&= \int_\Sigma (\nabla\phi_k \cdot \mathbf{n}_\Sigma)\, \rho_F \left(-\sum_n \phi_n \cdot \ddot{g}_n\right) \mathrm{d}\Sigma && (23)\\
&= \sum_n \left(-\rho_F \int_\Sigma (\nabla\phi_k \cdot \mathbf{n}_\Sigma)\, \phi_n \,\mathrm{d}\Sigma\right) \cdot \ddot{g}_n && (24)\\
&= \sum_n \mathcal{A}_{m;kn} \cdot \ddot{g}_n\,. && (25)
\end{aligned}$$

The term related to kinetic flow energy in equation (21) vanishes as the closed surface integral of the continuous integrand is zero. Expression (24) for the modal fluid-induced force is a function of the generalized structural acceleration $\ddot{g}_n$ of mode n. The final result (25) is achieved by introducing the *added mass* coefficient

$$\mathcal{A}_{m;kn} = -\rho_F \int_\Sigma (\nabla\phi_k \cdot \mathbf{n}_\Sigma)\phi_n \,\mathrm{d}\Sigma \tag{26}$$

of the k-th structural mode due to the action (fluid acceleration) done by the n-th mode. The added mass coefficients are symmetric

$$\mathcal{A}_{m;kn} = \mathcal{A}_{m;nk} . \tag{27}$$

In a similarly direct fashion the added mass coefficients for fluid-induced moments can be derived. As denoted in the equilibrium equation (7), the fluid force vector $\mathbf{f}_F$ points in the same direction as the vector of structural inertia forces.

- With the pressure work function on the fluid-structure interface

$$\begin{aligned} W &= \int_\Sigma (\mathbf{u}_S \cdot \mathbf{n}_\Sigma)\, p \,\mathrm{d}\Sigma & (28)\\ &= \int_\Sigma \left(\sum_k (\mathbf{v}_k \cdot \mathbf{n}_\Sigma)\, g_k \right) \rho_F \left(-\frac{\partial \phi}{\partial t} - \frac{1}{2} \|\nabla\phi\|_2^2 \right) \mathrm{d}\Sigma & (29)\\ &= \int_\Sigma \left(\sum_k (\nabla\phi_k \cdot \mathbf{n}_\Sigma)\, g_k \right) \rho_F \left(-\sum_n \phi_n \cdot \ddot{g}_n \right) \mathrm{d}\Sigma\,, & (30) \end{aligned}$$

the force component associated with mode k is determined by the partial derivative of the work function with respect to the k-th generalized coordinate

$$f_{F;k}(g_1, g_2, \dots) = \frac{\partial W}{\partial g_k} = \sum_n \mathcal{A}_{m;kn} \cdot \ddot{g}_n\,. \tag{31}$$

Again, the use of added mass coefficients allows for directly identifying modal interactions and inter-dependency in terms of fluid-induced inertial force contributions.

- The evaluation of kinetic fluid energy in the fluid domain,

$$T = \int_\Omega \frac{1}{2}\rho_F \|\mathbf{u}_F\|_2^2 \,\mathrm{d}\Omega = \frac{1}{2}\rho_F \int_\Omega \|\nabla\phi\|_2^2 \,\mathrm{d}\Omega\,, \tag{32}$$

induced by a specific structural deformation mode constitutes the third way of establishing added mass coefficients. Integration by parts of the above expression leads to

$$T = \frac{\rho_F}{2}\left[\int_\Omega \Delta\phi\,\phi\,\mathrm{d}\Omega - \int_\Sigma (\nabla\phi\cdot\mathbf{n}_\Sigma)\,\phi\,\mathrm{d}\Sigma - \int_\Gamma (\nabla\phi\cdot\mathbf{n}_\Gamma)\,\phi\,\mathrm{d}\Gamma \right] . \tag{33}$$

Taking into account the boundary conditions defined in equation (19) and the validity of equation (12) it follows, that

$$T = -\frac{1}{2}\rho_F \int\limits_{\Sigma} (\nabla\phi \cdot \mathbf{n}_\Sigma)\,\phi\,\mathrm{d}\Sigma \tag{34}$$

$$= \frac{1}{2}\int\limits_{\Sigma} -\rho_F \left(\sum_k (\nabla\phi_k \cdot \mathbf{n}_\Sigma)\,\dot{g}_k\right)\left(\sum_n \phi_n \cdot \dot{g}_n\right)\mathrm{d}\Sigma \tag{35}$$

$$= \frac{1}{2}\sum_k\sum_n \dot{g}_k \cdot \mathcal{A}_{m;kn} \cdot \dot{g}_n \tag{36}$$

where $\mathcal{A}_{m;kn}$ is defined and interpreted as in equation (26).

Added mass of circular beam structure immersed in potential flow. The computation of the added mass coefficients is demonstrated for a plane flexible ring structure (radius a) embedded in a fluid-containing cylinder (radius b) as shown in Figure 1.

For the structural component, small undamped in-plane motions of an elastic ring are considered. With the radial displacement $v(\theta)$, the angular displacement $u(\theta)$, and the assumption $\partial u/\partial\theta + v = 0$ (angular rigidity) the equation of radial motion is given by

$$\rho_S A\ddot{v} + \frac{EI}{a^4}\left(\frac{\partial^4 v}{\partial\theta^4} + 2\frac{\partial^2 v}{\partial\theta^2} + v\right) = f(\theta,t) \tag{37}$$

where A is section area, EI represents bending stiffness and $f(\theta,t)$ is time-dependent loading. The conditions on the periodicity of the displacement, the bending angle, the bending moment and the shear force: $\partial^m v(0,t)/\partial\theta^m = \partial^m v(2\pi,t)/\partial\theta^m$ for $m = 0,1,2,3$ constitute the boundary conditions. The eigensolution of the present radial vibration problem

$$v_{k1}(\theta) = \alpha_{k,1}\cos k\theta \quad \text{and} \quad v_{k2}(\theta) = \alpha_{k,2}\sin k\theta \quad \text{with}\, k = 0,1,2,\dots \tag{38}$$

proves to have double eigenvalues $\lambda_{k1,2}$ with two linearly independent phase-shifted eigenfunctions, $v_{k1,2}$, per wave number, k.

For potential flow (12), the Laplace equation in polar coordinates (r,θ)

$$\Delta\phi(r,\theta) = \frac{\partial^2\phi}{\partial r^2} + \frac{1}{r}\frac{\partial\phi}{\partial r} + \frac{1}{r^2}\frac{\partial^2\phi}{\partial\theta^2} = 0 \quad \text{for} \quad r = [a,b] \text{ and } \theta = [0,2\pi]\,, \tag{39}$$

is completed by the boundary condition at the interface (13) and outer fluid boundary (14)

$$\frac{\partial\phi_j(r=a,\theta)}{\partial r} = v_j(\theta) \qquad \text{and} \qquad \frac{\partial\phi_j(r=b,\theta)}{\partial r} = 0. \tag{40}$$

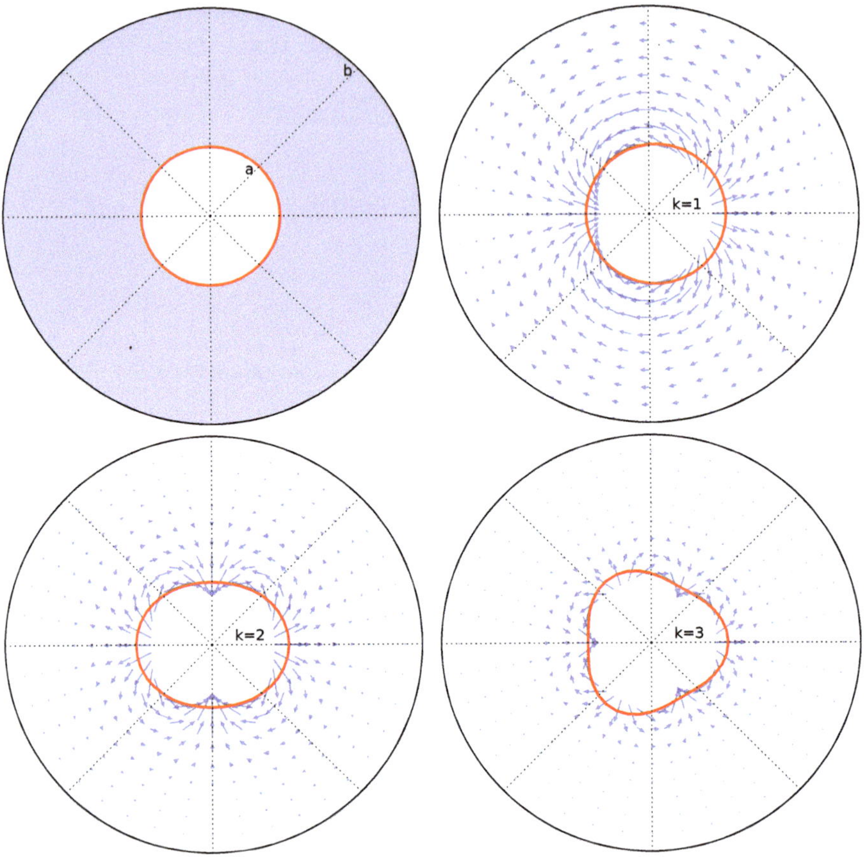

Figure 1. Circular beam embedded in circular fluid domain: structural eigensolution (deformation) and associated potential flow field (velocity).

An analytical solution to equation (39) is available through the separation ansatz $\phi(r,\theta) = \eta(r) \cdot \xi(\theta)$ which leads to

$$\phi(r,\theta) = \sum_{n=1}^{\infty} \eta_n(r) \cdot \xi_n(\theta) = \sum_{n=1}^{\infty} (A_n r^n + B_n r^{-n}) \cdot (\cos n\theta + \sin n\theta) \quad (41)$$

The motion of the ring structure in the k-th eigenmode results in an associated fluid velocity field whose potential can be expressed by (41) and for which the coefficients, A and B, can be determined from the boundary conditions defined in equation (40). Equating coefficients for the structural mode, k, and wave number, n, one obtains for $i = 1, 2$

$$A_n\Big|_{v_{k,i}} = -\frac{\alpha_{k,i}}{k}\frac{a^{n+1}}{b^{2n}-a^{2n}} \quad \text{and} \quad B_n\Big|_{v_{k,i}} = b^{2n}A_n\Big|_{v_{k,i}}. \tag{42}$$

The modal added mass coefficient, $\mathcal{A}_{m;kn}$, can now be derived according to equation (26) as

$$\begin{aligned}
\mathcal{A}_{m;kn} &= -\rho_F \int_\Sigma \frac{\partial\phi(r=a,\theta)}{\partial r}\phi(r=a,\theta)\,\mathrm{d}s & (43)\\
&= -\rho_F \int_0^{2\pi} (\alpha_k \cos k\theta)\cdot(A_n r^n + B_n r^{-n})(\cos n\theta)\cdot a\ \mathrm{d}\theta & (44)\\
&= \rho_F \frac{\alpha_k\alpha_n}{k}a^2\frac{b^{2k}+a^{2k}}{b^{2k}-a^{2k}}\int_0^{2\pi}\cos k\theta\cdot\cos n\theta\ \mathrm{d}\theta & (45)\\
&= \rho_F \frac{\alpha_k\alpha_n}{k}a^2\frac{b^{2k}+a^{2k}}{b^{2k}-a^{2k}}\pi\cdot\delta_{kn} \quad \text{with} \quad \delta_{kn} = \begin{cases} 1\,\forall k = n \\ 0\,\forall k \neq n \end{cases} & (46)
\end{aligned}$$

For this axially symmetric problem, the added mass coefficients are modally decoupled. The analytical expression for $\mathcal{A}_{m;kn}$ moreover shows the influence of open/enclosed flow situations on the modal added mass coefficient. Enclosed incompressible flow ($b/a \to 1$) usually generates increased added mass compared to open situations ($b/a \to \infty$).

2.3 Variable *added mass* and more advanced situations

The range of applicability of the analytical approach to determination of added mass is limited by the same simplifications, e.g. restriction to inviscid flows, the assumption of small structural motion or changes in the flow boundaries, that allow linear superposition and closed-form mathematical solution. In general, the effective added mass is a function of (1) the geometrical shape and deformability of the structure and its surface, (2) the geometry of the fluid domain and its boundaries, (3) the relative position/motion of the structure to the fluid (Konstantinidis (2013)) and adjacent fluid boundaries or free surfaces, (4) for compressible and free surface flows the propagation of pressure and gravity waves, respectively, giving rise to a phase shift between body motion and induced fluid force which reveals the possibility of added fluid damping even if the fluid is inviscid. Computational modelling and approximate numerical solution of fluid-structure interaction problems can help to overcome certain limitations of the analytic approach and allow consideration of the above mentioned dependencies not only for variable added fluid mass but also variable added fluid damping and stiffness.

3 Computational Modelling

Computational analysis of fluid-structure interaction requires the solution of a two-field problem consisting of model equations for structural and fluid dynamics. These two fields may be accompanied by additional descriptions for the coupling interface and geometrical changes of the fluid domain involved. The algorithmic treatment of the coupled problem needs to consider the dominating coupling characteristics between participating fields. For both, weakly and strongly coupled systems, very efficient numerical schemes have been developed over the last decades. Comprehensive overviews are given in the work of Wall (1999) and Steindorf (2002). More recent developments can be fond in Bungartz and Schäfer (2006), Bungartz et al. (2012) or Dettmer and Perić (2013).

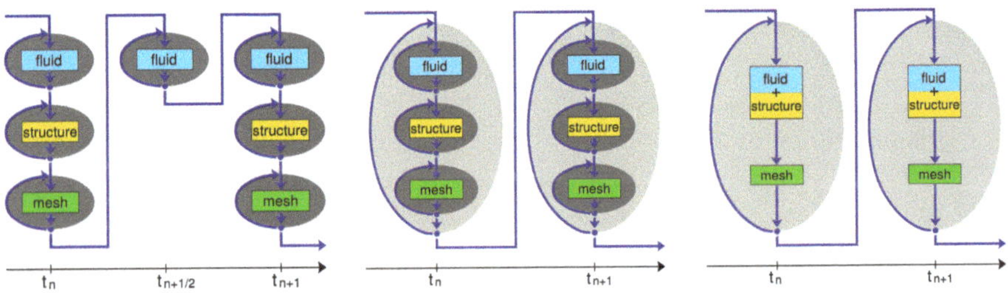

Figure 2. Coupling: partitioned(weak), partitioned(strong), simultaneous.

Coupling Strategies. *Partitioned approaches* solve the involved fields sequentially and are designed for highly modular application of specific discretisation and solution methods for each sub-problem. Interaction of the sub-systems fluid and structure is realized by communicating interface constraints as boundary conditions and leads to Dirichlet-Neumann methods: flow interface tractions acting as loads on the structure, structural velocities at the interface as Dirichlet conditions for the flow field, and structural deformations as conditions for mesh motion or re-meshing algorithms. Weakly coupled solutions in the time domain are carried out if exchange of information between the sub-systems takes place only once per time step and therefore the overall system will not be in a physically consistent state. While this very popular approach is acceptable for weakly coupled systems, it cannot be applied successfully for numerical investigations of strong couplings. Strongly coupled solutions are achieved in partitioned methods by iterated solution of the sub-systems. Different staggered schemes have been derived by choosing appropriate orderings in solving the sub-systems (including mesh motion etc.) with respect to specific temporal discretization methods of the solver modules for the structural and fluid part. The consistent state of the coupled system at a time instant is reached, often supported by selective numerical relaxation during the coupling iteration.

Simultaneous approaches do not apply a strict partitioning of the physical domain. Such an holistic view requires the solution of the discretised model equations for solid and fluid in a single algebraic system. The numerical properties of the resulting large system of equations may indicate the need for specialized preconditioning of iterative solvers. Monolithic schemes with identical time discretisation of both fields can represent an ideal basis for conservative and precise transient coupling. For physically strongly coupled systems, such schemes may provide improved convergence and numerical efficiency compared to partitioned approaches.

Deforming flow domain. In a fluid-structure system, structural motion causes geometrical changes of the flow domain. The discretisation of the varying fluid domain has to fulfil compatibility requirements along the interface. The model equations of the fluid are therefore defined and to be solved on a time-dependent deforming domain. Arbitrary Lagrange-Euler (ALE) formulations are often applied in semi-discrete methods and introduce the kinematics of an arbitrary reference configuration (e.g. the moving fluid mesh) to the model equations. Alternatively, uniform discretisation in space and time allows inherently consistent integration of the weak forms on the deforming space-time domain of the flow.

If the topology of the flow domain does not change, updated locations of internal grid nodes can be determined by using a pseudo-structure approach. The fluid mesh is then seen as a (hyper-)elastic solid subject to boundary displacements according to the current deformation of the physical structure. While mesh-moving schemes are sufficient to treat moderate displacements of involved structures, partial re-meshing of the flow discretisation or the application of overlapping chimera-techniques become necessary if topological changes occur. For rotating flow-immersed structures special clicking- and sliding-mesh techniques were developed that implement consistent coupling of adjacent but rotating fluid grids.

3.1 FSI with space-time finite elements

The following sections give the mathematical description of the boundary-coupled fluid-structure problem, involving an elastic structure at large deformations and a viscous incompressible fluid in terms of the strong form equations. The weak form is derived in context of the space-time finite element method and a monolithic solution approach. The final weak formulation is based on the velocity state of the involved continua, simplifying the treatment of coupling conditions considerably.

Strong form of governing equations for fluid and structure.

Structure. The conservation of momentum of a solid body or structure is described on the material configuration

$$\rho_{S0}\dot{\mathbf{v}} - \nabla_0 \cdot (\mathbf{FS}) - \mathbf{f}_0 = \mathbf{0} \quad \text{on} \quad Q_0 \tag{47}$$

with density, ρ_{S0}, and volume force, $\mathbf{f}_0$. The domain, $Q_0 = \Omega_0 \times I$, refers to the space-time continuum under consideration with the spatial reference domain, Ω_0, and time interval, $I = [t_a, t_e]$. Linear elastic material behaviour is assumed and given in rate form

$$\underline{\mathbf{C}}^{-1} : \dot{\mathbf{S}} - \dot{\mathbf{E}} = \mathbf{0}, \tag{48}$$

where $\mathbf{S}$, $\dot{\mathbf{E}}$ and $\underline{\mathbf{C}}$ are 2nd Piola-Kirchhoff stress, Green-Lagrange strain rate, and the fourth order elasticity tensor, respectively. The rate of strain at a material point is a function of deformation and velocity state

$$\dot{\mathbf{E}}(\mathbf{v}, \mathbf{u}) = \frac{1}{2}\left(\nabla_0\mathbf{v} + (\nabla_0\mathbf{v})^T + (\nabla_0\mathbf{u})^T\nabla_0\mathbf{v} + (\nabla_0\mathbf{v})^T\nabla_0\mathbf{u}\right). \tag{49}$$

Dirichlet and Neumann boundary conditions are defined on the outer space-time boundary $P_0 = \Gamma_0 \times I$ of the solid body

$$\mathbf{v} - \bar{\mathbf{v}} = \mathbf{0} \quad \text{on} \quad P_0^v \qquad \text{and} \qquad \mathbf{t}_0 - \bar{\mathbf{t}}_0 = \mathbf{0} \quad \text{on} \quad P_0^t, \tag{50}$$

where $\bar{\mathbf{v}}$ and $\bar{\mathbf{t}}_0$ are imposed boundary velocities and tractions, respectively. Moreover, the dynamic problem at hand requires the definition of initial values for the velocity state

$$\mathbf{v}(\mathbf{x}, t = 0) - \mathbf{v}_{t_a}(\mathbf{x}) = \mathbf{0} \quad \text{on} \quad \Omega_0. \tag{51}$$

Fluid. The incompressible Navier-Stokes equations are used to describe viscous flow at moderate speeds. Momentum balance is defined on the current configuration

$$\rho_F(\mathbf{v}_{,t} + \mathbf{v} \cdot \nabla\mathbf{v}) - \nabla \cdot \mathbf{T} - \mathbf{f} = \mathbf{0} \quad \text{on} \quad Q \tag{52}$$

and accompanied by the continuity equation or incompressibility constraint on the velocity field

$$\nabla \cdot \mathbf{v} = 0 \quad \text{on} \quad Q. \tag{53}$$

Assuming a Newtonian fluid, the constitutive relation,

$$\mathbf{T} = 2\mu\mathbf{D}(\mathbf{v}) - p\mathbf{I} \tag{54}$$

between Cauchy stress state $\mathbf{T}$ and the rate of strain

$$\mathbf{D}(\mathbf{v}) = \frac{1}{2}\left(\nabla_0\mathbf{v} + (\nabla_0\mathbf{v})^T\right) \tag{55}$$

introduces hydrostatic pressure p and viscosity μ of the fluid. Again, Dirichlet and Neumann boundary conditions are defined on the space-time boundary $P = \Gamma \times I$ of the fluid body

$$\mathbf{v} - \bar{\mathbf{v}} = \mathbf{0} \quad \text{on} \quad P^v \qquad \text{and} \qquad \mathbf{t} - \bar{\mathbf{t}} = \mathbf{0} \quad \text{on} \quad P^t, \tag{56}$$

where $\bar{\mathbf{v}}$ and $\bar{\mathbf{t}}$ are imposed boundary velocities and tractions, respectively. Compatible (divergence-free) initial values for the velocity state are to be defined

$$\mathbf{v}(\mathbf{x}, t = 0) - \mathbf{v}_{t_a}(\mathbf{x}) = \mathbf{0} \quad \text{on} \quad \Omega. \tag{57}$$

Coupling conditions. Flow domain and structural domain are coupled along the common space-time boundary $R = \Sigma \times I$ representing the fluid-structure interface. At the interface no-slip conditions are applied, requesting continuity of fluid and structural velocities

$$\mathbf{v}_F - \mathbf{v}_S = \mathbf{0} \quad \text{on} \quad R \tag{58}$$

and ensuring herewith geometrical conservation. Further, in order to fulfil momentum balance at the interface, fluid and solid interface tractions have to be of the same magnitude and opposite in direction

$$\mathbf{t}_F + \frac{d\Gamma_0}{d\Gamma}\mathbf{t}_S = \mathbf{0} \quad \text{on} \quad R. \tag{59}$$

The referential solid traction $\mathbf{t}_S$ is projected to the current frame.

Space-time weak form and discretisation. The weighted residual method is applied to the strong form equations of solid, fluid, and coupling conditions presented in the previous section. The resulting weak form of the whole coupled system and the space-time domain is then discretised using the space-time finite element method, see Argyris and Scharpf (1969), and a discontinuous Galerkin method for integration in time. The basic idea of a space-time approach is to include the temporal axis in the finite element discretisation. For numerical efficiency, the space-time domain Q

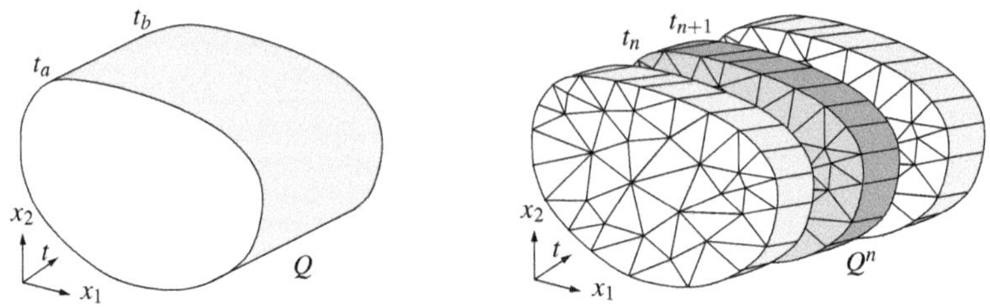

Figure 3. Discretisation of the continuous space-time domain using time-discontinuous space-time finite elements.

is divided into a sequence of N time slabs $Q^n = \Omega^n \times [t_n, t_{n+1}]$, as shown in Figure 3, which are solved successively.

At time instant, t_n, the energy of the discretised system at the end of the previous time slab, t_n^-, must be equal to the energy at the beginning of the next time step t_n^+. For time-discontinuous approximations of field unknowns, this leads to additional jump terms in the weak form. Moreover, spatial discretisations from t_n^- and t_n^+ do not need to be conforming. For first order ordinary differential equations, the resulting time integration scheme is A-stable and third-order accurate for linear temporal interpolation.

Structure. The stabilized space-time finite element formulation of the structural component (47)-(51) within the time slab, Q_0^n, including boundaries P_0^n reads

$$\int_{Q_0^n} \delta\mathbf{v} \cdot \rho_{S0}\, (\dot{\mathbf{v}} - \mathbf{f}_0)\, \mathrm{d}Q_0 + \int_{Q_0^n} \dot{\mathbf{E}}(\delta\mathbf{v}, \mathbf{u}) : \mathbf{S}\, \mathrm{d}Q_0 \tag{60a}$$

$$+ \sum_e \int_{Q_0^{n,e}} \delta\mathbf{S} : \big(\underline{\mathbf{C}}^{-1} : \dot{\mathbf{S}} - \dot{\mathbf{E}}(\mathbf{v}, \mathbf{u})\big)\, \mathrm{d}Q_0 \tag{60b}$$

$$+ \int_{\Omega_0} \delta\mathbf{v}(t_n^+) \cdot \rho_0 \big(\mathbf{v}(t_n^+) - \mathbf{v}(t_n^-)\big)\, \mathrm{d}\Omega_0 \tag{60c}$$

$$+ \sum_e \int_{\Omega_0^e} \delta\mathbf{S}(t_n^+) : \underline{\mathbf{C}}^{-1} : \big(\mathbf{S}(t_n^+) - \mathbf{S}(t_n^-)\big)\, \mathrm{d}\Omega_0 \tag{60d}$$

$$+ \text{terms stabilization of momentum equation} \tag{60e}$$

$$- \int_{P_0^{n,t}} \delta\mathbf{v} \cdot \bar{\mathbf{t}}_0\, \mathrm{d}P_0 \; = 0 \qquad \forall\ \delta\mathbf{v},\, \delta\mathbf{S}\,. \tag{60f}$$

In equation (60), line (a) represents the weak form of the momentum conservation and line (b) fulfils the constitutive law on element level, leading to a mixed-hybrid formulation as presented in (Knippers and Harbord, 1994), where only the velocities are global degrees of freedom. The jump terms for velocities (c) and stresses (d) satisfy the initial conditions of the time slab in integral form. The stabilization mentioned in line (e) is useful for wave propagation problems. The definition for the stabilization parameter given in (Hughes and Hulbert, 1988) is used. Interpolation functions for velocities are chosen to be multi-linear in space and linearly discontinuous in time, while the stress interpolation is discontinuous and incompletely linear in space and discontinuous linear in time, see (Hübner, 2003). Temporal integration of the velocities leads to the displacement field $\mathbf{u}$, which is required for computation of the rate of the Green-Lagrange strain tensor and for specifying the current position of the fluid-structure interface.

Fluid. The weighted residual formulation of the strong forms of the incompressible viscous fluid (52)-(57) in a space-time slab Q_t^n using the Galerkin method is

$$\int_{Q_t^n} \delta\mathbf{v} \cdot \rho_F(\mathbf{v}_{,t} + \mathbf{v} \cdot \nabla\mathbf{v} - \mathbf{f})\, \mathrm{d}Q_t +$$

$$\int_{Q_t^n} \mathbf{D}(\delta\mathbf{v}) : 2\mu\mathbf{D}\, \mathrm{d}Q_t - \int_{Q_t^n} \nabla \cdot (\delta\mathbf{v})\, p\, \mathrm{d}Q_t \tag{61a}$$

$$+ \int_{Q_t^n} \delta p\, \nabla \cdot \mathbf{v}\, \mathrm{d}Q_t \tag{61b}$$

$$+ \int_{\Omega_t^n} \delta\mathbf{v}(t_n^+) \cdot \rho(\mathbf{v}(t_n^+) - \mathbf{v}(t_n^-))\, \mathrm{d}\Omega_t \tag{61c}$$

$$+ \text{ terms Galerkin/least squares stabilization} \tag{61d}$$

$$- \int_{P_t^{n,h}} \delta\mathbf{v} \cdot \bar{\mathbf{t}}\, \mathrm{d}P_t = 0 \qquad \forall \delta\mathbf{v}, \delta p. \tag{61e}$$

Line (61a) represents the weak form of conservation of momentum, fulfilling the constitutive relation for the Newtonian fluid and the kinematics exactly. The incompressibility constraint is weighted with the variation of the pressure in line (61b). Line (61c) ensures the consistent transfer of kinetic energy from the previous time slab end at t_n^- to the current time slab at t_n^+. The weighted residual form is stabilized by a Galerkin/least-squares term (see (Tezduyar et al., 1992; Masud and Hughes, 1997)) of the momentum balance as denoted in line (61d). The Galerkin/least-squares stabilization suppresses numerical oscillations in solutions to hyperbolic differential equations by the introduction of additional numerical diffusion, allowing herewith the application of equal order approximations of velocities and pressure for the incompressible flow field. Neumann boundary conditions can be imposed in a weak sense by (61e), while (56) is treated a priori as an essential boundary condition.

Fluid-structure coupling. The space-time finite element formulation of the coupling conditions (58) and (59) uses boundary traction variables on the interface in order to fulfil momentum conservation and geometrical continuity between the structure and the fluid in integral form (see Hübner et al. (2004)). In particular, the equality of velocity values at the interface of the fluid and structure is enforced with the fluid interface traction

$$\int_R \delta \mathbf{t}_F \cdot (\mathbf{v}_F - \mathbf{v}_S)\, \mathrm{d}R \;-\; \int_R \delta \mathbf{v}_F \cdot \mathbf{t}_F \,\mathrm{d}R \;-\; \int_{R_0} \delta \mathbf{v}_S \cdot \left(-\frac{d\Sigma_t}{\mathrm{d}\Sigma_0}\, \mathbf{t}_F \right) dR_0 \,. \quad (62)$$

The tractions are projected onto the reference configuration and change their sign, since the unit outward normal vectors of fluid and structure are in opposite directions.

As an alternative, one may use a direct coupling of fluid and structural velocity degree of freedom in the case of fitting mesh discretisations along the common fluid-structure interface. This enables the feature of an unconstrained formulation without Lagrange multipliers (interface traction) and can improve efficiency in the numerical solution.

Simultaneous solution of the monolithic system of equations. The monolithic discrete form of fluid, structure, and coupling conditions of one space-time slab results in a single equation system, shown in Figure 4 (depicted without pressure).

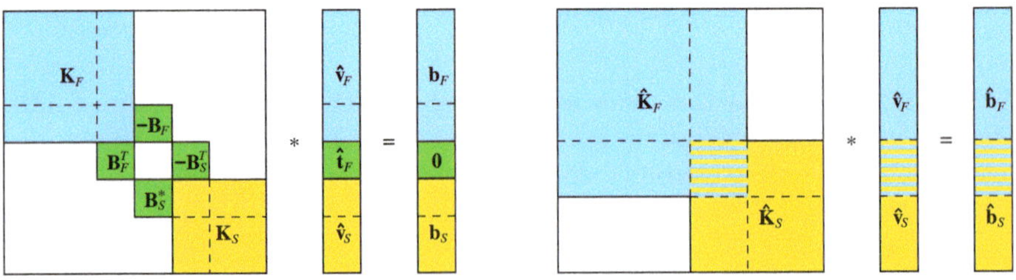

Figure 4. Structure of algebraic system: use of Lagrange multipliers (left) and native velocity coupling (right) in a monolithic approach.

The resulting highly non-linear system – describing the behaviour of both fluid and structure within the time slab n – is solved by the Newton-

Raphson iteration scheme:

$$\mathbf{A}(\mathbf{x}^n)\mathbf{x}^n = \mathbf{b}(\mathbf{x}^{n-1}) \quad \rightarrow \quad \begin{aligned} \mathbf{r}_i^n &= -\mathbf{A}(\mathbf{x}_i^n)\mathbf{x}_i^n + \mathbf{b}(\mathbf{x}^{n-1}) \\ \mathbf{A}_T(\mathbf{x}_{i-1}^n)\Delta\mathbf{x}^n &= \mathbf{r}_i^n \\ \mathbf{x}_{i+1} &= \mathbf{x}_i + \Delta\mathbf{x} \end{aligned}$$

Non-linearities in the coefficient matrix **A** are due to non-linear kinematics of the structure, the convection and stabilization terms of the fluid and the motion of the fluid space-time mesh. In the considered applications, the iteration scheme reaches the coupled monolithic solution to a time slab within three to four steps only. Within each non-linear iteration step, i, the monolithic linear system of equations is solved with an ILU(k)-preconditioned GMRES.

For an in-depth description of the design and implementation of the space-time finite element method for fluid-structure interaction, the interested reader is referred to the following publications: (Hübner et al., 2004; Walhorn et al., 2005; Hübner and Dinkler, 2005; Zilian and Legay, 2008), which demonstrate the monolithic space-time finite element method for strongly coupled fluid-structure interaction problems and validate the model for various applications in aero- and hydro-elasticity.

4 Applications

4.1 Bridge Aero-elasticity

Wind may induce high amplitude vibrations of long-span cable bridges. Mitigating the effects of winds must be regarded as a major objective in bridge design. Different excitation mechanisms are possible such as buffeting, vortex shedding, or self-excitation. In the case of self-excited bridge vibrations, which develop due to the interaction of wind flow and structural motion, amplitudes may increase up to the point of failure of the bridge. The dynamic system behaviour becomes unstable, if the energy transfer from the flow field to the structural oscillator is higher than the dissipation due to structural damping, averaged over a full period. The most popular example for such an aero-elastic instability, also called bridge flutter, is the failure of the Tacoma Narrows Bridge in 1940. In the case of the Tacoma bridge deck (H-shaped section), flow separation appears at the leading edges, and vortices, which increase up to the size of the profile height and move with the mean flow over the profile. At the trailing edges, vortex shedding reappears, leading to a vortex street in the wake of the profile. Large vortices, moving over the profile, cause strong pressure fluctuations and lead to high fluid force amplitudes.

In order to investigate the aero-elastic properties of the Tacoma Narrows cross section by numerical simulation (see the comprehensive study presented in Hübner et al. (2002) and Hübner (2003), which main results are reflected here), the bridge deck is modelled as plane H-section, fixed only in in the horizontal direction and spring supported with respect to vertical and angular motion. Velocity and pressure fields close to the profile are

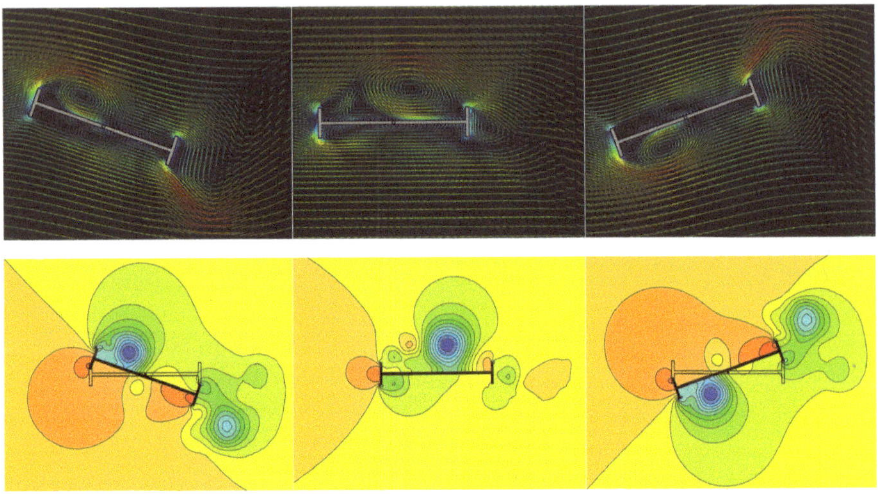

Figure 5. H-shaped bridge deck in cross flow: velocity and pressure fields.

shown in Figure 5 for half a period of the coupled motion. A vortex which develops due to flow separation at the upper leading edge of the H-section is moving over the profile. Thus, the vortex affects a fluctuating moment acting on the profile, since the distance between profile centre and vortex is changing with time. When the profile is in a straight position, the vortex is situated behind the profile centre and causes a moment in the direction of the rotation. This phase shift between vortex induced moment and profile rotation leads to energy transfer from the flow field into the structural system. Consequently the profile vibrations increase. The time histories in Figure 6 (upper left and right) represent the development of resulting fluid forces and moments acting on the profile in comparison with the respective spring forces and moments. Since the spring force values give a measure of the displacement and rotation of the profile, they represent the system response, while the fluid force values define the magnitude of the impact.

For an inflow velocity of $v_\infty = 5\text{m/s}$, vortex shedding causes forced vertical and torsional vibrations with the shedding frequency $f_{St} = 0.24\text{Hz}$.

Vertical and torsional motion are in phase and show a phase shift of $\phi = 180°$ compared to the excitation forces and moments, since the excitation frequency is higher than both natural frequencies and structural damping is not present. Maximum vertical profile displacement and rotation are very small, but in comparison with the excitation the rotation is clearly higher, since the related natural frequency is much nearer to the shedding frequency.

In the case of higher inflow velocities the initial system behaviour is dominated by vortex shedding in conjunction with small structural motions. The shedding frequency of the fixed profile is predominating, but the structural motions are unsteady and a periodical steady state does not appear. In fact, the torsional motion increases rapidly after some time of vortex excitation and the system behaviour becomes unstable. The instability caused by the interaction of structural motion and viscous fluid flow is characterized by exponentially increasing amplitudes of profile rotations and a frequency shift. The dominant frequency changes from the shedding frequency of the fixed profile, which depends on the inflow velocity, to a coupled motion frequency of $f_c = 0.19\text{Hz}$, which only depends on the structural properties. Due to a strong interaction of the fluid flow and structural motion, the coupled system frequency is valid for vertical and torsional motion as well as vortex shedding phenomena, which cause fluid forces and moments. The torsional motion is predominating, but the centre of rotation is shifted in upstream direction, leading to small in-phase oscillations in lift direction and a coupled system frequency, which is slightly smaller than the natural frequency of angular motion.

The dynamic system behaviour becomes unstable for all inflow velocities above a critical velocity ($v_\infty \approx 7.5\text{m/s}$ for the investigated model set-up). But for velocities not far above the critical velocity, the increasing torsional motion is limited, leading to stable limit cycle oscillations with moderate amplitudes. In the case of higher velocities, the simulation does not predict limit cycle oscillations in the regarded time domain. For an inflow velocity of $v_\infty = 10\text{m/s}$, rotation and displacement amplitudes increase further. The investigation of the excitation mechanisms and initiating instabilities is carried out for the inflow velocity, $v_\infty = 10\text{m/s}$. At first, the range of transition is investigated, where the system behaviour changes completely. The increasing torsional vibrations become unsteady until the frequency shift has completed and sinusoidal oscillations appear. The profile motion influences the fluid flow, and with a time delay, also fluid forces and moments become irregular and change the frequency. After transition of system properties has completed, runs of aero-elastic coefficients become more regular and exciting mechanisms may be identified.

The mechanism of energy transfer is visualized in the left diagram of Figure 6 (lower left and right), where the moment coefficient is displayed for a single period depending on the profile rotation. The enclosed area specifies an energy input in case of a clockwise direction and an energy output in case of a counter clockwise direction. The curve around the middle in the left diagram has a clockwise direction, leading to an excitation of profile vibrations, while the negative directions of the outer areas cause damping. But for a full period energy input is much higher than damping. The moment

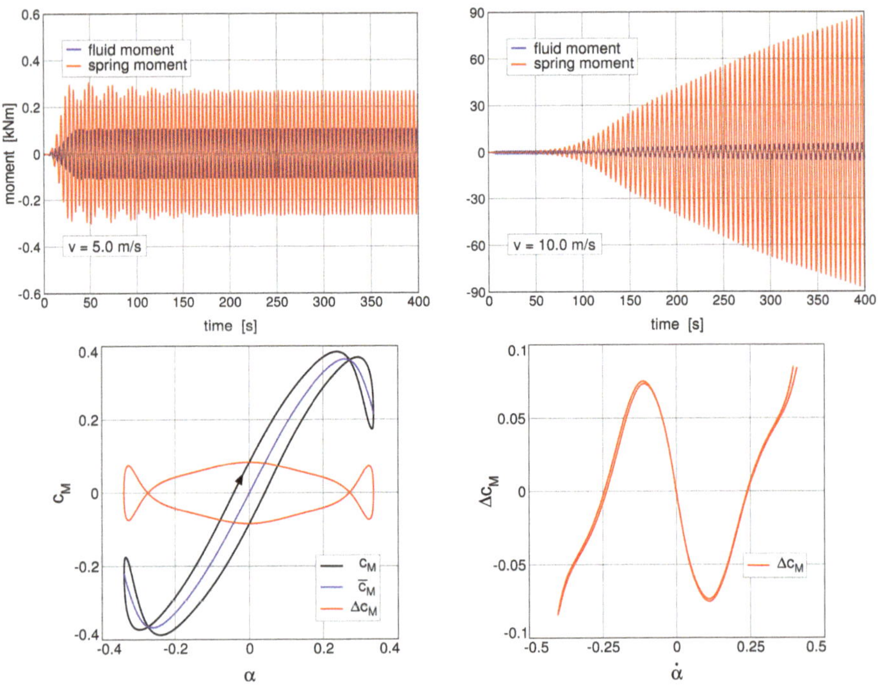

Figure 6. H-shaped bridge deck in cross flow: Time histories of resultant fluid and structure moment at $v_\infty = 5\text{m/s}$ and $v_\infty = 10\text{m/s}$ (upper row). Aero-elastic moment coefficients (lower row), see (Hübner et al., 2002).

coefficient characteristics can be separated into a mean value, which only depends on the rotation angle, and a deviating part, which is responsible for excitation and damping. The deviating part is displayed in the right diagram depending on the angular velocity. Here, the enclosed area is approximately zero, leading to a unique correlation between angular velocity and moment coefficient. Thus, if aero-elastic calculations of a simplified bridge model consider wind effects as deformation dependent impacts, nonlinear flow induced stiffness properties can be defined by the slope of the mean curve in the left diagram and non-linear damping properties by the slope of the curve in the right diagram of Figure 6 (lower row).

4.2 Wave Impact on Structures

Recent natural hazards emphasize the need for estimating effects generated by impacting water waves. The destructive power is not only related to peak forces but also mainly due to flash floods with sedimentation and debris transport. Since permanent retreat from populated regions that are at risk is not an option in most affected countries, civil engineers have the responsibility to take preventative action. In addition to the traditional coastal engineering solutions, mobile protection devices can provide rapidly deployable and customizable local safeguarding. An example of such a device is a modularized array of compliant structures that is able to dissipate wave energy and at the same time act as a trash-rack for floating debris.

The wave impact onto a deformable structure is an example of fluid-structure interactions involving a free-surface flow and a considerable amount of added mass. Characteristic of wave impact is a dominant unidirectional information transfer and therefore a weak coupling between fluid and structure. In addition, multiple deformable obstacles can considerably affect the unsteady flow field and beneficial interaction effects can be expected. Numerical simulations of wave impact on structures accompany experimental investigations and allow cost-effective studies of different system set-ups while providing detailed design quantities. The visualization of a model sit-

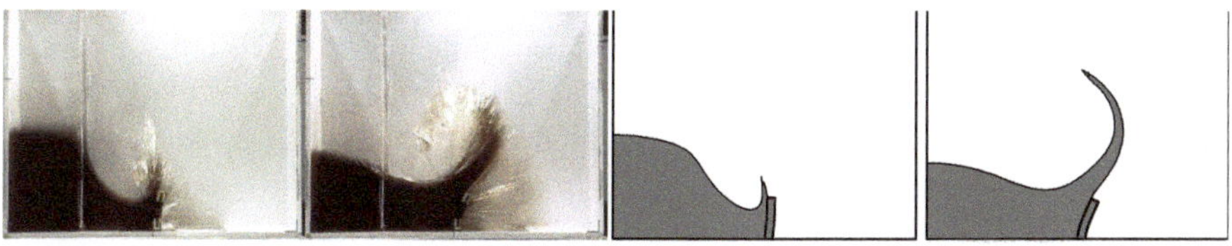

Figure 7. Collapse of a water column onto a compliant structure: Comparison of experimental and numerical results for the evolving free surface.

uation is shown in Figure 7 with experimental and numerical results (compare (Kölke, 2005) and extensions in (Pasenow et al., 2013)). A water packet (coloured fluid) is initially restrained in an open tank. In the middle of the tank a compliant cantilever structure (rubber-like material) is mounted. The water is released suddenly and collapses due to gravitational forces. While the fluid body deforms only slightly during the first instants of the process, it shows topological changes upon reaching the obstacle. The characteristics of such free-surface flows are considered in the simulation by using a two-fluid approach and a level-set representing the fluid-fluid interface, enabling the capturing of surface separation and merging.

The leading front of the collapsing water column reaches the obstacle and the elastic structure is deformed mainly by the developing pressure distribution on its left. The redirection angle of the water forefront is lower compared to results for a fixed obstacle. The structure reaches its maximum deflection before the wave hits the right wall of the tank and shows damped vibrations governed by unsteady dynamics of the viscous free-surface flow. Figure 8 shows close-up views of both, experimental and numerical investigations to the deformation of the structure. The above mentioned free-surface capturing technique is modelled on top of a mesh-moving approach.

Bibliography

C. Scruton. *Some considerations of wind effects on large structures.* In: Structures technology for large radio and radar telescope systems, Mar, Liebowitz (editors) MIT Press, 1969.

H. Sockel. *Aerodynamik der Bauwerke.* Vieweg & Sohn, 1984.

H. Ruscheweyh. Practical experiences with wind-induced vibrations. *Journal Wind Engineering and Industrial Aerodynamics*, 33, 493, 1988.

R. L. Bisplinghoff, H. Ashley, and R. L. Halfmann. *Aeroelasticity.* Addison-Wesley, 1955.

H. W. Försching. *Grundlagen der Aeroelastik.* Springer, 1974.

E. H. Dowell, H. C. Curtis, R. H. Scanlan, and F. Sisto. *A modern course in aeroelasticity.* Kluwer Academic Publishers, 1989.

R. Blevins. *Flow-induced Vibration.* Van Nostrand Reinhold, 1990.

E. Naudascher and D. Rockwell. *Flow-Induced Vibrations: An Engineering Guide.* A.A. Balkema, Rotterdam, 1994.

M. P. Paidoussis. Flow-induced instabilities of cylindrical structures. *Applied Mechanics Review*, 40:163, 1987.

G. V. Parkinson. Phenomena and modelling of flow-induced vibrations of bluff bodies. *Progress in Aerospace Sciences*, 26:169, 1989.

A.I. Korotkin. *Added Masses of Ship Structures.* Springer, Berlin, 2009.

E. Konstantinidis. Added mass of a circular cylinder oscillating in a free stream. *Proceedings of the Royal Society A: Mathematical, Physical and Engineering Science*, 469:2156, 2013.

W. A. Wall. *Fluid-Struktur-Interaktion mit stabiliserten Finiten Elementen.* PhD Thesis, Universität Stuttgart, 1999.

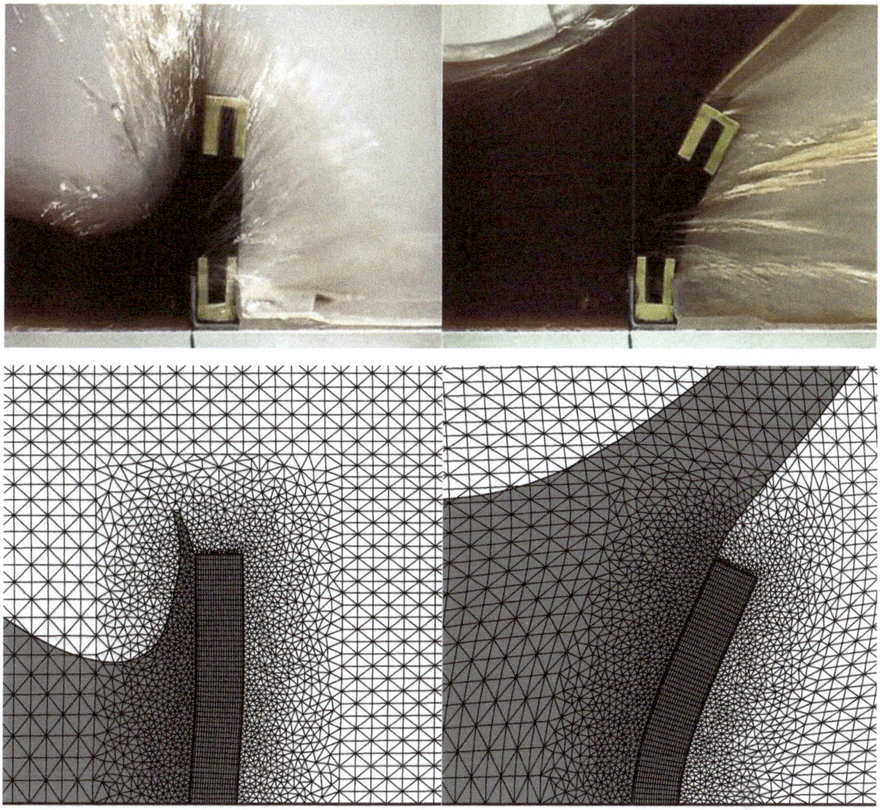

Figure 8. Wave-induced obstacle deformation: Experiment and simulation.

J. Steindorf. *Partitionierte Verfahren für Probleme der Fluid-Struktur-Wechselwirkung.* PhD Thesis, TU Braunschweig, 2002.

H.J. Bungartz and M. Schäfer. *Fluid–Structure Interaction Modelling, Simulation, Optimisation.* Springer, 2006.

H.J. Bungartz, M. Mehl, and M. Schäfer. *Fluid–Structure Interaction II Modelling, Simulation, Optimisation.* Springer, 2012.

W. G. Dettmer and D. Perić. A new staggered scheme for fluid-structure interaction. *International Journal for Numerical Methods in Engineering,* 93(1):1–22, 2013.

J. H. Argyris and D.W. Scharpf. Finite Elements in Space and Time. *Journal of Royal Aeronautical Society,* 73:1041–1044, 1969.

J. Knippers and R. Harbord. A Mixed Hybrid FE-Formulation for Solution of Elasto-Viscoplastic Problems. *Computational Mechanics*, 13:231–240, 1994.

TJR Hughes and GM Hulbert. Space-time finite element methods for elastodynamics: Formulations and error estimates. *Computer Methods in Applied Mechanics and Engineering*, 66(3):339–363, 1988.

B. Hübner. *Simultane Analyse von Bauwerk-Wind-Wechselwirkungen.* Dissertation, Technische Universität Braunschweig, Braunschweig, 2003.

T. E. Tezduyar, M. Behr, and J. Liou. A New Strategy for Finite Element Computations Involving Moving Boundaries and Interfaces - The Deforming-Spatial-Domain/Space-Time Procedure: I. The Concept and the Preliminary Numerical Tests. *Computer Methods in Applied Mechanics and Engineering*, 94:339–351, 1992.

A. Masud and T. J. R. Hughes. A Space-Time Galerkin/Least-Squares Finite Element Formulation of the Navier-Stokes Equations for Moving Domain Problems. *Computer Methods in Applied Mechanics and Engineering*, 146:91–126, 1997.

B. Hübner, E. Walhorn, and D. Dinkler. A Monolithic Approach to Fluid-Structure Interaction using Space-Time Finite Elements. *Computer Methods in Applied Mechanics and Engineering*, 193(23–26):2069–2086, 2004.

E. Walhorn, A. Kölke, B. Hübner, and D. Dinkler. Fluid-structure coupling within a monolithic model involving free surface flows. *Computers and Structures*, 83(25-26):2100–2111, 2005.

B. Hübner and D. Dinkler. A simultaneous solution procedure for strong interactions of generalized Newtonian fluids and viscoelastic solids at large strains. *International Journal for Numerical Methods in Engineering*, 64: 920–939, 2005.

A. Zilian and A. Legay. The enriched space-time finite element method (EST) for simultaneous solution of fluid-structure interaction. *International Journal for Numerical Methods in Engineering*, 75:305–334, 2008.

B. Hübner, E. Walhorn, and D. Dinkler. Numerical investigations to bridge aeroelasticity. In HA Mang, FG Rammerstorfer, and J Eberhardsteiner, editors, *Proceedings of the Fifth World Congress on Computational Mechanics, Vienna*, 2002.

A. Kölke. *Modellierung und Diskretisierung bewegter Diskontinuitäten in randgekoppelten Mehrfeldsystemen.* PhD thesis, Technische Universität Braunschweig, 2005.

F. Pasenow, A. Zilian, and D. Dinkler. Extended space–time finite elements for landslide dynamics. *International Journal for Numerical Methods in Engineering*, 93(3):329–354, 2013.

Dynamics and Stability of Engineering Systems with Moving Continua

Alexander K. Belyaev
Institute for Problems in Mechanical Engineering, Russian Academy of Sciences, St. Petersburg State Polytechnic University, Russia

Abstract The present chapter is concerned with a special case of dynamics of systems with variable mass, namely the engineering systems with moving continua. A characteristic feature of these systems is that a material enters the system and leaves the system. The material itself is assumed to be deformable and under some conditions the entire system exhibits unstable behaviour in the transverse direction. The intent of the present chapter is to demonstrate that the systems with axially moving material are inherently unstable. Dynamics and stability are known to be strongly related to each other, for this reason the study of dynamics and stability for each engineering system under consideration is carried out in the framework of the same approach. A number of special cases which are important for the mechanical engineering are considered. Some of these belong to the class of problems of fluid-structure interaction, in particular, dynamics and stability of the fluid conveying pipes and the shaft rotating in the oil film plain bearings. The dynamics and stability of the belts and chains are studied in detail, too.

1 Introduction

Dynamics and stability of systems with moving continua is a special case of the systems with variable mass. A characteristic feature of these systems is that a material enters the system and leaves the system. Some of these systems belong to the class of the problems referred to as the fluid-structure interaction. The purpose of this chapter is to demonstrate as to the axially moving continua affect the dynamics and stability of carrying structure. In the problems of unstable pipeline the part of the axially moving continua is played by fluids whereas the belts and chains members are moving in the axial direction and shown to be unstable themselves. In addition to this, a detailed analysis of the problem of vibration and stability of rotors caused by hydrodynamically induced bearing forces is performed in detail.

H. Irschik, A. K. Belyaev (Eds.), *Dynamics of Mechanical Systems with Variable Mass*,
CISM International Centre for Mechanical Sciences DOI 10.1007/ 978-3-7091-1809-2_6

2 Stability of non-conservative systems. General and specific features

2.1 Determining critical follower force

The objective of this part is determining the critical follower force. Nonetheless, for the tutorial purpose it is worth to start with considering a conservative system, namely a cantilever beam subjected to a dead load F. The beam is shown in Fig. 1. Let us determine the Euler critical force provided that force F is a dead load.

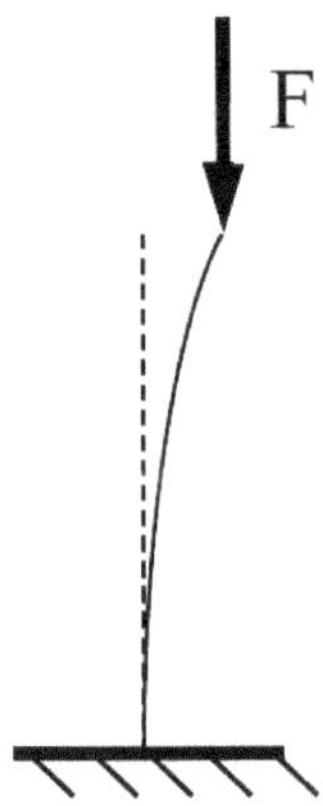

Figure 1. Cantilever beam subjected to a dead load F

The equilibrium equation for the beam deflection w is as follows

$$EI\frac{d^2w}{dx^2} + F\left(w - w\left(l\right)\right) = 0 \tag{1}$$

where EI is the bending rigidity and F is the axial compressive force. The boundary conditions for the cantilever beam under consideration are as follows

$$x = 0, \;\; w = \frac{dw}{dx} = 0; \quad x = l, \quad w = w\left(l\right) \tag{2}$$

The solution to the boundary-value problem is sought in the following form

$$w\left(x\right) = A\sin\lambda x + B\cos\lambda x + w\left(l\right), \;\; \lambda = \sqrt{F/EI}$$

Satisfying the boundary conditions we arrive at the following system of equations for the unknown coefficients A, B and $w(l)$

$$\begin{cases} B + w(l) = 0 \\ A\lambda = 0 \\ A \sin \lambda l + B \cos \lambda l = 0 \end{cases} \tag{3}$$

It is clear that the problem eigenvalue λ is obtained from the equation $\cos \lambda l = 0$, i.e. $\lambda l = \pi/2$ and it allows us to determine the first (Euler) critical force which is

$$F_c = \frac{\pi^2}{4l^2} EI \tag{4}$$

This classical result is generally accepted as the first critical force. However it turned out to be sensitive to the way of loading by force F. Let us consider a couple of possible technical realization of loading by means of force F.

The first way of loading is shown in Fig. 2. The transmission of the dead load F on the end of the cantilever beam is seen to be realised with the help of a connecting rod of length b.

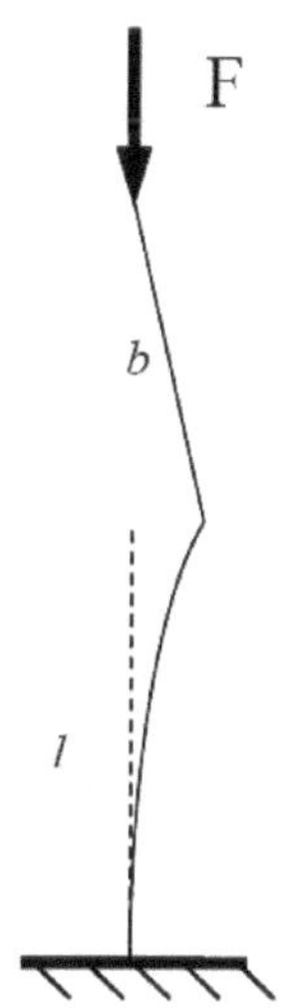

Figure 2. Cantilever beam subjected to dead load F via connecting rod b

The equilibrium equation (1) for small deflections w takes now another form

$$EI\frac{d^2w}{dx^2} = F\left(w\left(l\right) - w\right) + F\frac{w\left(l\right)}{b}\left(l - x\right) \tag{5}$$

The solution is now sought is the form

$$w\left(x\right) = A\sin\lambda x + B\cos\lambda x + w\left(l\right)\left(1 + \frac{l - x}{b}\right) \tag{6}$$

and the same boundary conditions (2) lead to the following system of equations

$$\begin{cases} B + w\left(l\right)\left(1 + l/b\right) = 0 \\ A\lambda - w\left(l\right)/b = 0 \\ A\sin\lambda l + B\cos\lambda l = 0 \end{cases} \tag{7}$$

The condition of vanishing determinant yields the following transcendental equation for λl

$$\tan\lambda l = \lambda l\left(1 + \frac{b}{l}\right) \tag{8}$$

Therefore, eigenvalue λ and thus the critical force depend on the ratio b/l. We will analyse only two limiting cases: $b/l = \infty$ and $b/l = 0$. In the first case ($b/l = \infty$) the result is $\lambda l = \pi/2$ and we have the same critical force (4) as in the previous analysis. However in the second case ($b/l = 0$) we have $\lambda l = 0$ and thus the critical force is equal to zero. The case of arbitrary length $0 < l < \infty$ yields the critical force having the value between the above limiting cases, i.e.

$$0 < F_c < \frac{\pi^2}{4l^2}EI$$

The conclusion which can be made from this example is that the way of imposing a dead load has a crucial influence on the value of the critical load. Therefore the system proved to be very sensitive to the particular way of loading. A number of various loading types different from the above ones are analysed in e.g. Volmir [1967] and Bolotin [1963].

Let us now proceed to the case of follower force depicted in Fig. 3.

The differential equation for small deflection $w\left(x\right)$ is now given by

$$EI\frac{d^2w}{dx^2} = F\left(w\left(l\right) - w\right) - F\varphi\left(l\right)\left(l - x\right),$$

where in addition to the previous denotations we introduced the angle of rotation of the cross-section $\varphi\left(x\right) = w'\left(x\right)$. The boundary conditions are

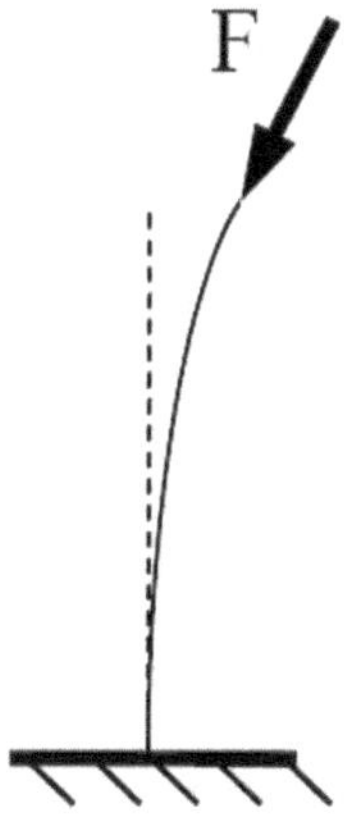

Figure 3. Cantilever beam loaded by a follower force

$$x = 0,\ \ w = \frac{dw}{dx} = 0; \qquad x = l,\ \ w = w\,(l)\,,\ w' = \varphi\,(l)\,.$$

We seek the solution in the form

$$w\,(x) = A \sin \lambda x + B \cos \lambda x + w\,(l) - \varphi\,(l)\,(l - x)$$

Satisfying the boundary conditions leads to the system of equations

$$\begin{cases} B + w\,(l) - \varphi\,(l)\,l = 0 \\ A\lambda + \varphi\,(l) = 0 \\ A \sin \lambda l + B \cos \lambda l = 0 \\ \lambda\,(A \cos \lambda l - B \sin \lambda l) = 0 \end{cases}$$

One can see that the determinant of the third and fourth equations is equal to unity for any non-zero force F $(\lambda \neq 0)$. Therefore $A = B = 0$ and from the remaining two equations we have $w\,(l) = 0,\ \varphi\,(l) = 0$ which means that the only solution to the problem is a trivial one, i.e. $w\,(x) \equiv 0$. Since $w\,(x) \equiv 0$ for any loading force one could come to the conclusion that the system was always stable. However it is a wrong conclusion. The right conclusion is based upon the observation that the follower force is a non-conservative force, that is, it has no potential. Thus the standard Euler-Lagrange approach is not applicable here, i.e. the system has no side static equilibrium state. This is the only conclusion which can be carried out from the condition of absence of any critical force. In other words, a dynamic analysis is needed for this case. About the relationships between the static and dynamic approaches to stability analysis see Bolotin [1963].

2.2 The Ziegler paradox

This paradox was first reported by Ziegler [1968]. Let us consider a double pendulum with a follower force acting on the lower pendulum, see Fig. 4. Each joint has a spring with the angular stiffness c and a dashpot with the angular resistance factor b. It is assumed that each pendulum has a lumped mass on its end, i.e. the pendulum rods themselves are assumed to be massless.

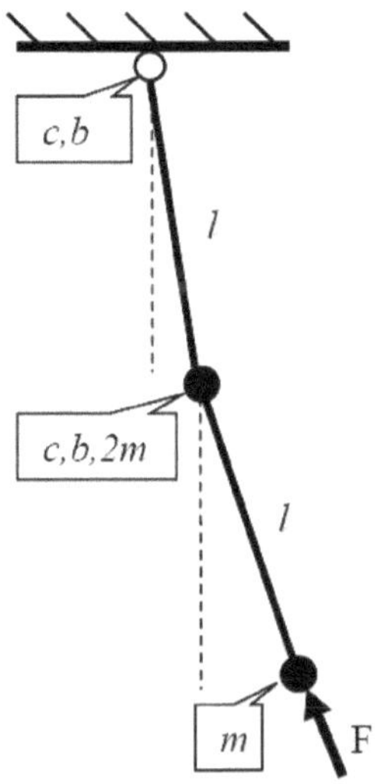

Figure 4. Double pendulum subjected to a follower force F

The Lagrange equation for the system is as follows

$$\frac{d}{dt}\left(\frac{\partial T}{\partial \dot{q}}\right) - \frac{\partial T}{\partial q} + \frac{\partial \Phi}{\partial \dot{q}} + \frac{\partial \Pi}{\partial q} = Q \tag{9}$$

where q denotes the vector of degrees of freedom, that is,

$$q = \begin{pmatrix} \varphi_1 \\ \varphi_2 \end{pmatrix} \tag{10}$$

Here φ_1, φ_2 designate the angles of the corresponding pendulum to the vertical. The assumption is made that the angles φ_1, φ_2 are small which implies small displacements of the system elements. In this case the kinetic energy of the system is given by

$$T = \frac{1}{2} 2m \left(l\dot{\varphi}_1\right)^2 + \frac{1}{2} m \left(l\dot{\varphi}_1 + l\dot{\varphi}_2\right)^2 .$$

The potential energy of the system is

$$\Pi = \frac{1}{2} c\varphi_1^2 + \frac{1}{2} c \left(\varphi_1 - \varphi_2\right)^2 .$$

The Rayleigh dissipative function is as follows

$$\Phi = \frac{1}{2} c\dot{\varphi}_1^2 + \frac{1}{2} c \left(\dot{\varphi}_1 - \dot{\varphi}_2\right)^2 .$$

In order to determine the generalized forces corresponding to the generalized coordinates (10) one has to calculate the virtual work due to the single external force F. To this aim one calculates the virtual displacement of the lower end of the second rod

$$\delta A = F \cdot \delta r = Fl(\varphi_1 - \varphi_2)\delta\varphi_1 = \begin{pmatrix} Q_1 \\ Q_2 \end{pmatrix} \cdot \begin{pmatrix} \delta\varphi_1 \\ \delta\varphi_2 \end{pmatrix} = Q_1\delta\varphi_1 + Q_2\delta\varphi_2$$

It allows one to determine the generalized forces

$$Q_1 = Fl(\varphi_1 - \varphi_2),\ Q_2 = 0. \tag{11}$$

Substituting the obtained equations in the Lagrange equations (9) yields the following system of two linear coupled differential equations for the generalized coordinates φ_1, φ_2

$$\begin{cases} 2ml^2\ddot{\varphi}_1 + ml^2\left(\ddot{\varphi}_1 + \ddot{\varphi}_2\right) + b(2\dot{\varphi}_1 - \dot{\varphi}_2) + c(2\varphi_1 - \varphi_2) - Fl(\varphi_1 - \varphi_2) = 0 \\ ml^2\left(\ddot{\varphi}_1 + \ddot{\varphi}_2\right) + b(\dot{\varphi}_2 - \dot{\varphi}_1) + c(\varphi_2 - \varphi_1) = 0 \end{cases} \tag{12}$$

The solution is sought in the standard form

$$q = \begin{pmatrix} \varphi_1 \\ \varphi_2 \end{pmatrix} = \begin{pmatrix} \Phi_1 \\ \Phi_2 \end{pmatrix} e^{\lambda t} \tag{13}$$

where Φ_1, Φ_2 are the eigenvector components and λ is the eigenvalue. Inserting the last equation into the equations of system (12) we arrive at the following system of two linear coupled algebraic equations for the eigenvector components Φ_1, Φ_2

$$\begin{cases} \left(3ml^2\lambda^2 + 2b\lambda + 2c - Fl\right)\Phi_1 + \left(ml^2\lambda^2 - b\lambda - c + Fl\right)\Phi_2 = 0 \\ \left(ml^2\lambda^2 - b\lambda - c\right)\Phi_1 + \left(ml^2\lambda^2 + b\lambda + c\right)\Phi_2 = 0 \end{cases} \tag{14}$$

The nontrivial solution of the system of homogeneous equations (14) is obtained from the condition that the determinant of the system vanishes, that is,

$$\begin{gathered} \left(3ml^2\lambda^2 + 2b\lambda + 2c - Fl\right)\left(ml^2\lambda^2 + b\lambda + c\right) - \\ - \left(ml^2\lambda^2 - b\lambda - c + Fl\right)\left(ml^2\lambda^2 - b\lambda - c\right) = 0 \end{gathered}$$

This polynomial of the fourth order is the characteristic equation which can be written in the form

$$a_0\lambda^4 + a_1\lambda^3 + a_2\lambda^2 + a_3\lambda + a_4 = 0 \tag{15}$$

where the coefficients are as follows

$$\begin{aligned} a_0 &= 2m^2l^4 \\ a_1 &= 7bml^2 \\ a_2 &= 7cml^2 + b^2 - 2l^3mF \\ a_3 &= 2bc \\ a_4 &= c^2 \end{aligned}$$

The Routh-Hurwitz criterion provides one with the necessary and sufficient conditions ensuring the stability. The necessary condition requires that all the above coefficients must be positive. This condition reduces to a single one and is satisfied when

$$7cml^2 + b^2 - 2l^3mF > 0 \;\Rightarrow F < \frac{7}{2}\frac{c}{l} + \frac{b^2}{2l^3m} \tag{16}$$

The sufficient condition operates with the Hurwitz matrix

$$\left\| \begin{matrix} a_0 & a_2 & a_4 & 0 \\ 0 & a_1 & a_3 & 0 \\ 0 & a_0 & a_2 & a_4 \\ 0 & 0 & a_1 & a_3 \end{matrix} \right\|$$

This condition guarantees the stability provided that all principal minors of the matrix are positive. Calculation leads to the following conditions

$$\begin{aligned} a_0 &> 0 \\ a_0 a_1 &> 0 \\ a_3 \Delta_3 &> 0 \\ a_3 &> 0 \end{aligned}$$

One can see that these four conditions can be reduced to a single condition $\Delta_3 > 0$ where Δ_3 denotes the third principal minor of the Hurwitz matrix. This minor can be finally rewritten in the form of the inequality for force F

$$F < \frac{41}{28}\frac{c}{l} + \frac{b^2}{2l^3 m} \tag{17}$$

Comparison of conditions (16) and (17) shows that the latter is stronger that condition (16). It allows one to introduce a critical force

$$F_c = \frac{41}{28}\frac{c}{l} \approx 1.464\frac{c}{l} \tag{18}$$

which ensures the system stability even for a vanishingly small damping b.

The latter result demonstrates that the critical force decreases with decreasing damping and the smallest force is obtained when the damping vanishes. This naturally leads to the idea that the damping should be set to zero from the very beginning and then the critical force should be determined. There is no need to derive again the characteristic equation since the equations of motion (12), ansatz (13) and the characteristic equation (14) are also valid under the additional condition of b=*0*.

The characteristic equation (14) now becomes a biquadratic equation

$$a_0 \lambda^4 + a_2 \lambda^2 + a_4 = 0$$

with the coefficients

$$a_0 = 2m^2 l^4,\ a_2 = 7cml^2 - 2l^3 mF,\ a_4 = c^2.$$

The solution is given by

$$\lambda_{1,2}^2 = \frac{-\left(7cml^2 - 2l^3 mF\right) + \sqrt{\Delta}}{4m^2 l^4};\ \lambda_{3,4}^2 = \frac{-\left(7cml^2 - 2l^3 mF\right) - \sqrt{\Delta}}{4m^2 l^4}$$

where the discriminant Δ is as follows

$$\Delta = \left(7cml^2 - 2l^3 mF\right)^2 - 8m^2 l^4 c^2.$$

If the discriminant Δ is negative then $\lambda_{1,2}^2$ and $\lambda_{3,4}^2$ are complex-valued numbers, and it means that there always exist two eigenvalues (say λ_1 and

λ_3) with positive real parts. In turn it implies that the motion is unstable since the complete solution has exponentially growing unbounded components. In other words, the stability condition is equivalent to the condition of positive coefficient a_2 and positive discriminant Δ, that is

$$a_2 = 7cml^2 - 2l^3mF > 0, \quad \Delta = \left(7cml^2 - 2l^3mF\right)^2 - 8m^2l^4c^2 > 0.$$

These conditions are reduced to a single one

$$F < \frac{7-\sqrt{8}}{2}\frac{c}{l}$$

that enables introduction of the critical force for the undamped system

$$F_{c(undamped)} = \frac{7-\sqrt{8}}{2}\frac{c}{l} \approx 2.086\frac{c}{l} \tag{19}$$

It is clearly seen that the latter critical force (19) (i.e. the critical force for undamped system) does not coincide with the former one (18) (i.e. the critical force for the system with a vanishingly small damping). This is the essence of the celebrated Ziegler paradox, cf. Ziegler [1968]. However one should understand that the very word "paradox" implies a "seeming" contradiction rather than a real contradiction. In other words, an additional reasoning is needed in order to remove this "seeming" contradiction. In the problem under consideration the observed mismatch is due to the fact that the critical force (18) ensures asymptotic stability whereas the critical force (19) ensures stability only. Therefore, both results are correct and the value of the critical force, either (18) or (19), is dependent upon what type of stability is considered.

3 Dynamics and stability of belts and chains

The systems with axially moving material represent a special class of systems with variable mass. Stability of motion of such systems is addressed below on example of timing belt and chains.

3.1 Dynamics and stability of timing belts

Literature on transverse vibrations and stability of axially moving beams and strings in the vicinity of a steady state motion is presented in a comprehensive review paper Chen [2005]. It is also worth mentioning the paper by Eliseev and Vetyukov [2012] in which the nonlinear dynamics of both

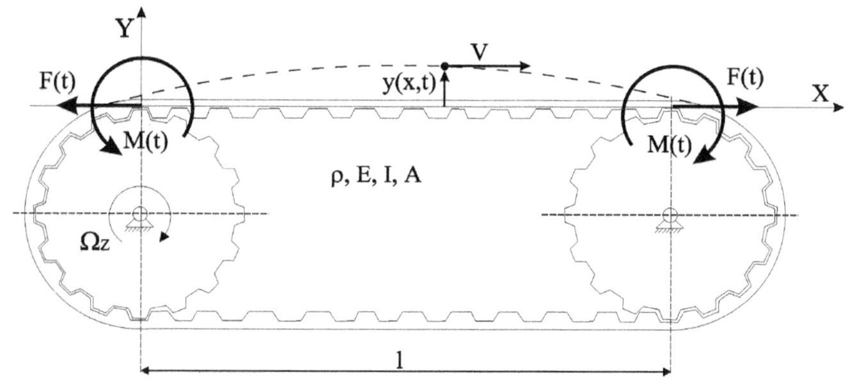

Figure 5. Timing belt and two gear wheels

synchronous and friction belts is addressed. The schematics of the drive system with a timing belt (e.g. to drive a camshaft) is shown in Fig. 5.

Since the belt possesses some bending rigidity it can be modeled by a beam whose parameters and deflection are shown in Fig. 5. In contrast to a classical beam the belt is moving in axial direction with velocity V and is pre-tensioned by a tension force $F(t)$. The tension force has two components (i) a constant pre-tension force and (ii) a time-dependent component due to the interaction of the gear teeth and belt teeth.

The Hamiltonian variational principle is applied for derivation of the equations of motion, cf. Hirmann and Belyaev [1997]. To this end, the expressions for the kinetic and bending strain energies are obtained

$$\begin{aligned} T &= \tfrac{1}{2}\int_0^l \left[V^2 + (\dot{y} + V\cdot\tan y')^2\right]\rho A dx = \\ &= \tfrac{1}{2}\int_0^l \left[V^2 + \dot{y}^2 + V^2 y'^2 + 2\dot{y}y'V\right]\rho A dx, \\ \Pi &= \tfrac{1}{2}\int_0^l EI(y'')^2 dx. \end{aligned} \tag{20}$$

While deriving these expressions we assumed that the deflection y is small. Work of the external forces and moments is given by

$$\begin{aligned} W &= -F\int_0^l (1-\cos y')dx + M(0,t)y'(0,t) + M(l,t)y'(l,t) = \\ &= -\tfrac{1}{2}F\int_0^l y'^2 dx + M(0,t)y'(0,t) + M(l,t)y'(l,t) \end{aligned} \tag{21}$$

Here the tension force $F(t)$ is assumed to have the same value in each cross-section. However this force is time-dependent because $F(t)$ describes the dynamic interaction between the belt and the teeth of the gear wheel. In other words, the axial wave processes in the belt are neglected here.

Making use of the Hamiltonian variational principle requires the Hamilton function $L = T - \Pi + W$ and the Hamilton action

$$S = \int_{t_0}^{t_1} L dt$$

whose variation must vanish, that is, $\delta S = 0$. Realisation of the variational procedure yields the following equation for the variation δS

$$\delta S = \int_{t_0}^{t_1} dt \left\{ \begin{array}{l} \int_0^l \left[-\rho A \ddot{y} - \rho A V^2 y'' - 2V\rho A \dot{y}' - EIy'''' + Fy'' \right] \delta y dx + \\ \delta y'(0,t) \left[EIy'' + M(0,t) \right] + \delta y'(l,t) \left[-EIy'' + M(0,t) \right] + \\ +\delta y(0,t) \left[-V^2 \rho A y' - V\rho A \dot{y} - EIy''' + Fy' \right] + \\ +\delta y(l,t) \left[V^2 \rho A y' + V\rho A \dot{y} + EIy''' - Fy' \right] \end{array} \right\}$$

The equation in variations $\delta S = 0$ is equivalent to the following differential equation of motion

$$0 < x < l, \qquad \rho A \ddot{y} + (\rho A V^2 - F) y'' + 2V\rho A \dot{y}' + EIy'''' = 0 \tag{22}$$

and the boundary conditions

$$EIy'' = -M(0,t), \qquad EIy'' = M(l,t) \tag{23}$$

The remaining two boundary conditions are identically satisfied because $y(0,t) = y(l,t) = 0$.

The obtained differential equation (22) can be solved by means of Galerkin approach. To this aim, we make the following substitution

$$y(x,t) = \varphi(x) q(t) + y_0(x,t)$$

where basic function $\varphi(x)$ satisfies the trivial boundary conditions

$$\varphi(0) = \varphi(l) = 0, \quad \varphi''(0) = \varphi''(l) = 0$$

The reason for introducing the additional function $y_0(x,t)$ is that it should meet the non-homogeneous boundary condition (23) such that function $\varphi(x)$

satisfies only the homogeneous boundary conditions. Following this strategy we seek the solution in the form

$$y(x,t) = q(t)\sin\frac{\pi x}{l} + \frac{M(t)}{6lEI}(2x^3 - 3lx^2 + l^2x) \tag{24}$$

where the only sought-for function is the generalized coordinate $q(t)$. One can prove that the boundary conditions (23) are met. Inserting eq. (24) in the equation of motion (22) we arrive at the following ordinary differential equation

$$\ddot{q}(t) + \frac{\pi^2}{l^2}\left(\frac{1}{\rho A}\left(F + EI\frac{\pi^2}{l^2}\right) - V^2\right)q(t) = \frac{4lV(12-\pi^2)}{3EI\pi^3}\dot{M}(t) \tag{25}$$

Let us now split the tension force $F(t)$ into two parts

$$F = F_0 + F_1\Phi(t) \tag{26}$$

where F_0 denotes the initial tension force which does not depend on time and $F_1\Phi(t)$ stands for the periodic component of the tension force. We also introduce the first Euler critical force $F_c = \pi^2 EI/l^2$ which enables us to rewrite eq. (25) in the following form

$$\ddot{q}(t) + \Omega^2(1 + 2\mu(t))q(t) = \frac{4lV(12-\pi^2)}{3EI\pi^2}\dot{M}(t) \tag{27}$$

Here Ω introduce the "eigenfrequency" of the bending vibration of the belt and μ is a non-dimensional magnitude of the time-dependent component of the axial excitation force

$$\Omega^2 = \frac{\pi^4}{l^4}\frac{EI}{\rho A}\frac{F_0 + F_c - V^2\rho A}{F_c}, \qquad \mu = \frac{F_1}{2\left(F_0 + F_c - V^2\rho A\right)} \tag{28}$$

Differential equation (27) is a kind of Hill's equation with periodic coefficient represented by function $\Phi(t)$. This function has a period $T = \pi r/Vz$, where r is the radius of the gear wheel and z denotes the number of teeth on the wheel.

Let us consider the homogeneous differential equation

$$\ddot{q}(t) + \Omega^2(1 + 2\mu(t))q(t) = 0 \tag{29}$$

Periodic function $\Phi(t)$ can be expanded in Fourier series in terms of the circular functions, that is,

$$\Phi(t) = \Phi_1\cos(\varpi t + \alpha_1) + \Phi_2\cos(2\varpi t + \alpha_2) + \ldots$$

where the frequency ϖ is related to period T by means of the equation $\varpi = 2\pi/T$. Besides, one can always choose the initial time instant such that $\alpha_1 = 0$. If we limit our consideration only to the first term in the Fourier series then Hill's equation (29) takes the form of the Mathieu equation

$$\ddot{q}(t) + \Omega^2(1 + 2\mu\Phi_1 \cos(\varpi t + \alpha_1))q(t) = 0 \tag{30}$$

It is known that this equation has unstable solutions which form the instability regions of the parametric resonance. The boundaries of these instability regions are obtained by the substitution

$$q(t) = a \cos \frac{\varpi t}{2} + b \sin \frac{\varpi t}{2} \tag{31}$$

Neglecting the terms with the higher harmonics we arrive at the following equation

$$a \cos \frac{\varpi t}{2} \left[-\left(\frac{\varpi}{2}\right)^2 + \Omega^2 (1 + \mu) \right] + b \sin \frac{\varpi t}{2} \left[-\left(\frac{\varpi}{2}\right)^2 + \Omega^2 (1 - \mu) \right] = 0$$

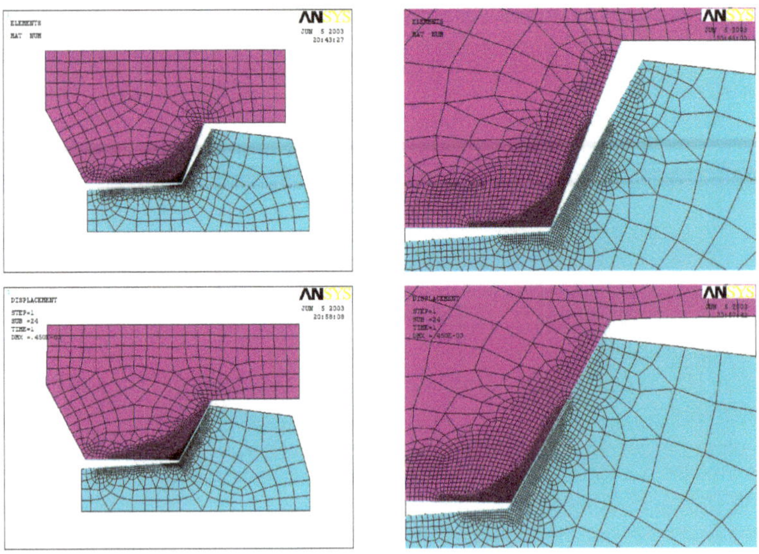

Figure 6. FE-mesh and the displacement field in the timing belt

which is equivalent to two algebraic equations

$$\begin{cases} -\varpi^2/4 + \Omega^2\,(1+\mu) = 0 \\ -\varpi^2/4 + \Omega^2\,(1-\mu) = 0 \end{cases}$$

since the sine and cosine functions are linearly independent functions. Therefore the boundaries of the instability region are given by

$$\varpi = 2\Omega\sqrt{1 \pm \mu} \quad . \tag{32}$$

The value of the contact force $\Phi(t)$ was calculated numerically by means of the finite element code. Figure 6 displays some details of the FE-mesh and the displacement field in the belt for some particular parameters of the belt and pulley.

The results of FE-analysis enables calculation of the contact pressure distribution and consequently the contact force in the belt which is shown in Fig. 7.

For this force we can calculate the magnitude of the first harmonics which allows us to plot the instability regions. The instability regions are displayed in Fig. 8 in the plane of parameters F_0, V. The region of the divergent instability is obtained from the condition that the square of "eigenfrequency" Ω^2 becomes negative. It follows from eq. (28) that it occurs if $V > V_{crit}$ where the critical velocity is given by the equation

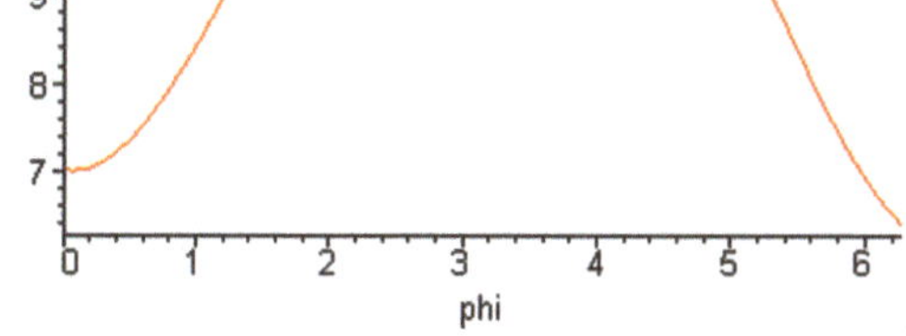

Figure 7. Contact force of the timing belt $\Phi(\varpi t)$, $0 \leq \varpi t \leq 2\pi$

$$V_{crit} = \sqrt{\frac{F_0 + F_c}{\rho A}} \quad .$$

In this case the solution growth exponentially which is typical for the divergence stability.

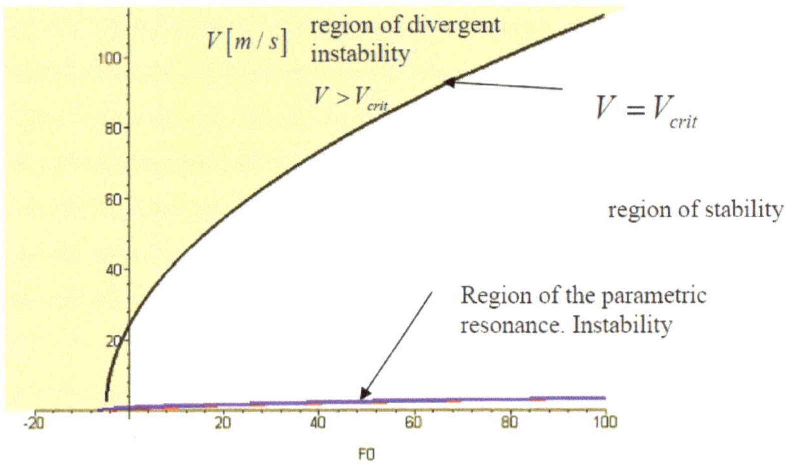

Figure 8. Instability regions for timing belt

3.2 Dynamics and stability of chains

Another interesting example of axially moving materials is a timing chain which is schematically shown in Fig. 9.

Stability of the chain drive is carried out in a similar manner in Eglseer and Belyaev [1997]. However there exists a principal difference between the excitation of belts and chains. The belt is excited dynamically in axial direction and the excitation is due to the contact interaction of the belt teeth and the pulley teeth. The belt possesses a low bending rigidity and is modeled as a beam in bending. Quite the contrary, the chain is excited kinematically because of the polygon effect. The latter is caused by the fact that any chain consists of links which are very stiff and "climb" of the sprocket-wheel. In addition to this, the chain has in fact zero bending rigidity, that is, a proper model for studying stability of a chain drive is a string excited by inertia forces due to the polygon effect on each sprocket.

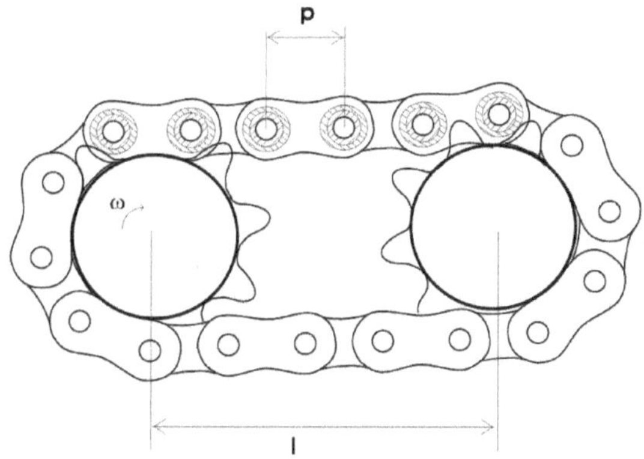

Figure 9. Schematic of a timing chain

Equation of the chain motion which is the string equation is as follows

$$P(t)\frac{\partial^2 W}{\partial x^2} - \rho A\left(\frac{\partial^2 W}{\partial t^2} + 2\frac{\partial^2 W}{\partial x \partial t}V + \frac{\partial^2 W}{\partial x^2}V^2 + \frac{\partial W}{\partial x}V\right) = 0\ .$$

Here W denotes the transverse displacement of the chain and V is the axial velocity of the chain. The tension force in the string $P(t)$ is given by

$$P(t) = F_0 + m_l\frac{\omega_1}{2}\left(V_{\max} - V_{\min}\right)\left[\sin\left(\omega_1 t\right) - \sin\left(\omega_1 t + \Phi\right)\right]$$

where F_0 denotes the pre-tension force, m_l denotes the mass of the chain link, ω_1 is the fundamental frequency corresponding to the process of the single link (similar to frequency ϖ for belt) and Φ stands for the phase shift between the sprocket-wheels.

The further analysis is similar to the belt analysis and is omitted because of the lack of space. The stability chart in the plane F_0, V is similar to that shown in Fig. 8, see Eglseer and Belyaev [1997] for detail.

4 Dynamics and stability of pipes conveying fluid

4.1 Dynamics of suspended pipe conveying heavy fluid

A pipe conveying a fluid is shown in Fig. 10. By analogy to the case of the timing belt we consider the pipe as a beam and derive the equation of motion which turns out to be very similar to eq. (22), namely

$$EI\frac{\partial^4 w}{\partial x^4}+\left((\rho A)_p+(\rho A)_f\right)\frac{\partial^2 w}{\partial t^2}+2v\,(\rho A)_f\frac{\partial^2 w}{\partial x\partial t}+v^2\,(\rho A)_f\frac{\partial^2 w}{\partial x^2}=0 \quad (33)$$

Here w denotes the transverse displacement of the pipe, the subscripts p and f are referred to the pipe and fluid, respectively, and v stands for the velocity of fluid.

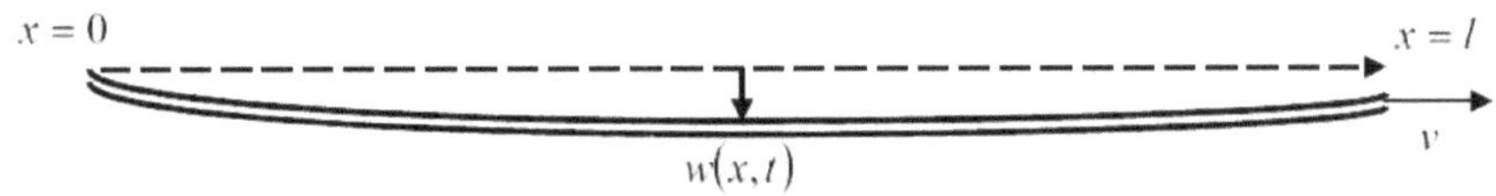

Figure 10. Pipe conveying a fluid

The stability analysis is carried out by means of the substitution $w = W(x)\,e^{st}$ which leads to the following ordinary differential equation for the deflection shape $W(x)$

$$W^{IV}+as^2W+bsW'+cW''=0 \quad (34)$$

where the newly introduced parameters are as follows

$$a=\frac{(\rho A)_p+(\rho A)_f}{EI},\quad b=\frac{2v\,(\rho A)_f}{EI},\quad c=\frac{v^2\,(\rho A)_f}{EI}.$$

The further step is looking for the basic functions. Let us assume that the pipe can be modeled by a simply supported beam. We make use of Galerkin approach, that is, we suggest the substitution

$$W(x)=A\sin\frac{\pi x}{l}+B\sin\frac{2\pi x}{l} \quad (35)$$

Inserting it into eq. (34), multiplying consequently by $\sin(\pi x/l)$, $\sin(2\pi x/l)$ and integrating over the beam length we arrive to the system of two equations for A and B

$$\begin{aligned}&A\left[\left(\frac{\pi}{l}\right)^4+as^2-c\left(\frac{\pi}{l}\right)^2\right]-\frac{8}{3}\frac{bs}{l}B=0\\&\frac{8}{3}\frac{bs}{l}A+B\left[16\left(\frac{\pi}{l}\right)^4+as^2-4c\left(\frac{\pi}{l}\right)^2\right]=0\end{aligned} \quad (36)$$

The nontrivial solution exists provided that the determinant of the system vanishes which results in the following biquadratic equation

$$
\begin{aligned}
& a^2 s^4 + s^2 \left[17a \left(\frac{\pi}{l}\right)^4 - 5ac \left(\frac{\pi}{l}\right)^2 + \left(\frac{8}{3}\frac{b}{l}\right)^2 \right] + \\
& +4 \left(\frac{\pi}{l}\right)^4 \left[\left(\frac{\pi}{l}\right)^2 - c \right] \left[4 \left(\frac{\pi}{l}\right)^2 - c \right] = 0
\end{aligned} \tag{37}
$$

If we introduce

$$
s \frac{l^2}{\pi^2} \sqrt{a} = \Omega, \quad \frac{cl^2}{\pi^2} = \alpha, \quad \frac{(\rho A)_f}{(\rho A)_f + (\rho A)_p} = \beta
$$

then eq. (37) takes the form

$$
\Omega^4 + \Omega^2 \left(17 - 5\alpha + \frac{256}{9\pi^2} \beta\alpha \right) + 4 (1 - \alpha) (4 - \alpha) = 0 \tag{38}
$$

It was proved numerically that for $0 \leq \alpha \leq 1$ the roots depend weakly on β which enables putting $\beta = 0$ in eq. (38) and representing the result in the form

$$
\Omega^4 + \Omega^2 (17 - 5\alpha) + 4 (1 - \alpha) (4 - \alpha) = \left[\Omega^2 + (1 - \alpha) \right] \left[\Omega^2 + 4 (4 - \alpha) \right] = 0 \tag{39}
$$

i.e. $\Omega_1 = i\sqrt{1 - \alpha}$ and $\Omega_2 = 2i\sqrt{4 - \alpha}$. It means that

$$
s_1 = i \frac{\pi^2}{l^2 \sqrt{a}} \sqrt{1 - v^2 \frac{l^2 (\rho A)_f}{\pi^2 EI}}, \quad s_2 = i \frac{2\pi^2}{l^2 \sqrt{a}} \sqrt{4 - v^2 \frac{l^2 (\rho A)_f}{\pi^2 EI}} \tag{40}
$$

hence, we obtain the critical velocity which is

$$
v_{crit} = \frac{\pi}{l} \sqrt{\frac{EI}{(\rho A)_f}} \tag{41}
$$

If this velocity is exceeded, the pipe becomes unstable.

4.2 Stability of suspended pipe conveying heavy fluid

The system is shown in Fig. 11.

The equation of the pipe bending coincides with that in eq. (33) in which the inertia term of the pipe is neglected in comparison with that of the fluid. Under zero initial conditions the Laplace transformation yields

$$
EI\bar{w}^{IV} + (\rho A)_f v^2 \bar{w}^{II} + 2v (\rho A)_f p\bar{w}^I + (\rho A)_f p^2 \bar{w} = 0 \tag{42}
$$

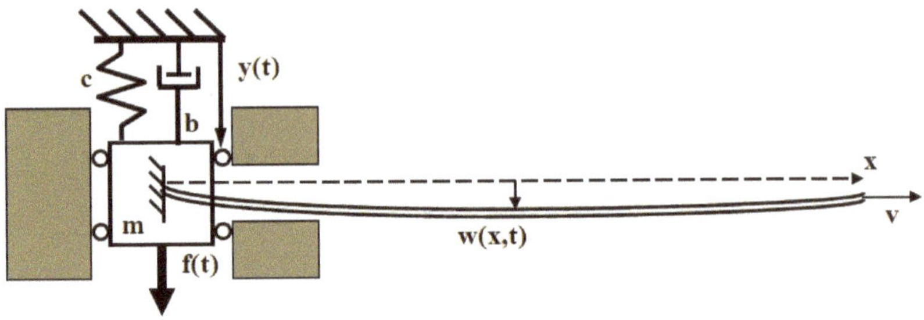

Figure 11. Schematic of suspended pipe conveying fluid

where a bar denotes the Laplace transform of the corresponding quantity.

The solution is sought in the form

$$\bar{w}\,(x,p) = A_1 \exp\left(\lambda_1 x\right) + A_2 \exp\left(\lambda_2 x\right) + A_3 \exp\left(\lambda_3 x\right) + A_4 \exp\left(\lambda_4 x\right) \quad (43)$$

where the eigenvalues λ_n are solutions of the equation

$$\lambda^4 + \beta^2 \left(2\lambda\sqrt{n} + \beta\, p\right)^2 = 0, \quad n = \frac{\beta^2 v^2}{4}, \quad \beta = \sqrt[4]{\frac{(\rho A)_f}{EI}} > 0, \quad (44)$$

One can prove that the transverse force in the beam at x=*0* is given by

$$Q|_{x=0} = EI\frac{d^3\bar{w}}{dx^3}\bigg|_{x=0} = -EI \cdot \left[\lambda_2^3 A_2 + \lambda_4^3 A_4\right]$$

or after some algebra we become asymptotically

$$\bar{Q}\big|_{x=0} = -EI\left[4i\beta^2 p \cdot \bar{w}\,(0,p) \cdot (-i\beta\sqrt{p})\right] = -4EI\beta^3 p^{3/2} \cdot \bar{w}\,(0,p) \quad (45)$$

Under zero initial conditions the equation of motion of mass m in the Laplace space is given by

$$mp^2\bar{y}\,(p) + pb\bar{y}\,(p) + c\bar{y}\,(p) = \bar{Q}\big|_{x=0} + \bar{f}\,(p) \quad (46)$$

Since $y\,(t) = w\,(0)$, i.e. $\bar{y}\,(p) = \bar{w}\,(0)$, we can substitute eq. (45) in eq. (46), to have

$$mp^2\bar{y}(p) + 4EI\beta^3 p^{3/2}\bar{w}(0,p) + (b - \rho Av)p\bar{w}(0,p) + c\bar{y}(p) = f(p) \quad (47)$$

which corresponds to the following ordinary differential equation with a fractional derivative of order 3/2

$$m\frac{d^2y}{dt^2} + 4EI\beta^3\frac{d^{3/2}y}{dt^{3/2}} + (b - \rho Av)\frac{dy}{dt} + cy = f(t) \quad (48)$$

This equation governs the motion of the pipe suspension. Introducing a non-dimensional time $\tau = kt, k = \sqrt{c/m}$ and the following non-dimensional system parameters

$$\delta = 2\frac{EI\beta^3}{m\sqrt{k}},\ \varepsilon = \frac{(\rho Av - b)}{mk}$$

we can set eq. (48) in the following form

$$D^2y + 4\delta D^{3/2}y + \varepsilon Dy + y = c^{-1}f(\tau),\ D = \frac{d}{d\tau}\ . \quad (49)$$

For handling this semi-differential equation we apply the method by Suarez and Shokooh [1997].

Eigenvector Expansion Method for solving differential equation with fractional derivative Equation (49) is an ordinary differential equations of second order with the derivatives of the order 3/2. For simplicity we consider eq. (49) in the case *f(t)=0*, that is,

$$D^2y + 4\delta D^{3/2}y + \varepsilon Dy + y = 0 \quad (50)$$

and solve it by means of the eigenvector expansion method suggested by Suarez and Shokooh [1997] for differential equations with fractional derivatives.

This equation can be represented in the normal form of four semi-differential equations by means of the substitution

$$z_1 = D^{3/2}y(t),\ z_2 = Dy(t),\ z_3 = D^{1/2}y(t),\ z_4 = y(t) \quad (51)$$

i.e. we can rewrite eq. (50) in the matrix form $\{A\}D^{1/2}\{z\} = \{B\}\{z\}$ where $\{z\}$ denotes the column composed of $z_n, n = 1, 2, 3, 4$ in the latter equation. Applying the standard methods of linear algebra yields the eigenvectors $\{\Psi\}_j$ and eigenvalues λ_j

$$\{A\}\{\Psi\}_j = \lambda_j \{B\}\{\Psi\}_j \tag{52}$$

where the eigenvectors are orthonormalized, i.e.

$$\{\Psi\}_i^T \{B\}\{\Psi\}_j = \delta_{ij}, \ \{\Psi\}_i^T \{A\}\{\Psi\}_j = \lambda_j \delta_{ij} \ . \tag{53}$$

Let us notice at this place that the eigenvalues λ_j of the matrix equation (51) have nothing in common with the eigenvalues of the differential equation (49). Namely the eigenvalues λ_j are solutions of equation $\lambda^4 + a\lambda + b = 0$ and are given by

$$\lambda_1 = \bar{\lambda}_2 = p + iq, \ \lambda_{3,4} = -p \pm is \tag{54}$$

where

$$p = \sqrt{\kappa}, \ q = \sqrt{\kappa + \frac{\delta}{2\sqrt{\kappa}}}, \ s = \sqrt{\kappa - \frac{\delta}{2\sqrt{\kappa}}} \tag{55}$$

$$\kappa = \frac{2^{1/3}}{4}\left[\left(\delta^2 + \sqrt{\delta^4 - \frac{16}{27}}\right)^{1/3} + \left(\delta^2 - \sqrt{\delta^4 - \frac{16}{27}}\right)^{1/3}\right]$$

By means of the substitution $\{z\} = \{\Psi\}\{h\}$ where matrix $\{\Psi\}$ is built from the eigenvectors columns $\{\Psi\}_j$ we arrive at the system of four uncoupled semi-differential equations

$$D^{1/2} h_j(t) - \lambda_j h_j(t) = 0, \ \ j = 1, 2, 3, 4 \ . \tag{56}$$

Solving these equations with the help of Laplace transformation, applying the inverse Laplace transformation and satisfying the initial conditions, we obtain the sought-for result, see Belyaev [2014] for detail.

Obtaining a closed form solution assumes the well-known property of the Laplace transformation, namely the Laplace transform $L[\ldots]$ of a fractional derivative of order α of function $x(t)$ is as follows

$$L[D^\alpha \varphi(t)] = p^\alpha L[\varphi(t)] - C \tag{57}$$

where C denotes an integration constant. This formula follows from the formal definition of a fractional derivative of order α which is given by

$$D^\alpha \{\varphi(t)\} = \frac{d}{dt}\left\{\frac{1}{\Gamma(1-\alpha)} \int_0^t (t-\tau)^{-\alpha} \varphi(\tau)\, d\tau\right\} \tag{58}$$

see Oldham and Spanier [1974]. Here the integration constant C is determined by the following condition

$$C = D^{\alpha-1}\varphi(t)\big|_{t=0} \ . \tag{59}$$

It is worth mentioning that the value of C is not necessarily equal to zero even for the zero initial conditions for the system. There exists a seeming discrepancy between the number of initial conditions in the system (two initial conditions in the initial-value problem) and the number of the integration constants in the system (56) of four uncoupled semi-differential equations (four integration constants). This discrepancy is easily removed since the general expressions for the displacement and velocity contains some functions which are unbounded at $t \to 0$. The requirement that these functions must vanish provide us with two additional conditions, see Suarez and Shokooh [1997] for detail. Satisfying four conditions (two initial conditions and two additional conditions of boundness) yields the sought-for integration constants for any initial conditions, which means that the solution to the considered problem is obtained.

Numerical analysis of the obtained solution shows that the stability border is described by the condition $\varepsilon = \delta$ (the details are omitted).

$$\delta = 2\frac{EI\beta^3}{m\sqrt{k}} \cong \varepsilon = \frac{(\rho A v_{crit} - b)}{mk}$$

Since parameter ε depends on the fluid velocity this condition yields the critical velocity of the flow that results in the suspension instability

$$v_{crit} = \frac{b}{\rho A} + \sqrt[4]{\frac{1}{8} \cdot \frac{EI}{\rho A} \cdot \frac{c}{m}} \ . \tag{60}$$

If this value is exceeded, i.e. $v > v_{crit}$, then the suspension and hence the pipe are unstable.

5 Dynamics and stability of the rapidly rotating shaft in floating bearing

A shaft of a rapidly rotating rotor on the plain floating bearing is schematically depicted in Fig. 12.

5.1 The Reynolds equation with account for the centrifugal force

The nonstationary momentum equation for the oil is given by

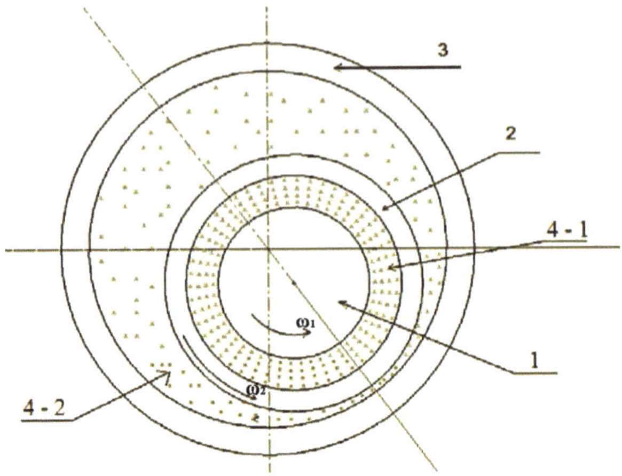

Figure 12. Schematic representation of the bearings with three cylinders: 1 denotes the floating rotor, 2 is the floating bushing (ring), 3 is a fixed cylinder (bearing housing), 4 stands for the incompressible oil (4-1 is the internal field and 4-2 is the external field) and ω_1, ω_2 denote the angular speed of rotation of the elements 1 and 2, respectively.

$$\nabla \cdot \tau - \rho \dot{\mathbf{v}} = 0, \tag{61}$$

where τ is the stress tensor, ρ is the mass density and $\mathbf{v}$ denotes the velocity vector

$$\mathbf{v} = \mathbf{e}_r u + \mathbf{e}_\varphi v + \mathbf{e}_z w.$$

where u,v,w are the velocity components in the corresponding direction. The Hamilton operator in the cylindrical coordinate system is defined as follows

$$\nabla = \mathbf{e}_r \frac{\partial}{\partial r} + \mathbf{e}_\varphi \frac{1}{r}\frac{\partial}{\partial \varphi} + \mathbf{e}_z \frac{\partial}{\partial z},$$

where r, φ, z are cylindrical coordinates.

The oil in the gap is assumed to be a Newtonian fluid, that is,

$$\tau = -p\mathbf{E} + \mathbf{s}, \;\; \mathbf{s} = 2\mu Dev(\nabla \mathbf{v})^S, \;\; p \geq 0,$$

where p and $\mathbf{s}$ imply the oil pressure and the shear stress tensor correspondingly. Next, μ is the dynamic viscosity and $Dev()$ denotes the deviator. It is assumed that the deviator contains only the non-diagonal elements. Besides, we assume the incompressible oil and a constant oil viscosity.

The material time-derivative of the velocity is

$$\dot{\mathbf{v}} = \frac{d\mathbf{v}}{dt} = \frac{\partial \mathbf{v}}{\partial t} + \mathbf{v} \cdot (\nabla \mathbf{v}) = \mathbf{v} \cdot (\nabla \mathbf{v}),$$

i.e. a quasi-stationary flow is assumed.

In what follows we make the following assumptions which are common in the bearing theory:

1. The gap flow is laminar;
2. $u = 0$, i.e. the velocity components in the radial direction are neglected since the gap is very narrow;
3. $|\partial v/\partial r| >> \left|r^{-1}\partial v/\partial \varphi\right|, |\partial v/\partial z|$ since the gap is very narrow whereas the velocity in the gap experiences drastic changing. For the same reason $|\partial w/\partial r| >> \left|r^{-1}\partial w/\partial \varphi\right|, |\partial w/\partial z|$.

Under these assumptions one obtains one vectorial equation which is equivalent to three scalar equations governing the balance of momentum in three directions:

$$\begin{aligned}
r: \quad & \frac{\partial p}{\partial r} = \rho \frac{v^2}{r}, \\
\varphi: \quad & \frac{\partial p}{\partial \varphi} = \mu \frac{\partial}{\partial r}\left(r \frac{\partial v}{\partial r}\right), \\
z: \quad & \frac{\partial p}{\partial z} = \mu \frac{\partial^2 w}{\partial r^2}.
\end{aligned} \tag{62}$$

The radial term $\rho v^2/r$ which is referred to as the centrifugal force represents a considerable generalisation of the existing theories of the flow in gaps. In the case of high velocities and small gaps this generalisation turns out to be of crucial importance.

The next step is formulation of the boundary conditions. We assume that the velocity of flow coincides with velocity of the corresponding boundary surface, that is,

surface of the shaft $(r = r_0)$: $\quad u = 0, \;\; v = V = \Omega r_0, \;\; w = 0;$

surface of the floating ring: $(r = r_i = r_0 + h)$: $\quad u = \dot{h}, v = \omega r_i, w = 0.$

Here V is the tangential velocity of the shaft, h is the gap height, ω and Ω denote the angular velocity of the ring and shaft respectively, r stands for the inner radius of the ring. Integration of the first equation in (62) yields

$$p = \rho \int v^2 \frac{dr}{r} + P(\varphi, z)\,, \tag{63}$$

whereas integrating the second equation in eq. (62) results in the expression for v

$$v = \frac{1}{\mu}\frac{\partial P}{\partial \varphi}\left[r + C_1 \ln r + C_2\right] \quad .$$

The integration constants are determined from the boundary conditions for v, which ensure the coincidence of the velocities of the flow and the wall

$$v(r = r_0) = \frac{1}{\mu}\frac{\partial P}{\partial \varphi}\left[r_0 + C_1 \ln r_0 + C_2\right] = V\,,$$
$$v(r = r_0 + h) = \frac{1}{\mu}\frac{\partial P}{\partial \varphi}\left[r_0 + h + C_1 \ln\left(r_0 + h\right) + C_2\right] = \omega r_i \ .$$

The solution is as follows

$$v = \frac{1}{\mu}\frac{\partial P}{\partial \varphi} f(r) + g(r), \tag{64}$$

where

$$f(r) = r + \frac{h \ln r + r_0 \ln(r_0 + h) - (r_0 + h)\ln r_0}{\ln \frac{r_0}{(r_0+h)}}\,,$$
$$g(r) = \frac{(V - \omega r_i)\ln r + \omega r_i \ln r_0 - V \ln(r_0 + h)}{\ln \frac{r_0}{(r_0+h)}}\,.$$

The integration of the third equation in (62) along with the boundary conditions yields

$$w(r) = \frac{1}{2\mu}\frac{\partial P}{\partial z}(r - r_0)(r - r_0 - h). \tag{65}$$

The continuity equation for the fluid with account for stationary flow is as follows

$$\dot{\rho} = \frac{\partial \rho}{\partial t} + \nabla \cdot (\rho \mathbf{v}) = 0 \ \rightarrow \ \nabla \cdot (\rho \mathbf{v}) = 0$$

Since we consider the incompressible fluid the latter equations reduces to

$$\nabla \cdot \mathbf{v} = 0.$$

In order to obtain the equation that governs the dynamics of fluid in the plain bearing we consider the balance of the mass in the control volume, that is, in the gap $r_0 \leq r \leq r_0 + h$. A control volume approach is an approach to study systems with variable mass describing the fluid that enters and leaves the control volume. In order to apply this approach we introduce the fluid rate in the gap in directions φ and z :

$$Q_\varphi = \int_{r_0}^{r_0+h} v dr, \quad Q_z = \int_{r_0}^{r_0+h} w dr.$$

Then the law of mass conservation for the gap which is, in fact, the continuity equation for the gap takes the following form:

$$\frac{1}{r}\frac{\partial Q_\varphi}{\partial \varphi} r d\varphi dr dz + \frac{\partial Q_z}{\partial z} r d\varphi dr dz + \frac{\partial h}{\partial t} r d\varphi dr dz = 0.$$

It can be rewritten as follows:

$$\frac{1}{r}\frac{\partial Q_\varphi}{\partial \varphi} + \frac{\partial Q_z}{\partial z} + \frac{\partial h}{\partial t} = 0. \tag{66}$$

Inserting the obtained formulae for v and w, eqs. (64) and (65), into eq. (66) yields

$$Q_z = \int_{r_0}^{r_0+h} w dr = \frac{1}{2\mu}\frac{\partial P}{\partial z}\int_{r_0}^{r_0+h} (r - r_0)(r - r_0 - h) dr = -\frac{h^3}{12\mu}\frac{\partial P}{\partial z}$$

and

$$Q_\varphi = \int_{r_0}^{r_0+h} v dr = \frac{1}{\mu}\frac{\partial P}{\partial \varphi}\Phi(h) + \Psi(h),$$

where the closed form expressions for $\Phi(h)$ and $\Psi(h)$ are as follows

$$\Phi(h) = \int_{r_0}^{r_0+h} f(r) dr =$$

$$= \int_{r_0}^{r_0+h} \left[r + \frac{h \ln r + r_0 \ln(r_0 + h) - (r_0 + h) \ln r_0}{\ln \frac{r_0}{(r_0+h)}} \right] dr =$$

$$= \left[\frac{1}{2} r^2 + r(\ln r - 1)\frac{h}{\ln \frac{r_0}{(r_0+h)}} + \frac{r_0 \ln(r_0 + h) - (r_0 + h) \ln r_0}{\ln \frac{r_0}{(r_0+h)}} r \right] \Bigg|_{r=r_0}^{r=r_0+h}$$

$$
\begin{aligned}
\Psi(h) &= \int_{r_0}^{r_0+h} g(r)dr = \\
&= \int_{r_0}^{r_0+h} \frac{(V-\omega r_i)\ln r + \omega r_i \ln r_0 - V\ln(r_0+h)}{\ln \frac{r_0}{(r_0+h)}} dr = \\
&= \left[\frac{(V-\omega r_i)\, r\,(\ln r - 1) + \{\omega r_i \ln r_0 - V \ln(r_0+h)\}\, r}{\ln \frac{r_0}{(r_0+h)}} \right] \Bigg|_{r=r_0}^{r=r_0+h} .
\end{aligned}
$$

Substitution of this result in the equation for the mass conservation leads to the following equation

$$
\frac{h^3}{6}\frac{\partial^2 P}{\partial z^2} - \frac{1}{r}\frac{\partial}{\partial \varphi}\left[\Phi(h)\frac{\partial P}{\partial \varphi}\right] = 2\mu\left[\frac{\partial h}{\partial t} + \frac{1}{r}\frac{\partial \Psi(h)}{\partial \varphi}\right] = 2\mu\left[\frac{\partial h}{\partial t} + \frac{1}{r}\frac{\partial \Psi(h)}{\partial h}\frac{\partial h}{\partial \varphi}\right]
$$

This equation is a generalisation of the celebrated Reynolds equation for the case of considerable centrifugal forces. Since

$$
\frac{\partial \Psi(h)}{\partial h} = g(r_0 + h)
$$

we arrive at the following form of the generalised Reynolds equation

$$
\frac{h^3}{6}\frac{\partial^2 P}{\partial z^2} - \frac{1}{r}\frac{\partial}{\partial \varphi}\left[\Phi(h)\frac{\partial P}{\partial \varphi}\right] = 2\mu\left[\frac{\partial h}{\partial t} + \frac{g(r_0+h)}{r}\frac{\partial h}{\partial \varphi}\right]. \tag{67}
$$

5.2 Generalised Reynolds equation in the case of short bearing

This is the case in which the change of the pressure in the circumferential direction (φ) is much smaller than that in the axial direction (z), i.e.

$$
\left|\frac{\partial P}{\partial \varphi}\right| << \left|\frac{\partial P}{\partial z}\right|. \tag{68}
$$

This assumption simplifies the Reynolds equation (67) to the following form

$$
\frac{\partial^2 P}{\partial z^2} = \frac{12\mu}{h^3}\left[\frac{\partial h}{\partial t} + \frac{g(r_0+h)}{r}\frac{\partial h}{\partial \varphi}\right]. \tag{69}
$$

Under the assumption that the shaft and the floating ring have circular forms we can express $h(\varphi, t)$ in terms of the normal gap height $h_0 = r_i - R$, the eccentricity $e(t)$, coordinate $\varphi(t)$ and the angle $\gamma(t)$, the latter representing the angle between the connecting line to the actual center of the shaft and the ring.

The boundary conditions for a symmetric bearing are given by

$$z = 0, \quad p = p_{in},$$
$$P(z) = P(-z)$$

where L stands for the length of the bearing and p_{in} denotes the input pressure of the oil in the middel of the bearing. Taking into account equation (63) for p and equation (64) for v we can satisfy the above boundary conditions by means of the equation

$$P = \frac{12\mu}{h^3}\left[\frac{\partial h}{\partial t} + \frac{g(r_0+h)}{r}\frac{\partial h}{\partial \varphi}\right]\left(\frac{1}{2}z^2 + C\right). \tag{70}$$

Since the pressure in the gap is given by the formula

$$p = P + \rho\int v^2\frac{dr}{r} \tag{71}$$

we obtain the following equation for the integration constant C :

$$\frac{12\mu}{h^3}\left[\frac{\partial h}{\partial t} + \frac{g(r_0+h)}{r}\frac{\partial h}{\partial \varphi}\right]C + \rho\int_{r_0}^{r_0+h} v^2\frac{dr}{r} = p_{in} \quad .$$

In turn this yields the following equation for the pressure in the gap

$$p(r,\varphi,z) = p_{in} - \rho\int_{r}^{r_0+h} v^2\frac{dr}{r} + \frac{6\mu}{h^3}\left[\frac{\partial h}{\partial t} + \frac{g(r_0+h)}{r}\frac{\partial h}{\partial \varphi}\right]z^2. \tag{72}$$

In the zones when the pressure is negative it is conventionally assumed that $p = 0$.

Integration over z yields the force per length unit. For example, the force acting on the ring $r = r_0 + h$ is as follows

$$q(h,\varphi)\big|_{r=r_0+h} = \int_{-L/2}^{L/2} p\,(r,\varphi,z)\big|_{r=r_0+h}\,dz =$$
$$= p_{in}L + \frac{\mu}{2h^3}\left[\frac{\partial h}{\partial t} + \frac{g(r_0+h)}{r}\frac{\partial h}{\partial \varphi}\right]L^3, \; p > 0.$$

The force on the shaft $r = r_0$ is given by

$$q(h,\varphi)|_{r=r_0} = \int_{-L/2}^{L/2} p\,(r,\varphi,z)|_{r=r_0}\, dz =$$

$$\left[p_{in} - \rho \int_{r_0}^{r_0+h} v^2 \frac{dr}{r}\right] L + \frac{\mu}{2h^3}\left[\frac{\partial h}{\partial t} + \frac{g(r_0+h)}{r}\frac{\partial h}{\partial \varphi}\right] L^3,\ p > 0\ ,$$

however with a good degree of accuracy we can neglect the first term in eq. (64), that is, we can take $v = g\,(r)$. Besides, the input pressure p_{in} should be sufficiently high in order to avoid negative pressure which inevitably leads to cavitation and the fluid model becomes no longer valid.

5.3 Determining angular velocity of the floating bushing in the bearing

In the framework of the classical theory of short bearing we consider a rigid rotor rotating in the housing that is fixed in the space. The gap between the solids is assumed to be filled with incompressible oil, cf. Dubois and Ocvirk [1953]. We explore a more complicated case of thrust bearings, which consists of three parts (see Fig. 12): a fixed housing 3, a floating bushing 2, rotating with angular velocity ω_2 and floating shaft (rotor) 1 rotating with an angular velocity speed of ω_1. The gaps between the solid bodies are filled with an incompressible lubricating oil, 4, with 4-1 and 4-2 denoting respectively the internal field and external field of lubricating oil.

We first consider a system of "floating bushing - rotor." The coordinate system is taken to be fixed in the center of the floating bushing. Then in this system the incompressible oil 4-1 is located in the gap between the floating bushing (O_2, R_2) (its rotation axis is fixed in space) and the floating rotating rotor (O_1, R_1), see Fig. 13. The floating bushing rotates with angular velocity ω_2 , while the angular velocity of the rotor is ω_1.

Let us introduce the following notation: $h_{01} = R_2 - R_1$ is the nominal gap, $e_1 = e_{P1}(t)$ denotes the eccentricity of the center of the floating rotor and $\gamma_1 = \gamma_1(t)$ is the angle describing the position of the line between the centers of the floating rotor and the rotating floating bushing. The motion of the rotor in the lubricating layer is time-dependent, i.e. the position and velocity of the center depends on time since the external load and the response of the lubricating layer depend on time. Let us suppose that at the time instant $t = t_0$ the center O_1 of the rotor is in the position corresponding to the eccentricity $e_{01} = e_1(t_0)$ and the angle $\gamma_{01} = \gamma_{P1}(t_0)$ describing the position of the line between the centers of the floating rotor and the bushing. Then

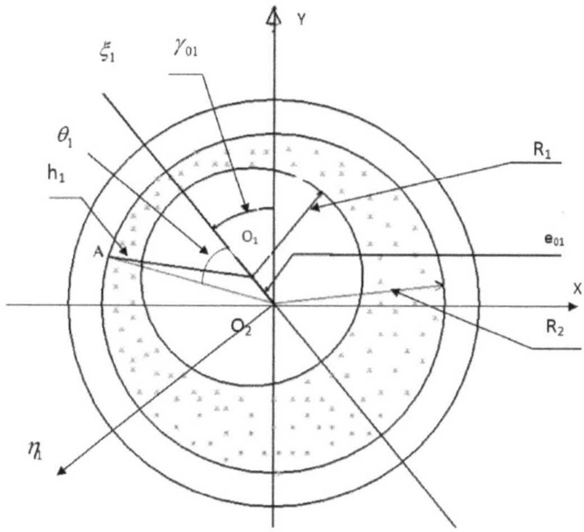

Figure 13. The notation for the system "floating bushing - rotor"

$$\theta_1 = \theta_1(t_0, \varphi) = \varphi - \gamma_{P1}(t_0) = \varphi - \gamma_{01}.$$

Denoting for the width of the gap (film thickness) as

$$h_1(\theta_1, t) = h_{01} - e_1(t)\cos(\theta_1)$$

we introduce the initial gap at $t = t_0$:

$$h_1(\theta_1, t_0) = h_{01} - e_{01}\cos(\theta_1) \quad .$$

Then the force acting on the rotor per length unit is given by

$$q_{01} = \frac{1}{L}\int_{-L/2}^{L/2} (p_{01} - \widetilde{p_{01}})\, dz = \frac{\mu L^2(\omega_1 + \omega_2)}{2h_{01}^2}\,\overline{q}_{01} \tag{73}$$

where

$$\overline{q}_{01} = \frac{\left(\frac{2\dot{\gamma}_1}{\omega_1+\omega_2} - 1\right)\varepsilon_1 \sin\theta_1 + \frac{2\dot{\varepsilon}_1}{\omega_1+\omega_2}\cos\theta_1}{(1-\varepsilon_1\cos\theta_1)^3}, \quad \varepsilon_1 = \frac{e_1}{h_{01}}. \tag{74}$$

and $p_{01} = p_{01}(r, \varphi, z, t)$ stands for the global pressure in the gap 4-1, $\widetilde{p_{01}} = \widetilde{p_{01}}(r, \varphi, t)$ denotes the pressure at the ends of the bearing and μ, L are respectively the dynamic viscosity of lubricant and the bearing length.

In a local coordinate system $(O_2\xi_1, O_2\eta_1)$, where the direction $O_2\xi_1$ corresponds to $\theta_1 = 0$ $(\varphi = \gamma_{P1})$, the projections of the force F_1^P acting on the rotor from the lubricant layer, are as follows:

$$F_{1\xi}^P = L \int_{\overline{\theta_1}}^{\overline{\theta_2}} (R_2 - h_1) q_{01} \cos\theta_1 d\theta_1 \tag{75}$$

$$F_{1\eta}^P = L \int_{\overline{\theta_1}}^{\overline{\theta_2}} (R_2 - h_1) q_{01} \sin\theta_1 d\theta_1 \tag{76}$$

The question of the boundaries of the lubricating layer is not yet resolved, cf. Hatakenaka et al. [2012]. The theory of dynamically loaded bearings usually adopts one of the following two hypotheses:

1. For the angles that determine the beginning and the end of the lubricating layer, the researchers accept the values, where the excessive pressure is equal to zero, i.e. at those places where the gap is the narrowest one and the widest one. Most often is $\overline{\theta}_1 = 0$, $\overline{\theta_2} = \pi$, that is, only the half of the gap is considered.

2. The value of the angles is $\overline{\theta_1} = 0$, $\overline{\theta_2} = 2\pi$, i.e. the lubricating layer embraces the entire rotor. According to this hypothesis, there is negative pressure in the gap which is comparable with or even equals in magnitude the positive one.

In what follows we use the second hypothesis, i.e. the lubricating layer fills the entire gap, however a comparison with the results from the first hypothesis will be given, too.

5.4 Distribution of the rotor equilibria in the lubricant layer

Substituting eqs. (73), (74) into eq. (75) we obtain the expression of the projection of force F_1^P in the direction $O_2\xi_1$

$$F_{1\xi}^P = \frac{\mu L^3(\omega_1 + \omega_2)}{2h_{01}^2} \left[\int_0^{2\pi} \frac{R_1 B \cos^2\theta_1}{(1 - \varepsilon_1 \cos\theta_1)^3} d\theta_1 + \int_0^{2\pi} \frac{e_1 B \cos^3\theta_1}{(1 - \varepsilon_1 \cos\theta_1)^3} d\theta_1 \right] \tag{77}$$

where

$$A = \left(\frac{2\dot{\gamma}_1}{\omega_1 + \omega_2} - 1 \right) \varepsilon_1, \quad B = \frac{2\dot{\varepsilon}_1}{\omega_1 + \omega_2} \quad .$$

While deriving eq. (77) we omitted some transformations. In particular, two integrals (of four ones) vanish since their integrands are odd functions of the argument. It is seen from the above formula that the expression of the projection of force F_1^P on direction $O_2\xi_1$ is $F_{1\xi}^P$ and does not contain the angular velocity $\overset{\bullet}{\gamma}_1$. If we would use the first hypothesis, this angular velocity appeared in the expression for $F_{1\xi}^P$, then the result was

$$F_{1\xi}^P = \frac{2\mu L^3 \overset{\bullet}{\varepsilon}_1}{h_{01}^2}\left[\int_0^\pi \frac{R_1 . \cos^2\theta_1}{(1-\varepsilon_1\cos\theta_1)^3}d\theta_1 + \int_0^\pi \frac{e_1 . \cos^3\theta_1}{(1-\varepsilon_1\cos\theta_1)^3}d\theta_1\right]. \quad (78)$$

Similarly by substituting eqs. (73) and (74) into eq. (76) we obtain the following expression for the projection of force F_1^P in the direction of $O_2\eta_1$:

$$F_{1\eta}^P = \frac{\mu L^3(\omega_1+\omega_2)\varepsilon_1}{h_{01}^2}\left(\frac{2\overset{\bullet}{\gamma}_1}{\omega_1+\omega_2}-1\right)\times$$

$$\left[\int_0^\pi \frac{R_1\sin^2\theta_1}{(1-\varepsilon_1\cos\theta_1)^3}d\theta_1 + \int_0^\pi \frac{e_1\cos\theta_1\sin^2\theta_1}{(1-\varepsilon_1\cos\theta_1)^3}d\theta_1\right]. \quad (79)$$

Let us suppose that the rotor (O_1, R_1) is loaded by a constant force. The coordinates of the equilibrium position of the bearing are denoted by (e_1^*, γ_1^*) and $(\varepsilon_1^*, \gamma_1^*)$, then $\overset{\bullet}{\varepsilon}_1^* = 0$, $\overset{\bullet}{\gamma}_1^* = 0$. It follows from eqs. (78) and (5.4):

$$F_{1\xi}^P(\varepsilon_1^*, \gamma_1^*) = 0, \quad F_{1\eta}^P(\varepsilon_1^*, \gamma_1^*) = -\frac{\mu L^3(\omega_1+\omega_2)\varepsilon_1^*}{h_{01}^2}\times$$

$$\left[\int_0^\pi \frac{R_1\sin^2\theta_1}{(1-\varepsilon_1^*\cos\theta_1)^3}d\theta_1 + \int_0^\pi \frac{e_1^*\cos\theta_1\sin^2\theta_1}{(1-\varepsilon_1^*\cos\theta_1)^3}d\theta_1\right].$$

Here $(\varepsilon_1^*, \gamma_1^*)$ are obtained from the equilibrium condition:

$$\left|\overrightarrow{Q^P}\right| = \left|\overrightarrow{F_1^P(\varepsilon_1^*, \gamma_1^*)}\right| = \sqrt{\left(F_{1\xi}^P(\varepsilon_1^*, \gamma_1^*)\right)^2 + \left(F_{1\eta}^P(\varepsilon_1^*, \gamma_1^*)\right)^2} =$$

$$= \frac{\mu L^3(\omega_1+\omega_2)\varepsilon_1^*}{h_{01}^2}\left[\int_0^\pi \frac{R_1 . \sin^2\theta_1}{(1-\varepsilon_1^*\cos\theta_1)^3}d\theta_1 + \int_0^\pi \frac{e_1^* . \cos\theta_1 . \sin^2\theta_1}{(1-\varepsilon_1^*\cos\theta_1)^3}d\theta_1\right] \quad (80)$$

$$\tan\widetilde{\theta}_1 = \frac{F_{1\eta}^P(\varepsilon_1^*, \gamma_1^*)}{F_{1\xi}^P(\varepsilon_1^*, \gamma_1^*)} = -\infty, \quad (81)$$

where $\widetilde{\theta_1} = (\overrightarrow{F_1^P}, \overrightarrow{O_2\xi_1})$. It is easy to see from eq. (80) that if the rotor velocity ω_1 is constant then unbounded increase of the external force $\overrightarrow{Q^P}$ leads to that the relative eccentricity ε_1^* tends to unity. If we assume that the external force $\overrightarrow{Q^P}$ is directed vertically downwards, then it follows from eq. (80) that $\left|\overrightarrow{Q^P}\right|$ is proportional to the angular velocity ω_1 of the rotor. The external load can be, for example, the gravity force of the rotor. The rotor is in a state of equilibrium $(\varepsilon_1^*, \gamma_1^*)$, that is, the force $\overrightarrow{F_1^P}$ must be directed vertically upwards, i.e.

$$\widetilde{\theta_1} = (\overrightarrow{F_1^P}, \overrightarrow{O_2\xi_1}) = -\gamma_1^*. \tag{82}$$

It follows from eqs. (81) and (82) that $\gamma_1^* = \pi/2$.

The set of equilibrium positions of the center of the rotor in the lubricating layer is a horizontal segment O_2M (this interval is shown in Fig. 14 in bold), respectively, $\varepsilon_1^* \in [0, 1]$ depends on the external load $\overrightarrow{Q^P}$ on the rotor. If one uses the first hypothesis, the curve of the equilibrium positions of the center of the rotor O_1 is the semicircle which is shown in Fig. 14 by dotted lines. Note that $O_2M = O_2M' = h_{01} = R_2 - R_1$, cf. Fig. 14.

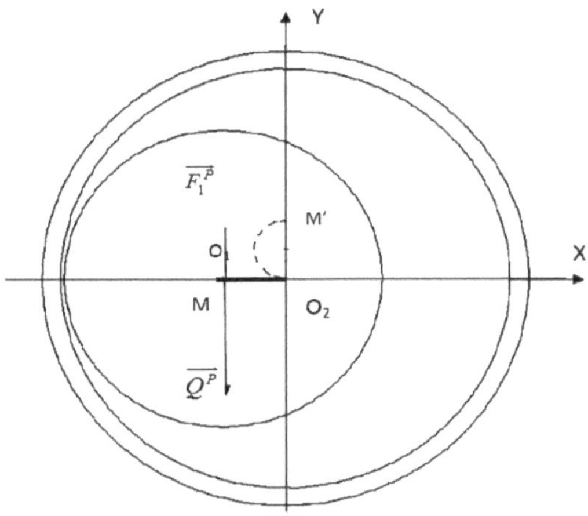

Figure 14. The locus of the equilibrium position of the center of rotor

5.5 The distribution of equilibria of the bushing in lubricating layer in the gap between the bearing housing and the rotor

So far we have only considered the system of "bushing - rotor" and have not yet taken into account the effect of an external field of lubrication 4-2. Now we consider the entire bearing, which consists of three rigid bodies, as shown in Fig. 15. In order to find the force F_2^B acting on the floating bushing (O_2, R_2) from the outer layer of lubricant, we consider a system of "bearing housing - bushing." By analogy, we obtain the projection of this force in the local coordinate system $(O\xi_2, O\eta_2)$

$$F_{2\xi}^B = L\int_0^{2\pi}(R-h_2)q_{02}\cos\theta_2 d\theta_2,$$

$$F_{2\eta}^B = L\int_0^{2\pi}(R-h_2)q_{02}\sin\theta_2 d\theta_2.$$

in which direction $O\xi_2$ corresponds to $\theta_2 = 0$ $(\varphi = \gamma_{B2}(t_0))$, see Fig. 15.

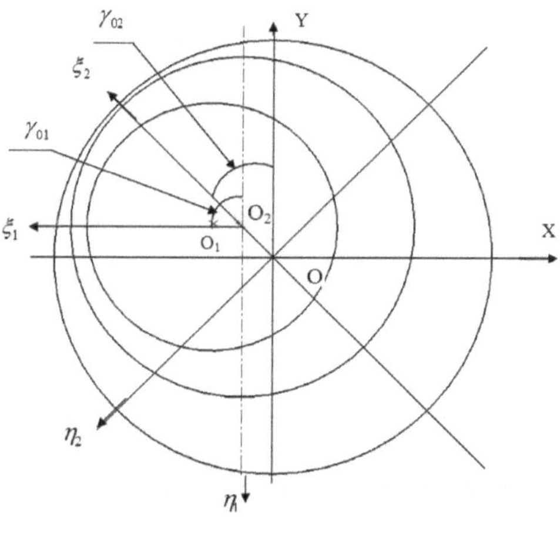

Figure 15. The notion for the bearing with the bushing

Note that the bearing housing *(O,R)* is fixed which yields the expression for these projections by analogy:

$$F_{2\xi}^B = \frac{2\mu L^3 \dot{\varepsilon}_2}{h_{02}^2}\int_0^{\pi}\frac{(R_2+\varepsilon_2\cos\theta_2)\cos^2\theta_2}{(1-\varepsilon_2\cos\theta_2)^3}d\theta_2, \qquad (83)$$

$$F_{2\eta}^B = \frac{\mu L^3 \omega_2 \varepsilon_2}{h_{02}^2} \left(\frac{2\dot{\gamma}_2}{\omega_2} - 1 \right) \int_0^\pi \frac{(R_2 + \varepsilon_2 \cos\theta_2)\sin^2\theta_2}{(1-\varepsilon_2\cos\theta_2)^3} d\theta_2, \tag{84}$$

where $h_{02} = R - R_2$ is the nominal gap of the outer layer of lubricant, $e_2 = e_{B2}(t)$ denotes the eccentricity of the center of the floating bushing, $\gamma_2 = \gamma_{B2}(t)$ is the angle describing the position of the line between the centers of the fixed-floating bushing and bearing housing. The motion of the floating bushing in the lubricating layer is time-dependent, i.e. the position and the velocity of the center depend on the time since external load and the response from the outer lubricant layer depends on the time. Suppose that at the time instant $t = t_0$ the center of the bushing O_2 is in the position corresponding to the eccentricity $e_{02} = e_{B2}(t_0)$ and the angle describing the position of the line between the centers of the fixed-floating bushing and bearing housing: $\gamma_{02} = \gamma_{B2}(t_0)$, then $\theta_2 = \theta_2(t_0, \varphi) = \varphi - \gamma_{B2}(t_0) = \varphi - \gamma_{02}$.

The projections of the force F_1^B acting on the floating bushing from the inner layer of lubricant in the local coordinate system $(O\xi_2, O\eta_2)$ are given by

$$F_{1\xi}^B = LR_2 \int_0^{2\pi} q_{01} \cos(\theta_1 + \gamma_{01} - \gamma_{02})\, d\theta_1, \tag{85}$$

$$F_{1\eta}^B = LR_2 \int_0^{2\pi} q_{01} \sin(\theta_1 + \gamma_{01} - \gamma_{02})\, d\theta_1. \tag{86}$$

Inserting eqs. (73), (74) into eqs. (85), (86) we obtain the expression for the projections of this force F_1^B:

$$F_{1\xi}^B = \frac{\mu L^3(\omega_1+\omega_2)R_2}{2h_{01}^2} \int_0^{2\pi} \frac{(A\sin\theta_1 + B\cos\theta_1)\cos(\theta_1+\gamma_{01}-\gamma_{02})}{(1-\varepsilon_1\cos\theta_1)^3} d\theta_1 \tag{87}$$

$$F_{1\eta}^B = \frac{\mu L^3(\omega_1+\omega_2)R_2}{2h_{01}^2} \int_0^{2\pi} \frac{(A\sin\theta_1 + B\cos\theta_1)\sin(\theta_1+\gamma_{01}-\gamma_{02})}{(1-\varepsilon_1\cos\theta_1)^3} d\theta_1 \tag{88}$$

Let the bushing (O_2, R_2) be loaded by the constant force $\overrightarrow{Q^B}$. Denoting the coordinates of the equilibrium position of the bearing by $(\varepsilon_2^*, \gamma_2^*)$ we have $\dot{\varepsilon}_2^* = 0$, $\dot{\gamma}_2^* = 0$. It follows from equations (83), (84):

$$F^B_{2\xi}(\varepsilon_2^*, \gamma_2^*) = 0$$

$$F^B_{2\eta}(\varepsilon_2^*, \gamma_2^*) = -\frac{\mu L^3 \omega_2 \varepsilon_2^*}{h_{02}^2}\left[\int_0^\pi \frac{R_2 \sin^2\theta_2}{(1-\varepsilon_2^* \cos\theta_2)^3} d\theta_2 + \int_0^\pi \frac{e_2^* \cos\theta_2 . \sin^2\theta_2}{(1-\varepsilon_2^*\cos\theta_2)^3} d\theta_2\right]$$

Let $(\varepsilon_1^*, \gamma_1^*)$ denote the coordinates of the equilibrium position of the center of the rotor in the bearing, then $\overset{\bullet}{\varepsilon}{}_1^* = 0$, $\overset{\bullet}{\gamma}{}_1^* = 0$. One obtains from eqs. (87) and (88):

$$F^B_{1\xi}(\varepsilon_1^*, \gamma_1^*) = -\frac{\mu L^3(\omega_1+\omega_2)R_2}{2h_{01}^2}\int_0^{2\pi} \frac{\sin\theta_1 \cos(\theta_1 + \gamma_1^* - \gamma_2^*)}{(1-\varepsilon_1^*\cos\theta_1)^3} d\theta_1,$$

$$F^B_{1\eta}(\varepsilon_1^*, \gamma_1^*) = -\frac{\mu L^3(\omega_1+\omega_2)R_2}{2h_{01}^2}\int_0^{2\pi} \frac{\sin\theta_1 \sin(\theta_1 + \gamma_1^* - \gamma_2^*)}{(1-\varepsilon_1^*\cos\theta_1)^3} d\theta_1.$$

The total force acting on the floating bushing from the inner and outer fields of lubrication is denoted as $\overrightarrow{F^B_{1,2}}$, i.e. $\overrightarrow{F^B_{1,2}} = \overrightarrow{F^B_1} + \overrightarrow{F^B_2}$. The values of $(\varepsilon_2^*, \gamma_2^*)$ are determined from the equilibrium conditions:

$$\left|\overrightarrow{Q^B}\right| = \left|\overrightarrow{F^B_{1,2}}\right| , \quad \tan\widetilde{\theta}_2 = \frac{F^B_{1\eta}(\varepsilon_1^*, \gamma_1^*) + F^B_{2\eta}(\varepsilon_2^*, \gamma_2^*)}{F^B_{1\xi}(\varepsilon_1^*, \gamma_1^*) + F^B_{2\xi}(\varepsilon_2^*, \gamma_2^*)} \tag{89}$$

where $\widetilde{\theta}_2 = (\overrightarrow{F^B_{1,2}}, \overrightarrow{O\xi_2})$ and

$$\left|\overrightarrow{F^B_{1,2}}\right| = \sqrt{\left(F^B_{1\xi}(\varepsilon_1^*, \gamma_1^*) + F^B_{2\xi}(\varepsilon_2^*, \gamma_2^*)\right)^2 + \left(F^B_{1\eta}(\varepsilon_1^*, \gamma_1^*) + F^B_{2\eta}(\varepsilon_2^*, \gamma_2^*)\right)^2}$$

If we assume that the external force $\overrightarrow{Q^B}$ is also directed vertically downwards, then it follows from eq. (89) that $\overrightarrow{Q^B}$ is proportional to the rotor angular velocity ω_1 and the angular velocity of the bushing ω_2. The bushing is in equilibrium $(\varepsilon_2^*, \gamma_2^*)$, i.e. in addition to eq. (89), force $\overrightarrow{F^B_{1,2}}$ must be directed vertically upwards, i.e.:

$$\widetilde{\theta}_2 = (\overrightarrow{F^B_{1,2}}, \overrightarrow{O\xi_2}) = -\gamma_2^*. \tag{90}$$

It follows from eqs. (89) and (90):

$$\tan \gamma_2^* = -\tan \widetilde{\theta}_2 = -\frac{F_{1\eta}^B(\varepsilon_1^*, \gamma_1^*) + F_{2\eta}^B(\varepsilon_2^*, \gamma_2^*)}{F_{1\xi}^B(\varepsilon_1^*, \gamma_1^*) + F_{2\xi}^B(\varepsilon_2^*, \gamma_2^*)}.$$

When the rotor and bushing are in equilibrium, then $\gamma_1^* = \pi/2$, $\varepsilon_1^* \in [0, 1]$. Each equilibrium position of the rotor in the gap $(\varepsilon_1^*, \gamma_1^*)$ corresponds to one curve of the equilibrium positions of the center bushing, which is determined by the equation:

$$\tan \gamma_2^* = -\frac{F_{1\eta}^B(\varepsilon_1^*, \frac{\pi}{2}) + F_{2\eta}^B(\varepsilon_2^*, \gamma_2^*)}{F_{1\xi}^B(\varepsilon_1^*, \frac{\pi}{2}) + F_{2\xi}^B(\varepsilon_2^*, \gamma_2^*)} \quad . \tag{91}$$

5.6 Determination of the constant speed of the bushing rotation in a plain bearing

Suppose that the rotor (O_1, R_1) rotates at a given constant angular velocity ω_1, whereas the busing (O_2, R_2) rotates with yet unknown constant angular velocity ω_2. This angular velocity ω_2 appears in the expressions for all the forces acting on the rotor and bushing. In order to determine the motion of the bearing parts, it is necessary to derive the dependence of the angular velocity of the floating bushing ω_2 on the rotor velocity ω_1. Up to now, the velocity of rotation of the bushing ω_2 has been determined only by means of experimental methods, cf. Lang and Steinhilper [1978], Boyaci et al. [2009].

In what follows we will obtain an analytic dependence of ω_2 on ω_1. We note that the friction moment exists only in the region of the positive pressure in the lubricant. Since $\overset{\bullet}{\varepsilon_1^*} = 0$, $\overset{\bullet}{\gamma_1^*} = 0$, $\overset{\bullet}{\varepsilon_2^*} = 0$, $\overset{\bullet}{\gamma_2^*} = 0$ in the equilibrium position we obtain from eqs. (75) and (76) that the pressure is positive in the domain $\theta_1 \in [\pi, 2\pi]$, and similarly in the domain $\theta_2 \in [\pi, 2\pi]$.

In order to derive the equation for the moments acting on the floating bushing we need the expression for the shear stress in the case of short journal bearings:

$$\tau_{r\varphi} = \mu \left(\frac{\partial v}{\partial r} + \frac{1}{r} (\frac{\partial u}{\partial \varphi} - v) \right) \cong \mu \left(\frac{\partial v}{\partial r} - \frac{v}{r} \right)$$

It allows us to determine the moments from the internal and external fields of lubricant, respectively

$$M_1^B = 2\mu R_1 L (\omega_1 - \omega_2) \int_{\pi}^{2\pi} \frac{d\theta_1}{1 - \left(\frac{R_1}{R_1 + h_1^*(\theta_1)} \right)^2} \quad ,$$

$$M_2^B = -\ 2\mu R_2 L\omega_2 \int_{\pi}^{2\pi} \frac{d\theta_2}{1-\left(\frac{R_2}{R_2+h_2^*(\theta_2)}\right)^2} \ .$$

The explicit expressions for the shear stresses and the moments can be found in e.g. Belyaev et al. [2008].

In a steady motion the floating bushing rotates with constant angular velocity ω_2, which implies that the moments acting on the bushing from both sides are self-equilibrated, that is, $M_1^B + M_2^B = 0$. The latter condition is understood as an equation from which we obtain the expression for the ratio of the angular velocities in closed form

$$\omega_2/\omega_1 = \left[1 + \frac{R_2}{R_1} \frac{\int_{\pi}^{2\pi} \left[1-\left(\frac{R_2}{R_2+h_2^*(\theta_2)}\right)^2\right]^{-1} d\theta_2}{\int_{\pi}^{2\pi} \left[1-\left(\frac{R_1}{R_1+h_1^*(\theta_1)}\right)^2\right]^{-1} d\theta_1}\right]^{-1} . \tag{92}$$

For numerical calculations we assume the following values of the bearing parameters: Bearing *(O, R)*: $R = 0.05$ (m), length $L = 0.05$ (m). Bushing (O_2, R_2): R_2=0.048 (m), this implies the following nominal gap of the outer layer of lubricant $h_{02} = R - R_2$=0.002 (m). Rotor (O_1, R_1): R_1=0.046(m). We took the nominal gap of the inner layer of lubricant to be $h_{01} = R_2 - R_1$=0.002(m) and the dynamic viscosity be $\mu = 1.754 \cdot 10^{-5}(N \cdot s \cdot m^{-2})$.

The result of calculation displayed in Fig. 16 shows that the ratio ω_2/ω_1 varies between 0.3 and 0.67 depending upon the relationship between the eccentricities $\varepsilon_1^*, \varepsilon_2^*$. It is generally assumed in most studies that the angular velocity of the floating bushing is equal to or slightly less than half the angular velocity of the rotor, as it was, for example, established experimentally in Lang and Steinhilper [1978], Boyaci et al. [2009]. This generally accepted rule is usually substantiated by a simple reasoning that the floating bushing is located in the middle of the gap, one wall of which rotates with angular velocity ω_1 while the second one is motionless. However, the present study shows that this is true only in some special cases, cf. Fig 16.

5.7 Dynamics and stability of the rotor rotation

In the Cartesian frame the governing equations for the rotor dynamics are as follows

$$m_1\ddot{x} = m_1 g + F_{1\xi}^P \cos\phi - F_{1\eta}^P \sin\phi \ ,$$
$$m_1\ddot{y} = F_{1\xi}^P \sin\phi + F_{1\eta}^P \cos\phi$$

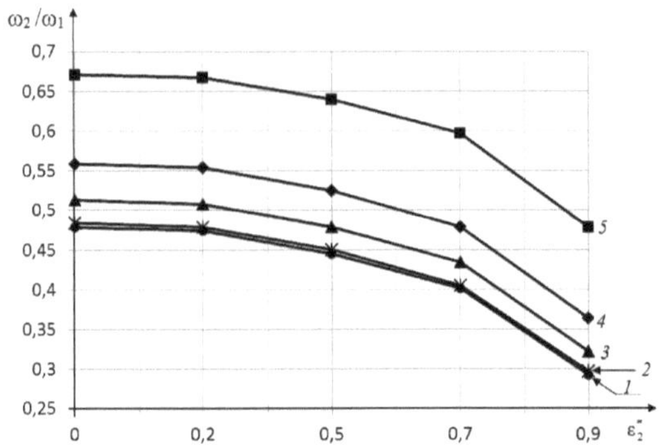

Figure 16. Relative velocity of the bushing for $\varepsilon_1^* = 0(1)$; 0,2(*2*); 0,5(*3*); 0,7(*4*); 0,9(*5*)

where $(\bar{x}, \bar{y})$ denote the position of the center of gravity G of the rotor, cf. Fig. 17. An analogy with the governing equations of motion for a pendulum is evident.

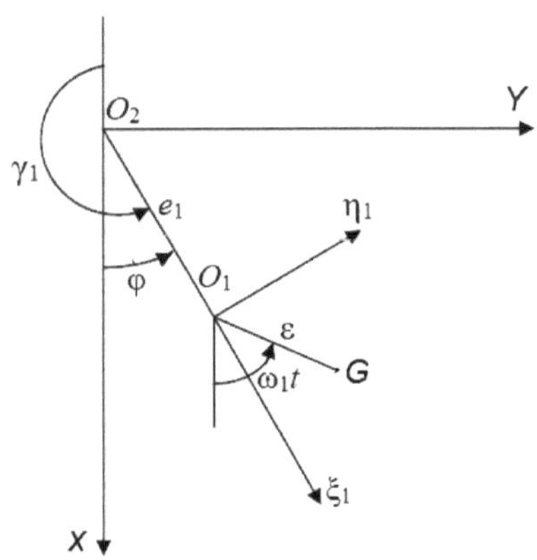

Figure 17. Position of the rotor center in the housing

In the polar system of coordinates the above equations take the following form

$$\ddot{e}_1 - \dot{\gamma}_1^2 e_1 = -g\cos\gamma_1 + m^{-1}F^p_{1\xi} - \varepsilon\omega_1^2\left(\cos\omega_1 t\cos\gamma_1 + \sin\omega_1 t\sin\gamma_1\right),$$

$$\ddot{\gamma}_1 e_1 - 2\dot{e}_1\dot{\gamma}_1 = g\sin\gamma_1 + m^{-1}F^p_{1\eta} + \varepsilon\omega_1^2\left(\cos\omega_1 t\sin\gamma_1 - \sin\omega_1 t\cos\gamma_1\right).$$

We now introducing the small parameters $\varepsilon_{11}(t)$, $\gamma_{11}(t)$ which describe the motion in the vicinity of the equilibrium position ε_1^*, γ_1^*

$$\varepsilon_1(t) = \varepsilon_1^* + \varepsilon_{11}(t);\quad \gamma_1(t) = \gamma_1^* + \gamma_{11}(t)\quad . \tag{93}$$

Performing the linearization with respect to these small parameters yields the following perturbation equations

$$\begin{aligned}
\ddot{\varepsilon}_{11} &= -\frac{k}{m_1 h_{01}}\frac{2T_1(\varepsilon_1^*)}{\omega_1+\omega_2}\dot{\varepsilon}_{11} - \frac{g}{h_{01}}\gamma_{11},\\
\ddot{\gamma}_{11}\varepsilon_1^* &= -\frac{k}{m_1 h_{01}}\frac{2T_2(\varepsilon_1^*)}{\omega_1+\omega_2}\dot{\gamma}_{11} + \frac{k}{m_1 h_{01}}T_2'(\varepsilon_1^*)\,\varepsilon_{11}
\end{aligned} \tag{94}$$

where the coefficients T_1, T_2 etc. are obtained from eq. (94) by substituting eq. (93) and removing the higher order terms.

The characteristic equation for this system of equations is as follows

$$\begin{aligned}
&\varepsilon_1^*\lambda^4 + \frac{2k}{m_1 h_{01}(\omega_1+\omega_2)}\left[\varepsilon_1^* T_1(\varepsilon_1^*) + T_2(\varepsilon_1^*)\right]\lambda^3 +\\
&+\left[\frac{2k}{m_1 h_{01}(\omega_1+\omega_2)}\right]^2 T_1(\varepsilon_1^*)\,T_2(\varepsilon_1^*)\,\lambda^2 + \frac{kg}{m_1 h_{01}^2}T_2'(\varepsilon_1^*) = 0
\end{aligned}$$

It can be written in the form of the polynomial of fourth order

$$a_0\lambda^4 + a_1\lambda^3 + a_2\lambda^2 + a_3\lambda + a_4 = 0$$

where the coefficients are as follows

$$\begin{aligned}
a_0 &= \varepsilon_1^*;\ a_1 = \frac{2k}{m_1 h_{01}(\omega_1+\omega_2)}\left[\varepsilon_1^* T_1(\varepsilon_1^*) + T_2(\varepsilon_1^*)\right];\\
a_2 &= \left[\frac{2k}{m_1 h_{01}(\omega_1+\omega_2)}\right]^2 T_1(\varepsilon_1^*)\,T_2(\varepsilon_1^*);\ a_3 = 0;\ a_4 = \frac{kg}{m_1 h_{01}^2}T_2'(\varepsilon_1^*).
\end{aligned}$$

The Routh-Hurwitz criterion states that the motion is unstable if

$$\Delta_1 = a_1 > 0; \Delta_2 = a_1 a_2 - a_0 a_3 > 0; \Delta_3 = a_3\Delta_2 - a_1^2 a_4 > 0; \Delta_4 = a_4\Delta_3 > 0.$$

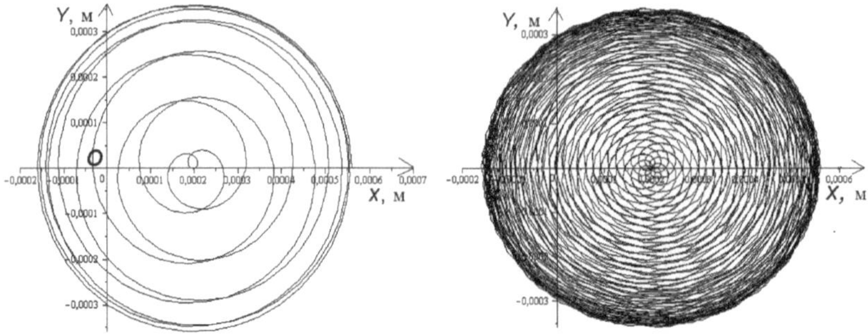

Figure 18. First type of unstable motion

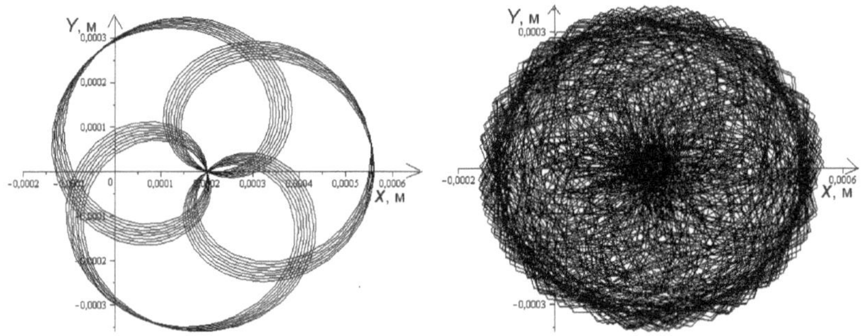

Figure 19. Second type of unstable motion

Applying the Routh-Hurwitz criterion convinces that in the case under consideration the motion of the rotor is unstable since the criterion conditions are not satisfied since $a_3 = 0$. It means that

$$\Delta_3 = -a_1^2 a_4 < 0, \quad \Delta_4 = a_4 \Delta_3 < 0 \ .$$

Some characteristic types of unstable motion are calculated by means of the above equations and shown in Figs. 18–21, cf. Nguyen et al. [2012]. The first plot in each Figure demonstrates a scenario of developing the self-excited oscillation after only a few rotations whereas the second plot displays fully developed oscillations of the rotor.

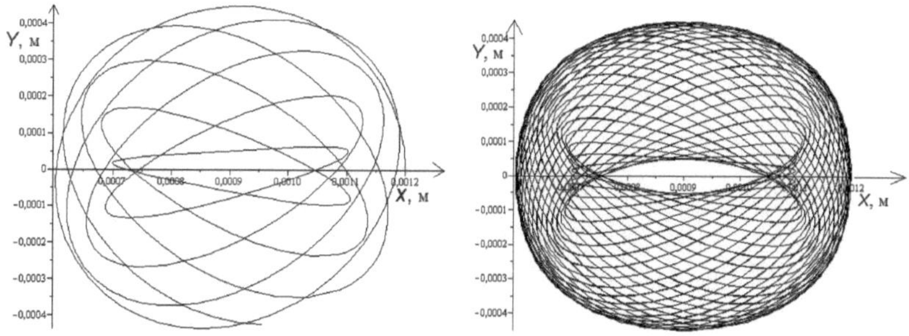

Figure 20. Third type of unstable motion

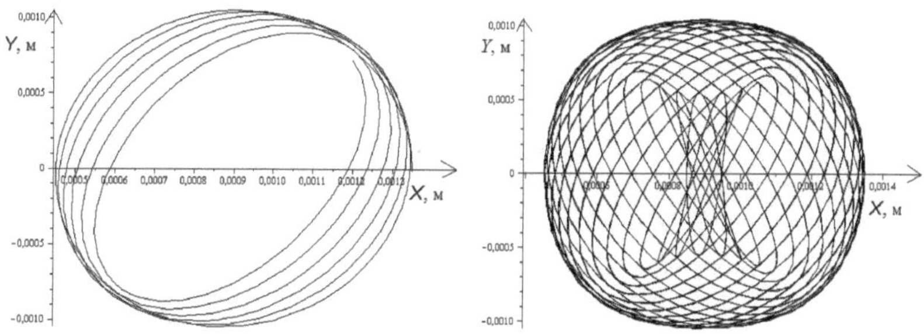

Figure 21. Fourth type of unstable motion

6 Conclusions

The intent of the present chapter was to demonstrate that the systems with axially moving material are inherently unstable. Dynamics and stability are known to be strongly related to each other, for this reason the study of dynamics and stability for each engineering system under consideration was carried out in the framework of the same approach.

Acknowledgement. The author acknowledges the support by the Russian Foundation for Basic Research, grants 14-08-00646-a and 12-01-92000-HHC-a.

Bibliography

A.S. Volmir. *Stability of Deformable Systems (in Russian)*. Nauka, 1967.

V.V. Bolotin. *Nonconservative Problems of the Theory of Elastic Stability.* Pergamon Press, 1963.

H. Ziegler. *Principles of Structural Stability.* Blaisdel Publishing Company, 1968.

L.-Q. Chen. Analysis and control of transverse vibrations of axially moving strings. *ASME Applied Mechanics Reviews*, 58:91–116, 2005.

V.V. Eliseev and Y.M. Vetyukov. Effects of deformation in the dynamics of belt drive. *Acta Mechanica*, 223(8):1657–1667, 2012.

G. Hirmann and A.K. Belyaev. Stabilitätsverhalten eines schnellaufenden synchronriemens. *Antriebstechnik*, 36(6):64–66, 1997.

R. Eglseer and A.K. Belyaev. Instabilitätbereiche eines kettentriebes. *Antriebstechnik*, 36(11):61–63, 1997.

L.E. Suarez and A. Shokooh. An eigenvector expansion method for the solution of motion containing fractional derivatives. *ASME Journal Applied Mechanics*, 64:64–66, 1997.

A.K. Belyaev. Fractional derivatives in some mechanical systems. In A.K. Belyaev, H. Irschik, and M. Krommer, editors, *Mechanics and Model-Based Control of Advanced Engineering Systems.* Springer WienNewYork, 2014.

K.B. Oldham and J. Spanier. *The fractional calculus.* Academic Press, 1974.

G.B. Dubois and F.W. Ocvirk. Analytical derivation and experimental evaluation of short-bearing approximation for full journal bearing. *Cornell University Report*, 1157:119–127, 1953.

K. Hatakenaka, K. Tanaka, and M. Suzuki. A theoretical analysis of floating bush journal bearing with axial oil film ruptures being considered. *Journal of tribology*, 124(3):494–505, 2012.

O.R. Lang and W. Steinhilper. *Gleitlager.* Springer, 1978.

A. Boyaci, H. Hetzler, W. Seemann, C. Proppe, and J. Wauer. Analytical bifurcation analysis of a rotor supported by floating ring bearings. *Nonlinear Dynamics*, 57:497–507, 2009.

A.K. Belyaev, M. Krommer, H. Irschik, and V.T. Nguyen. Forces and moments acting on the rapidly rotating floating bearing. In D.A. Indeitsev and A.M. Krivtsov, editors, *Advanced Problems in Mechanics (APM2008).* Russian Academy of Sciences, 2008.

V.T. Nguyen, D.G. Arseniev, and A.K. Belyaev. Distribution of the equilibrium positions of a shaft and defining the angular speed of the ring in a floating ring bearing. In D.A. Indeitsev and A.M. Krivtsov, editors, *Advanced Problems in Mechanics (APM2012).* Russian Academy of Sciences, 2012.

FSC
www.fsc.org
MIX
Papier | Fördert gute Waldnutzung
FSC® C083411

Kolibri 360 GmbH
Ferdinand-Jühlke-Straße 7
99095 Erfurt, Deutschland
produktsicherheit@kolibri360.de